Optical Systems and Wireless Communications

Optical Systems and Wireless Communications

Edited by
Josephine Winston

Willford Press
www.willfordpress.com

Published by Willford Press,
118-35 Queens Blvd., Suite 400,
Forest Hills, NY 11375, USA

ISBN: 978-1-68285-674-1

Cataloging-in-Publication Data

Optical systems and wireless communications / edited by Josephine Winston.
p. cm.
Includes bibliographical references and index.
ISBN 978-1-68285-674-1
1. Optical communications. 2. Wireless communication systems.
I. Winston, Josephine.

TK5103.59 .O58 2019
621.382 7--dc23

For information on all Willford Press publications
visit our website at www.willfordpress.com

Willford Press

Printed in China.

Contents

Preface

This book has been an outcome of determined endeavour from a group of educationists in the field. The primary objective was to involve a broad spectrum of professionals from diverse cultural background involved in the field for developing new researches. The book not only targets students but also scholars pursuing higher research for further enhancement of the theoretical and practical applications of the subject.

The branch of physics that studies the properties of light and its interactions with matter is known as optics. It is also concerned with the construction of instruments that use and detect light. Lenses, mirrors, fiber optics, lasers, telescopes and microscopes are some of the common applications of optical systems. Optical wireless communications is a form of optical communication. It uses unguided visible, ultraviolet and infrared light for carrying a signal. The wireless communication that operates in the visible region of the electromagnetic spectrum is known as visible light communication. This form of communication has applications in wireless local area networks, chip-to-chip communications, underwater communications, etc. At near infrared frequencies, communication is achieved using free space optical systems. The ever-growing need of advanced technology is the reason that has fueled the research in the fields of optical systems and wireless communications in recent times. This book is a valuable compilation of topics, ranging from the basic to the most complex advancements in these fields. The extensive content of this book will provide the readers with a thorough understanding of the subject.

It was an honour to edit such a profound book and also a challenging task to compile and examine all the relevant data for accuracy and originality. I wish to acknowledge the efforts of the contributors for submitting such brilliant and diverse chapters in the field and for endlessly working for the completion of the book. Last, but not the least; I thank my family for being a constant source of support in all my research endeavours.

Editor

1

Ultra-Fast All-Optical Memory based on Quantum Dot Semiconductor Optical Amplifiers (QD-SOA)

Yossef Ben Ezra and Boris I. Lembrikov

Additional information is available at the end of the chapter

Abstract

All-optical signal processing is characterized by high bit-rate, power efficiency, high bandwidth, and transparency. All-optical logic gates are basic logic units for the all-optical signal processing implementation. Typically, all-optical gates are based on strong optical nonlinearities related in particular to semiconductor optical amplifiers (SOA). We briefly review the state of art in the field of all-optical logic gates and all-optical memory. In the original part, we discuss the ultrafast all-optical memory loop based on the Mach-Zehnder interferometer (MZI) with quantum dot (QD) semiconductor optical amplifier (SOA) in each arm.

Keywords: all-optical signal processing, quantum dots (QD), semiconductor optical amplifier (SOA)

1. Introduction

Optical signal processing is based on the using of linear and nonlinear optical techniques in order to manipulate and process digital, analogue, and quantum information [1]. Optical signal processing increases the processing speed of devices and reduces the energy consumption and latency of communication systems [1]. In particular, ultrafast optical nonlinearities provide a substantial speed advantage as compared to electronic techniques for simple logic: switching, regeneration, wavelength conversion (WC), performance monitoring, and analog-to-digital conversion (ADC) [1]. Silicon photonics and highly nonlinear nanophotonic devices are providing strong optical nonlinearities for ultrafast processing on millimeter length scales [1].

The recent progress in optical signal processing is based on the combination of the advanced modulation techniques, coherent detection, and digital signal processing (DSP) [2]. The interface of all-optical techniques and advanced DSP will enhance electronic processing capabilities [1]. Optical signal processing is essentially based on the following advanced technologies: coherent detection, high-speed electronics for DSP, advances in strongly nonlinear materials and devices, photonic integrated circuits (PIC), and access to four optical domains of amplitude, phase, polarization, and wavelength [2]. A simple digital modulation scheme is the on-off keying (OOK) referred to as intensity modulation with direct detection (IM/DD) [3]. In such a case, an electrical binary bit stream modulates the intensity of an optical carrier inside the optical transmitter, and the resulting optical signal is converted to the original signal in the electrical domain in an optical receiver [4]. The phase modulation combined with the coherent detection increases the spectral efficiency (SE) of optical communication systems and improves the sensitivity of optical receivers [4]. In general case, amplitude-shift keying (ASK), phase-shift keying (PSK) or M-ary quadrature amplitude modulation (QAM) can be realized [3, 4]. Polarization-division multiplexing (PDM), advanced multilevel modulation formats such as M-ary QAM, digital spectral shaping at the transmitter, coherent detection and advanced forward error correction (FEC) can increase SE of the communication system [1]. Typically, DSP must overcome deterministic signal distortions, while FEC overcomes stochastic impairments caused by noise and interference [1]. At the transmitter, DSP together with digital-to-analog converters (DAC) and FEC converts the incoming data bits into a set of analogue signals [1]. An optical coherent receiver recovers the amplitude and phase of the signal by mixing it with the local oscillator (LO) which is typically a continuous-wave (CW) laser [3, 4]. DSP, ADC, and FEC recover the data from the set of analogue electrical signals [1]. The main functions of the receiver-based DSP are equalization and synchronization [1]. Equalization must realize the polarization rotation tracking and dispersion compensation including both the chromatic dispersion and polarization-mode dispersion (PMD) [1]. Synchronization carries out the transmitter and receiver electrical and optical signal frequency and phase matching [1].

The all-optical signal processing is implemented by using the nonlinear optical phenomena such as self-phase modulation (SPM), cross-phase modulation (XPM), four-wave mixing (FWM) related to the third-order susceptibility and sum frequency, difference frequency, second harmonic generation (SHG) related the second-order susceptibility [1–4]. The typical nonlinear elements used in optical communication systems are highly nonlinear optical fibers (HNLF), silicon waveguides, chalcogenide waveguides, photonic crystals, nonlinear optical loop mirrors (NOLM), parametric amplifiers, and semiconductor optical amplifiers (SOA) [2, 3]. SOA are characterized by the extremely strong third-order optical nonlinearity and fast response and can be integrated monolithically with other devices on the same chip [3].

Optically assisted signal processing combines optics and electronics for what each one of them does best [2]. Optical components can perform some functions very fast, while electronic components carry out complex computations with buffers and memory [2]. For instance, optically assisted network routing technique uses optical correlation on headers of Internet data packets [2]. Optically assisted signal processing can be also used for a target pattern search in large amounts of data [2]. In such cases, the data information is encoded on an optical carrier at Tb/s speed and sent to an optical correlator for pattern recognition [2]. The output at Gb/s speed is searched and processed electronically with high accuracy before being sent to the user [2].

In optical networks, the bandwidth mismatch between optical transmission and electronic routers results in the development of different optical signal processing and the investigation of optical packet switching (OPS) [5]. Some applications require selective switching of one or more bits to a different port [3]. The packet switching takes place when a packet of tens or hundreds of bits is selected from a bit stream [3]. The flip-flop memory is an essential component of the packet switch [3]. Typically, such a memory is implemented using two coupled lasers switching the output signal between two wavelengths λ_1 and λ_2 [3]. Recently, we proposed a novel architecture of an all-optical memory loop combining the ultrafast all-optical signal processor based on the Mach-Zehnder interferometer (MZI) with quantum dot (QD) SOA and the DSP block for the mitigation of dispersion and nonlinearity impairments [6].

The chapter is organized as follows. OPS in optical communication systems is discussed in Section 2. The different types of all-optical logic gates used in OPS are briefly reviewed in Section 3. The operation principle of the novel all-optical memory is described in Section 4. The QD SOA theoretical model is briefly discussed in Section 5. The numerical simulation results and conclusions are presented in Sections 6 and 7, respectively.

2. Optical packet switching (OPS)

OPS process requires many components for buffering, header processing, and switching [3]. Each packet begins with a header containing the destination information [3]. When a packet arrives at a node, a router reads the header and sends it toward its destination [3]. The basic element of an optical router is a packet switch directing incoming packets to the corresponding output ports depending on the information in the header [3].

Consider the architecture and operation principle of the all-optical packet switch. The scheme of the 1 × 2 all-optical packet switch is shown in **Figure 1** [3, 5, 7]. The all-optical packet switch consists of three functional blocks: the all-optical header-processing block, the all-optical flip-flop memory block, and the WC block [7]. All-optical header processing can be realized by using the different methods such as tunable Bragg gratings, FWM in a SOA, terahertz optical asymmetric demultiplexers (TOAD), two-pulse correlation in a semiconductor laser amplifier in a loop optical mirror (SLALOM) [5, 7].

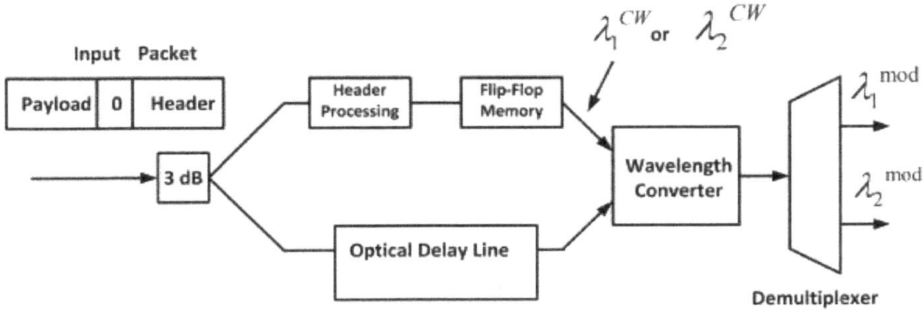

Figure 1. System concept for 1 × 2 all-optical packet switches.

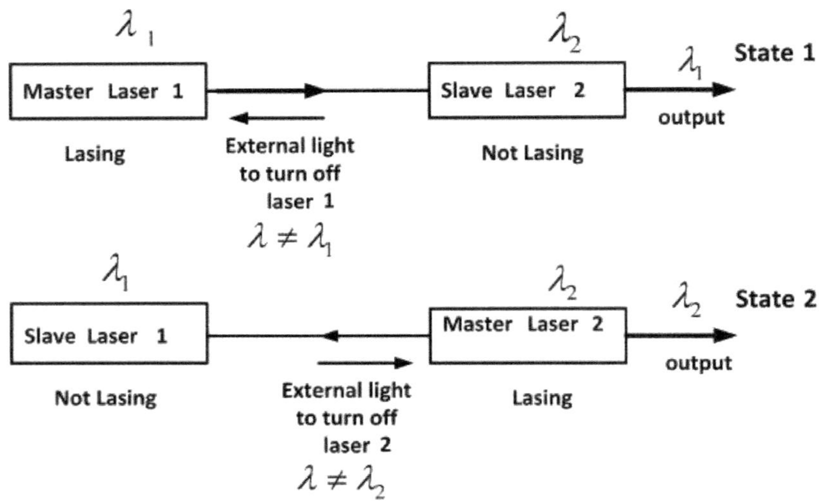

Figure 2. The all-optical flip-flop memory based on two coupled lasers.

The high-speed memory is necessary for OPS networks in order to avoid the packet collisions during packet routing [8]. The all-optical flip-flop memory is based on two coupled lasers with separate laser cavities and can have two states [7]. It is shown in **Figure 2** [7].

In state 1, light from laser 1 suppresses lasing in laser 2 emitting CW light at wavelength λ_1, while in state 2, light from laser 2 suppresses lasing in laser 1 emitting CW light at wavelength λ_2 [7]. The output pulse of the optical header processor is used to set the optical flip-flop memory into the desired wavelength [7]. The amount of light P_{sw} which is necessary for the change of states, the threshold carrier number N_{th}, and photon lifetime τ_p are given by, respectively [7]:

$$P_{sw} = E\frac{v_g R \tau_p}{L(1-R)}\ln\left(\frac{1}{R}\right)\left(1-\frac{2R}{\delta(1-R)}\right)\left(\frac{I}{q}-\frac{N_{th}}{\tau_e}\right) \tag{1}$$

$$N_{th} = \frac{V}{\tau_p \Gamma v_g a} + N_0;\ \frac{1}{\tau_p} = v_g\left(\alpha_{\text{int}} + \frac{1}{L}\ln\left(\frac{1}{R}\right)\right) \tag{2}$$

Here, E is the photon energy, R is the reflectivity at the end facets of lasers, δ is the coupling constant between the two laser cavities, v_g is the group light velocity, L is the length of the active region in the laser, I is the injection current, q is the electron charge, τ_e is the carrier lifetime, V is the volume of the laser cavity active region, Γ is the confinement factor, a is the gain factor, N_0 is the carrier number at transparency, and α_{int} is the internal laser cavity losses factor. Note that the outputs of the lasers on the left side are defined by the reflectivity at the end facets of lasers R and do not influence the memory states.

WC component converts the incoming data packet wavelength to the output wavelength of the flip-flop memory [3]. The demultiplexer directs output at different wavelengths to different ports depending on the header information [3].

3. All-optical logic gates with SOA

In this section, we briefly discuss the scheme and operation principle of all-optical gates which are core logic units for the all-optical signal processing system implementation [9]. All-optical gates may be divided into two groups: without SOA and with SOA [9]. The all-optical gates without SOA are based on the change in nonlinear refraction index in silica fiber [9]. The intensity-dependent refractive index of silica results in the following nonlinear optical effects: SPM, cross gain modulation (XGM), and FWM [2, 9]. The all-optical gates without SOA based on these nonlinear optical phenomena can be realized in the following configurations: dispersion shifted fiber/high nonlinear fiber (DSF/HNLF) configuration; waveguide configuration; circular configuration; optical channel-dropping (C/D) filter configuration; multilayer waveguide configuration; double heterostructure optical thyristor (DHOT) configuration; and acousto-optical tunable filter (AOTF) configuration [9]. The detailed description and comparison between non-SOA gates are presented in Ref. [9]. For instance, DSF/HNLF, waveguide, circular, and AOTF configurations are polarization sensitive; DSF/HNLF, waveguide, circular configurations are characterized by bad or moderate integration capacity [9].

On the other hand, the SOA-based devices are mainly polarization non-sensitive and possess compact integration capacity [9]. They are highly competitive due to the high nonlinearity, low switching power, wide gain bandwidth, and compact size [8]. Recently, novel two inputs optical logic gates (NOT, AND, OR and NOR) based on a traveling wave SOA (TW-SOA) operating at 40 Gb/s had been demonstrated [10].

The implementation of all-optical gates with SOA is based on the different interferometer techniques such as ultra-high nonlinear interferometer (UNI), Sagnac interferometer (SI), MZI, and delay interferometer (DI) [9]. In these techniques, the XPM-induced phase shift is used for optical switching [4]. Typically, a weak signal pulse is divided equally between two arms of the interferometer and is undergoing identical phase shift in each arm [4]. In such a case, it is transmitted through constructive interference [4]. Consider now the situation when a pump pulse at a different wavelength as compared to the signal pulse is injected into one arm of the interferometer. As a result, the signal phase in that arm would be changed due to XPM. If the XPM-induced phase shift is close to π, the signal pulse will not be transmitted due to the destructive interference at the interferometer output [4]. An intense pulse pump can switch the signal pulse through the XPM-induced phase shift [4].

In particular, all-optical gate based on SOA-MZI can be realized with the copropagating, counterpropagating, and copropagating push-pull configurations [9]. Copropagation MZI operates on the principle of phase change caused by the light propagating through the 3 dB coupler [9]. MZI copropagating gates consist of a symmetrical MZI with two SOA placed in the upper and lower arm of the interferometer [9]. Data and clock pulses of different wavelengths are inserted into SOA operated under the gain saturation condition, where the optical gain is distributed between wavelengths according to their relative photon densities [9]. The data are transferred in the clock pulse in the inverted form in both arms of MZI [9]. After passing through the first 3 dB coupler, the phase difference $\pi/2$ is created between the upper and lower arms of clock pulse, after passing through the second 3 dB coupler the total phase shift becomes π [9]. Then, if both data have the same value, they will cancel, and at the T-port 0 will appear, if data have different value, then it will not cancel and 1 will appear at the T-port [9]. In MZI counter-propagating gates, the clock and data pulse propagate in opposite directions through MZI [9]. If any of the data is 1, then XPM between the clock and data pulse inside SOA creates the differential phase shift between the two clock components, MZI becomes unbalanced, and the clock pulse exits at T-port [9]. If both the data are the same, the total phase shift will become π, and the clock pulse is cancelled at T-port [9].

Consider now a typical all-optical logic element based on transforming of XPM into an intensity modulation and implemented as the MZI copropagating push-pull gate with SOA in the two arms shown in **Figure 3** [9]. The optical fibers are used as interconnects. The SOA-based MZI with couplers at the input and output is shown in **Figure 4**. At the SOA-based MZI block output, there is a coupler shown in **Figure 4**. The outputs at the right side of this coupler are connected to the T-port and R-port shown in **Figure 3**.

Note that the co-propagating data streams configuration permits to avoid the SOA length restriction, and the MZI with push-pull configuration allows increasing the memory bit-rate beyond the limitation of the SOA carrier recovery time [8]. The copropagating data streams A and B of the same wavelengths are inserted into upper and lower arm of MZI shown in **Figure 4**.

The data A in the upper arm is ahead of one bit period to data B traveling in the lower arm of MZI, and the lower arm data B is one bit period ahead to upper arm data A [9]. As a result, a

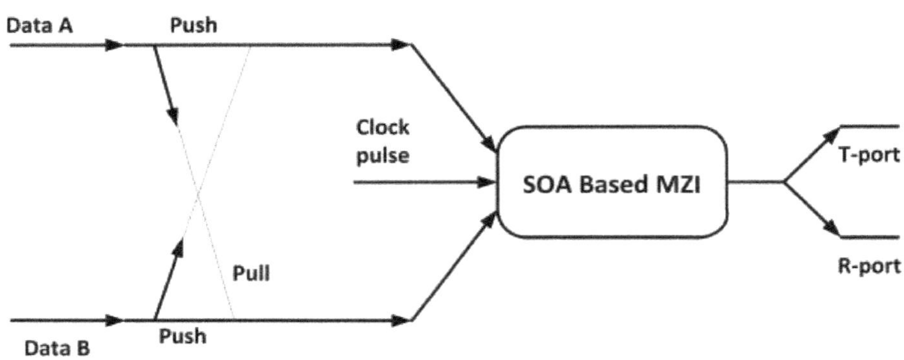

Figure 3. MZI with push-pull configuration.

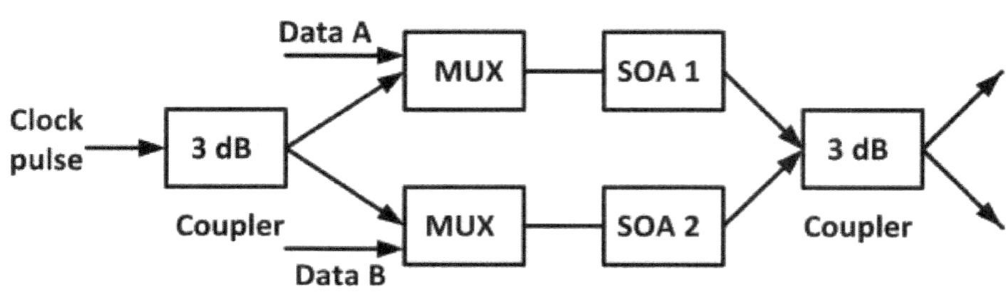

Figure 4. SOA–based MZI.

switching window for data streams occurs [9]. The clock pulse copropagating with the data streams A and B is inserted into the 3 dB coupler. Assume that the data A is 1 and data B is 0. Then, the pulse from data A splits into two parts in such a way that one pulse is pushed to the upper SOA 1 and other is delayed by the switching window. Consequently, the upper SOA 1 is switched before the lower SOA 2 [9]. The MZI is unbalanced, and clock wave is switched to the T-port. In the opposite case, when data A is 0 and data B is 1, then lower SOA 2 is switched and wave also appears at T-port [9]. Assume now that data A and data B are the same. As a result, SOA 1 and SOA 2 are equally influenced by the injected pulse. The respective push and pull pulse temporarily coincide with each other, the phase difference between the two arms of MZI equals to zero, and no switching occurs at T-port [9].

The disadvantages of the all-optical logic gates discussed above are twofold: (i) the operating speed of SOA is limited by the carrier recovery time of the order of magnitude of 100 ps; (ii) the scheme can be used only for OOK modulation format [8]. However, the SOA operation rate can be significantly increased up to 100 Gb/s by using the quantum dot semiconductor optical amplifiers (QD-SOA) [8, 11]. A theoretical model of an ultrafast all-optical signal processor based on QD-SOA MZI has been developed with limiting bit rates of 100 and 200 Gb/s at the injection currents of I = 30 mA and I = 50 mA, respectively [12].

4. All-optical memory loop based on SOA MZI

Consider now an all-optical memory loop consisting of an AND gate and a regenerator based on the two push-pull co-propagating MZI with a SOA in each arm and couplers at the input and the output shown in **Figure 4** and discussed in the previous section [8]. The scheme of the all-optical memory loop is shown in **Figure 5** [8]. The input data burst at the wavelength λ_2 is inserted into the memory through the AND gate is converted into the output data burst at the wavelength λ_1 [8]. A regenerator is used into the loop in order to improve the quality of the data burst at the wavelength λ_1 eliminating the signal degradation caused by dispersion, nonlinearity, and other physical impacts [8]. The output of the regenerator is fed back to the AND gate as a pump at the wavelength λ_3 [8]. The data can be stored in the loop for a long term [8]. The length of the data burst nT must be less than or equal to the length of the memory loop NT where T is the bit period [8].

The data format of the all-optical memory shown in **Figure 5** is OOK, while for the PSK modulation format the memory should be modified [8].

Recently, we proposed a novel architecture of an all-optical memory loop combining the ultrafast all-optical signal processor based on QD-SOA MZI [12] and the DSP block for the mitigation of the dispersion and nonlinearity impairments [6]. The proposed all-optical memory loop is shown in **Figure 6**. The advantages of the novel all-optical memory loop are following [6].

1. It includes only one MZI with two QD-SOA reducing the complexity of the electronic synchronization scheme.
2. It can operate at the bit rates up to 100 Gb/s due to the fast gain recovery time of QD-SOA.

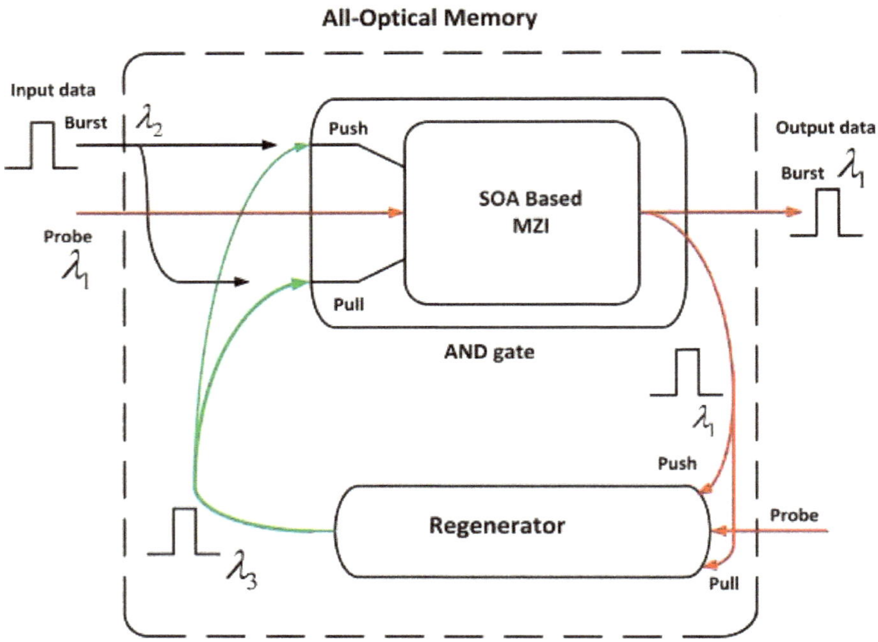

Figure 5. Schematic setup of the all-optical memory based on SOA MZI.

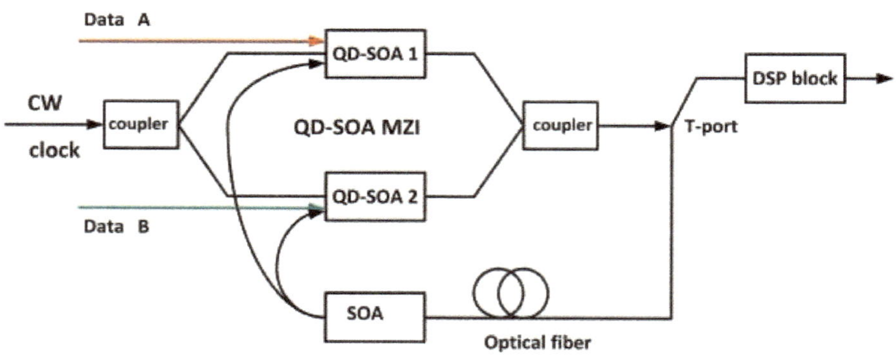

Figure 6. The architecture of the all-optical memory loop based on QD-SOA–based MZI.

3. DSP block can improve the signal quality as compared to an MZI-based regenerator.
4. An additional SOA is inserted into the loop in order to compensate the optical fiber losses and to increase the loop length.

In our case, the phase difference at the output of the QD-SOA–based MZI is caused by the signal power difference in the upper and lower MZI arms, unlike the MZI copropagating push-pull gate mentioned above [9]. Typically, 80 and 20% of the input signal power were fed through the coupler into the upper and lower arm of the MZI, respectively [6]. As a result, dynamic processes in QD-SOA placed into the upper and lower arms of MZI are characterized by different carrier relaxation time and gain recovery time, and the flatness of the switching window is significantly improving [6].

The QD-SOA MZI output light intensity P_{out} and phase difference $\varnothing_1(t) - \varnothing_2(t)$ are given by Refs. [12–14]:

$$P_{out} = \frac{P_{in}}{4}\{G_1(t) + G_2(t) - 2\sqrt{G_1(t)G_2(t)}\cos[\phi_1(t) - \phi_2(t)]\} \tag{3}$$

$$\phi_1(t) - \phi_2(t) = -\frac{\alpha}{2}\ln\left(\frac{G_1(t)}{G_2(t)}\right) \tag{4}$$

where the P_{in} is continuous wave (CW) clock stream optical signal divided and introduced into the two QD-SOA, $G_{1,2} = exp(g_{1,2}L)$ and $\varnothing_{1,2}(t)$ are the gain and phase shift, respectively, in the two arms of QD-SOA MZI, α is the line width enhancement factor (LEF), $g_{1,2}$ is the SOA gain, and L is the active medium length. In order to evaluate the QD-SOA gain, we must use the theoretical model of the QD-SOA, which will be briefly discussed in Section 5.

5. The theoretical model of the QD-SOA

QD-SOA had been thoroughly investigated both theoretically and experimentally (see, for example, [15, 16] and references therein). The promising features of QD-SOA such as high saturation power, broad gain bandwidth, pattern-free amplification of single- and multi-channel signals, efficient WC can provide high-performance amplifiers and all-optical switches for optical networks [16].

QD is a nanostructure where the electron and hole movement is confined in the three dimensions, and these dimensions are of a few nanometers [15]. In QD, the charge carriers occupy only a restricted number of discrete energy levels like the electrons in an atom [15]. The density of states in QD is quantized, and the number of carriers necessary to fill these states decreases as compared to the structures with higher dimensionality. Consequently, the threshold current density in QD lasers substantially reduces, while the transparency and the population inversion necessary for the optical gain can be achieved more easily [15]. The QD grown by using the Stranski-Krastanov technology typically has a pyramidal shape with a base of about 15–20 nm and a height of about 5 nm [15]. The QD has a significant size dispersion, which results in the inhomogeneous broadening of the QD laser and SOA spectrum [15]. QD structures exhibit ultrafast gain recovery time which results in the ultrafast carrier dynamics [15]. The active layer of a QD-SOA contains one or several quantum wells (QW) referred to as a wetting layer (WL) with a continuous carrier energy band [15]. The electrons in the Stranski-Krastanov grown QD are typically characterized by two energy levels: the ground state (GS) situated about 100 meV below the band gap of WL and the excited state (ES), or the first excited level [15]. The energy level structure of a QD laser or QD-SOA is presented in **Figure 7**.

The carrier dynamics in QD is described by the system of the rate equations taking into account the following electron transitions: the fast electron transitions from WL to ES with the relaxation time $\tau_{wE} \sim 3$ ps; the fast electron transitions between ES and GS with the corresponding relaxation times$\tau_{EG} \approx 0.16$ ps; $\tau_{GE} \sim 1.2$ ps; the slow electron transitions from ES to WL with the escape time $\tau_{Ew} \sim 1$ ns. The QD-SOA rate equations have the form [12, 17]:

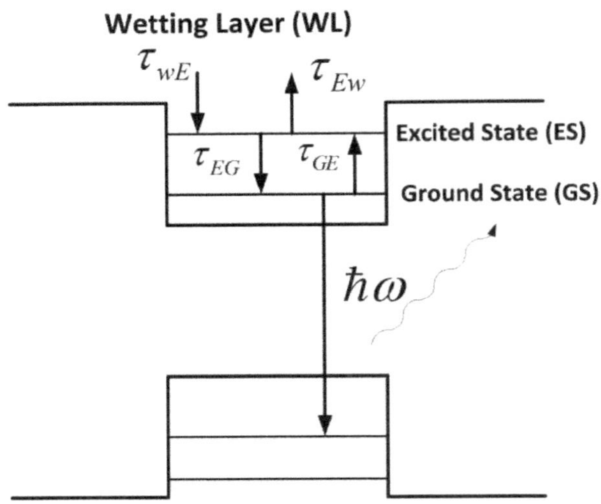

Figure 7. The energy level structure and the electron transitions in a QD.

$$\frac{\partial N_w}{\partial t} = \frac{J}{eL_w} - \frac{N_w(1-h)}{\tau_{wE}} + \frac{N_w}{\tau_{Ew}} - \frac{N_w}{\tau_{wR}} \tag{5}$$

$$\frac{\partial h}{\partial t} = \frac{N_w L_w(1-h)}{N_Q \tau_{wE}} - \frac{N_w L_w h}{N_Q \tau_{Ew}} - \frac{(1-f)h}{\tau_{EG}} + \frac{f(1-h)}{\tau_{GE}} \tag{6}$$

$$\begin{aligned}\frac{\partial f}{\partial t} = {} & \frac{(1-f)h}{\tau_{EG}} - \frac{f(1-h)}{\tau_{GE}} - \frac{f^2}{\tau_R} - \frac{g_p L}{N_Q}(2f-1)S_p \frac{c}{\sqrt{\varepsilon_r}} \\ & - \frac{g_s L}{N_Q}(2f-1)S_s \frac{c}{\sqrt{\varepsilon_r}}\end{aligned} \tag{7}$$

where J is injection current density, N_w is the WL carrier density per unit volume, f is the electron occupation probability of GS, h is the electron occupation probability of ES, $S_{p,s}$ are the pump (data A or data B) and signal (clock stream) wave photon densities averaged over the length of SOA L, $g_{p,s}$ are the pump and signal wave modal gains, respectively, e is the electron charge, $N_Q \sim (10^{10} - 10^{12})$ cm^{-2} is the QD density per unit area, L_w is the effective thickness of the active layer, ε_r is the SOA material permittivity, and c is the free space light velocity. The average photon densities $S_{p,s}$ are given by Ben Ezra [12]:

$$S_{p,s}(\tau) = \frac{[S_{p,s}(\tau)]_{in}}{L} \int_0^L dz \exp\left(\int_0^z (g_{p,s} - \alpha_{\text{int}})dz'\right) \tag{8}$$

where $\tau = t - \left(\frac{z}{v_g}\right)$, t, z, $v_g$are time, coordinate, and the optical wave group velocity, respectively. The pump and signal wave phase $\varnothing_{p,s}$ and modal gain $g_{p,s}$ are given by, respectively [12]:

$$\theta_{p,s}(\tau) = -\frac{\alpha}{2}\int_{0}^{L} g_{p,s}dz \tag{9}$$

$$g_{p,s}(\omega) = \frac{2\Gamma N_Q}{a}\int d\omega F(\omega)\sigma(\omega_0)(2f-1) \tag{10}$$

where Γ is the confinement factor, l is the number of QD layers, a is the mean size of QD, $\sigma(\omega_0)$ is the cross section of interaction of photons of frequency ω_0 with carriers in QD at the transition frequency ω including the homogeneous broadening factor, $F(\omega)$ is the Gaussian distribution of the transition frequency in the QD ensemble related to the inhomogeneous broadening. The inhomogeneous broadening is caused by the QD shape and size variations as it was mentioned above. It is given by Ben Ezra [12]:

$$F(\omega) = \frac{1}{\Delta\omega\sqrt{\pi}}\exp\left[-\frac{(\omega-\overline{\omega})^2}{(\Delta\omega)^2}\right] \tag{11}$$

where $\Delta\omega$ is related to the inhomogeneous bandwidth, $\gamma_{inhom} = 2\Delta\omega\sqrt{ln2}$, and $\overline{\omega}$ is the average transition frequency.

6. The simulation results

We solved numerically Eqs. (3)–(11) for the typical values of parameters [12] for the QD-SOA–based all-optical loop shown in **Figure 6** [6]. The numerical simulations have been carried out for the OOK and pulse amplitude modulation 4 (4-PAM) formats [6]. We used the MATLAB environment. Simulation results for the eye diagrams in the case of the OOK modulation format are shown in **Figures 8** and **9**.

We used the QD-SOA–based all-optical memory model with the loop length of L = 2 km and the input OOK modulated signal with the bit rate of 50 Gb/s and the quality factor at the input Q = 15.8932. The comparison of the eye diagrams presented in **Figures 8** and **9** shows that after 4 rounds in the loop, the quality factor of the signal decreases by approximately 18%.

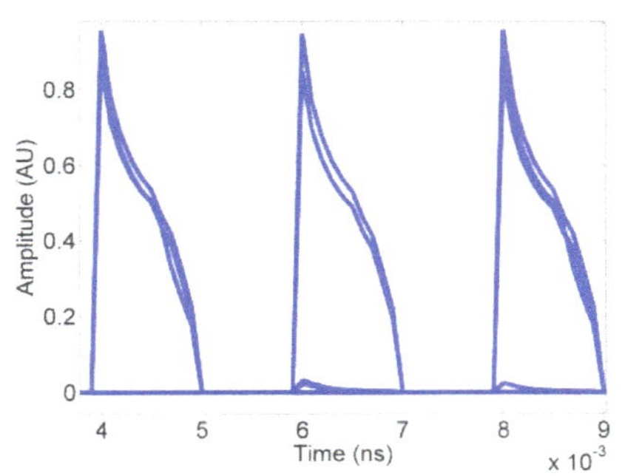

Figure 8. The eye diagram for the OOK modulation format, a bit rate of 50 Gb/s after one round in the QD-SOA–based all-optical memory loop. The memory loop length L = 2 km, T = 10 μs, the quality factor Q = 13.3156.

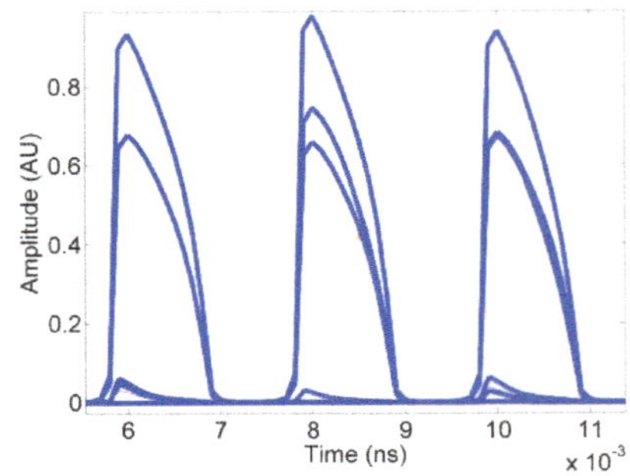

Figure 9. The eye diagram for the OOK modulation format, a bit rate of 50 Gb/s after four rounds in the QD-SOA–based all-optical memory loop. The memory loop length L = 2 km, $T = 40$ μs, the quality factor $Q = 13.0327$.

The simulations results for the 4-PAM modulation format, QD SOA–based all-optical memory loop with the length of L = 1 km and bit rates of 50 and 100 Gb/s are shown in **Figures 10–13**, respectively.

Eye diagrams in **Figures 10** and **11** clearly show that for the bit rate of 50 Gb/s, the patterning effect is negligible after two rounds in the memory loop. In the case of the bit rate of 100 Gb/s, the patterning effect is slightly pronounced after the two rounds in the memory loop as it is

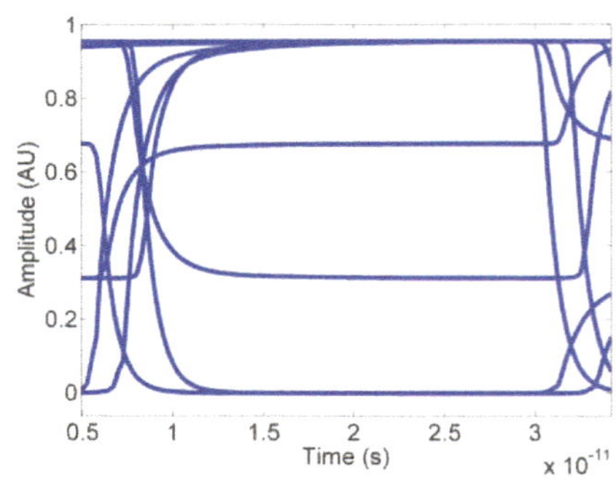

Figure 10. The eye diagram for 4-PAM modulation format and a bit rate of 50 Gb/s, the memory loop length L = 1 km. The signal is at the input of the all-optical memory loop.

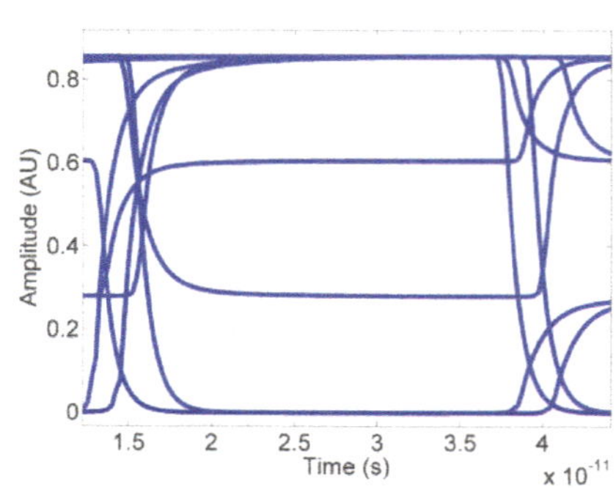

Figure 11. The eye diagram for 4-PAM modulation format and a bit rate of 50 Gb/s, the memory loop length L = 1 km. The signal is after two rounds in the all-optical memory loop.

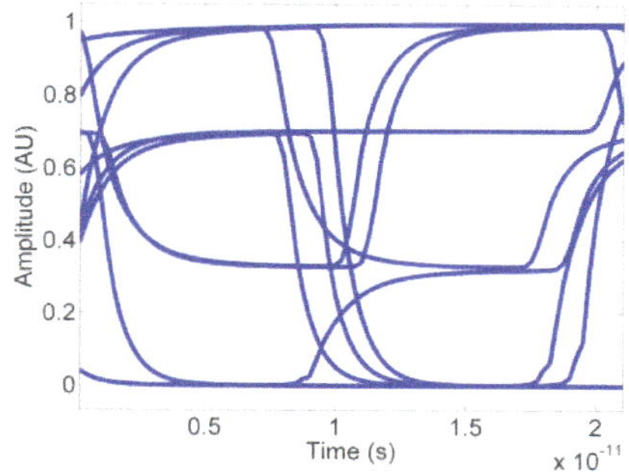

Figure 12. The eye diagram for 4-PAM modulation format and a bit rate of 100 Gb/s, the memory loop length L = 1 km. The signal is at the input of the all-optical memory loop.

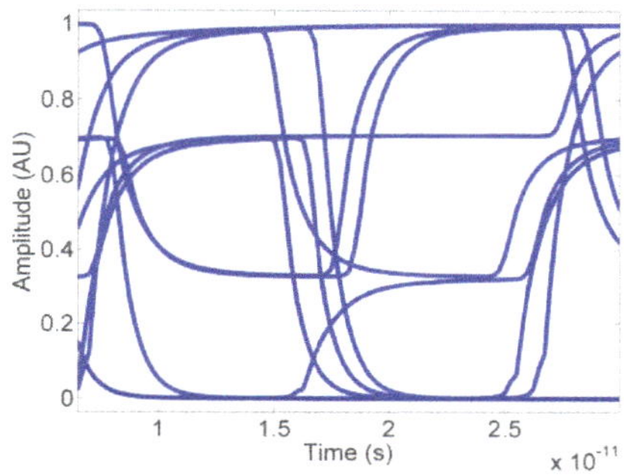

Figure 13. The eye diagram for 4-PAM modulation format and a bit rate of 100 Gb/s, the memory loop length L = 1 km. The signal is after two rounds in the all-optical memory loop.

seen from **Figures 12** and **13**. The QD-SOA–based all-optical memory performance does not deteriorate up to the bit rate of 100 Gb/s.

The numerical estimations show that for loop length $L = 1$ km, the light velocity in the optical fiber $v \approx 2 \times 10^8$ m/s and the bit rates of 50 and 100 Gb/s the all-optical memory storage values are of 0.25 and 0.5 Mb, the storage times are 5 and 10 μs, respectively [6].

7. Conclusions

Optical signal processing substantially increases the device processing speed, provides an alternative to electronic techniques, and, at the same time, can enhance the processing capabilities of electronics when the optical and electronic signal processing techniques are combined. Optical packet-switched networks are promising candidates for the advanced optical telecommunication systems. All-optical signal processing is essential for ultrafast OPS in such networks. Flip-flop memory is a basic component of an all-optical packet switch. We discussed the all-optical memory based on SOA-MZI. It is characterized by the high performance due to the SOA high nonlinearity, low switching power, wide gain bandwidth, compact size, and integration capability with other photonic devices. Unfortunately, the SOA operation rate is limited by its comparatively slow gain recovery time.

We proposed a novel architecture of the ultrafast all-optical memory based on MZI with two QD-SOA. QD-SOA is characterized by high operation rate and low threshold current caused by the 3D carrier confinement and fast electron transitions in QD. For this reason, the operation rate of the proposed all-optical memory loop increases up to 100 Gb/s. The time delay in the typical push-pull scheme of the memory loop is replaced with the output signal phase difference caused by the signal power difference in the MZI arms due to the highly efficient XPM in QD-SOA. The DSP block is inserted for the dispersion and nonlinearity impairment mitigation. The proposed all-optical memory loop includes only one MZI with two QD-SOA and reduces the complexity of electronic synchronization scheme. We carried out numerical simulations of the proposed all-optical memory loop based on the QD-SOA rate equations and the expression for the MZI output power for the OOK and 4-PAMmodulation formats. The simulation results show that the proposed all-optical memory exhibits a high performance up to the bit rate of 100 Gb/s and the corresponding memory storage and storage time values of 0.5 Mb and 10 μs, respectively.

Author details

Yossef Ben Ezra[1,2] and Boris I. Lembrikov[1*]

*Address all correspondence to: borisle@hit.ac.il

1 Department of Electrical Engineering and Electronics, Holon Institute of Technology (HIT), Holon, Israel

2 MER Cellos, Holon, Israel

References

[1] Agrell E, Karlsson M, Chraplyvy AR, et al. Roadmap of optical communications. Journal of Optics. 2016;**18**:1–40. DOI: 10.1088/2040-8978/18/6/063002

[2] Willner AE, Khaleghi S,Chitgarha MR, Yilmaz OF. All-optical signal processing. Journal of Lightwave Technology. 2014;**32**:660–680. DOI: 10.1109/JLT.2013.2287219

[3] Agrawal GP. Fiber-Optic Communication Systems. 4th ed. New York: Wiley; 2010. ISBN: 978-0-470-50511-3

[4] Agrawal GP. Nonlinear Fiber Optics. 5th ed. New York: Academic Press; 2013. ISBN: 978-0-12397-023-7

[5] Liu Y, Hill MT, Calabretta N, Tangdiongga E, Geldenhuys R, Zhang S, Li Z, De Waardt H, Khoe GD, Dorren HJS. All-optical signal processing for optical packet switching networks. In: Proceedings of SPIE; September 15, 2005; San Diego, California, USA; 2005. 59070J-1-12. DOI: 10.1117/12.621346

[6] Ben Ezra Y, Lembrikov BI. All-optical memory based on quantum dot semiconductor optical amplifiers (QD-SOAs) for advanced modulation formats. In: Proceedings of 18th International Conference on Transparent Optical Networks (ICTON 2016); July 10-14, 2016; Trento, Italy; Tu. A5.3 1-3. ISBN: 978-1-5090-1466-8

[7] Dorren HJS, Hill MT, Liu Y, Calabretta N, Srivatsa A, Huijskens FM, De Waardt H, Khoe GD. Optical packet switching and buffering by using all-optical signal processing methods. Journal of Lightwave Technology. 2003;**21**:2–12. DOI: 10.1109/JLT.2002.803062

[8] Xuelin Y, Qiwei W, Weisheng H. High-speed all-optical long-term memory using SOA MZIs: Simulation and experiment. Optics Communications. 2012;**285**:4043-4047. DOI: 10.1016/j.optcom.2012.06.027

[9] Singh P, Tripathi D Kr, Jaiswal S, Dixit HK. All-optical logic gates: Designs, classification and comparison. Advances in Optical Technologies. 2014;**2014**:1–13. DOI: 10.1155/2014/275083

[10] Shehata MI, Mohammed NA. Design and optimization of novel two inputs logic gates (NOT, AND, OR and NOR) based on single commercial TW-SOA operating at 40 Gbit/s. Optical and Quantum Electronics. 2016;**48**:336(1–16). DOI: 10.1007/s1 1082-016-0602-2

[11] Stubkjaer KE. Semiconductor optical amplifier-based all-optical gates for high-speed optical processing. IEEE Journal of Selected Topics in Quantum Electronics. 2000;**6**:1428–1435. DOI: 10.1109/2944.902198

[12] Ben Ezra Y, Lembrikov BI, Haridim M. Ultra-fast all-optical processor based on quantum dot semiconductor optical amplifiers (QD-SOA). IEEE Journal of Quantum Electronics. 2009;**45**:34-41. DOI: 10.1109/JQE.2008.2003497

[13] Sun H, Wang Q, Dong H, Dutta NK. XOR performance of a quantum dot semiconductor optical amplifier based Mach-Zender interferometer. Optics Express. 2005;**13**:1892-1899. ISSN: 1094-4087

[14] Wang Q, Zhu G, Chen H, Jaques J, Leuthold J, Picirilli AB, Dutta NK. Study of all-optical XOR using Mach-Zender interferometer and differential scheme. IEEE Journal of Quantum Electronics. 2004;**40**:703-710. DOI: 10.1109/JQE.2004.828261

[15] Rafailov EU, Cataluna MA, Avrutin EA. Ultrafast Lasers Based on Quantum Dot Structures. Weinheim: Wiley-VCH; 2011. p. 250. ISBN: 978-3-527-40928-0

[16] Sugawara M, Ebe H, Hatori N, et al. Theory of optical signal amplification and processing by quantum-dot semiconductor optical amplifiers. Physical Review B. 2004;**69**:235332-1-2 35332-39. DOI: 10.1103/PhysRevB.69.235332

[17] Qasaimeh O. Optical gain and saturation characteristics of quantum-dot semiconductor optical amplifiers. IEEE Journal of Quantum Electronics. 2003;**39**:793-798. DOI: 10.1109/JQE.2003.810770

2

Improvement of the Quantum Dot-in-a-Well (QDWELL) Laser and Amplifier Performance under the Optical Injection

Yossef Ben Ezra and Boris I. Lembrikov

Additional information is available at the end of the chapter

Abstract

Quantum dot (QD) laser devices can be successfully used in optical communications due to their unique properties caused by the carrier localization in three dimensions. In particular, quantum dot-in-a-well (QDWELL) lasers are characterized by an extremely low threshold current density and the high modulation frequency. However, their operation rate is limited by the strongly nonlinear electron and hole scattering rates in and out of QD. We investigated theoretically the nonlinear optical phenomena in QDWELL lasers and amplifiers under the optical injection. We have shown that the synchronization of the carrier dynamics in QD and quantum well (QW) caused by the optical injection improves the QDWELL laser performance and, in particular, enhances the relaxation oscillation (RO) frequency. As a result, the QDWELL laser performance in the analogous optical link (AOL) is significantly improving. The optical injection also improves the performance of the QDWELL-based semiconductor optical amplifiers (SOA).

Keywords: quantum dot (QD), quantum well (QW), laser, semiconductor optical amplifier (SOA), optical communications

1. Introduction

Advanced high-capacity communication systems are necessary for different applications, such as medical diagnosis, traffic safety, Internet, data services, etc. [1]. These new applications generate a giant data traffic which requires the time sensitive analysis and data processing at high-performance computing infrastructures (HPC) and the data storage, transport, and exchange in datacenters (DC) [1]. Recently, all-optical architecture of DC has been proposed.

It is based on the switching of all data in the optical domain [1]. This approach requires the development of new photonic devices [1]. One of the most efficient technologies for the realization of such devices is Silicon Photonics. Silicon Photonics can resolve the so-called bandwidth bottleneck by the integration of photonic integrated circuits (PIC) and electronic integrated circuits [1]. The active components such as lasers are made of the III-V compositions. There are two techniques of the III-V laser integration with PIC: (i) the III-V laser die can be butt-coupled to the silicon photonic chip using active alignment and (ii) III-V materials are wafer bonded to the silicon photonic chip in order to fabricate lasers lithographically aligned to the silicon waveguide circuit [1].

A generic optical communication system consists of an optical transmitter, optical communication channel, and an optical receiver [2]. A block diagram of such a system is shown in **Figure 1** [2].

The information highways providing these services are based on optical fibers [1]. Typically, silica optical fibers are used as the communication channel due to their low losses of about 0.2 dB/km [2]. In such a case, the losses are 20 dB after the propagation distance of 100 km, and the optical power decrease by 100 times defines the amplifier spacing in the long-haul lightwave systems [2, 3]. In long-haul networks extended over thousands of kilometers fiber losses are compensated by using a chain of amplifiers boosting the signal power periodically to its original value [1, 2]. Erbium-doped fiber amplifiers (EDFA) are widely used in optical communication systems due to their compatibility with transmission fibers, energy efficiency, and low cost [1, 2]. Raman amplifier operation is based on the stimulated Raman scattering (SRS) in silica fibers [2]. The advantages of the Raman amplifiers are the using of the fiber itself as an active medium and the large bandwidth [2]. However, the Raman amplifiers require a comparatively large pumping power [2].

The optical communication system performance is limited by the fiber dispersion leading to the optical pulse broadening with propagation through the channel [2, 3]. As a result, the original signal recovery with the high enough accuracy may be impossible [2]. The dispersion influence is strongly manifested in multimode fibers (MMFs) where different fiber modes propagate with different velocities. For this reason, optical communication systems are mainly based on single-mode fibers (SMFs) [2]. In such a case, the intermodal dispersion vanishes because the pulse energy is carried by a single mode [2]. The main types of the SMF dispersion are as follows: (i) group velocity dispersion (GVD) is caused by group velocity v_g frequency dependence (chromatic dispersion); (ii) material dispersion is caused by the silica refractive index n dependence on the optical frequency; (iii) waveguide dispersion is defined by the following parameters: the fiber

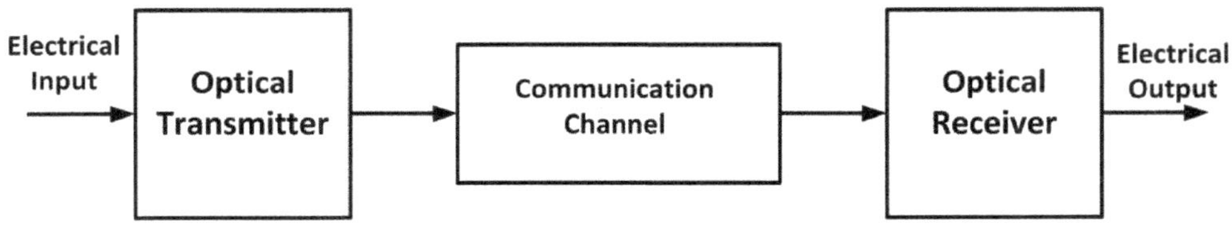

Figure 1. A block diagram of a generic optical communication system.

core radius r_0, the fiber core refractive index n_1 and the cladding refractive index n_2; and (iv) polarization-mode dispersion (PMD) is caused by a modal birefringence [2, 3]. SMF material dispersion is small enough, and it actually limits the optical communication system bit rate and transmission distance under the condition that other types of dispersion are mitigated [2].

Optical transmitter converts the electrical signal into the optical signal and launches it into the optical fiber [2]. An optical transmitter consists of an optical source, a modulator, and a channel coupler [2]. An optical source is usually a semiconductor laser or light-emitting diode (LED) compatible with the optical communication channel [2]. In LED, population inversion is not realized, and the light is generated through spontaneous emission caused by the radiative recombination of electron-hole pairs in the active layer [2]. In semiconductor lasers, the stimulated emission of light is the dominating operation mechanism [2, 4]. For this reason, the LED radiation is rather weak as compared to the semiconductor laser light. The typical values of the launched power are less than 100 μW (−10 dBm) for LED and about 1 mW (10 dBm) for semiconductor lasers [2]. The optical signal is generated by the direct or external modulation of the optical carrier wave radiated by a semiconductor laser [2]. Generally, time, quadrature, polarization, and frequency are widely used in optical networking technologies for complex quadrature modulation formats, polarization multiplexing, digital pulse shaping, and coherent detection [1]. Recently, space as a physical dimension for modulation and multiplexing in communication systems attracted interest for fiber capacity scaling [1]. Space-division multiplexing (SDM) uses multiplicity of space channels, or spatial parallelism, in order to increase capacity of the optical communication system [1]. For instance, fiber bundles 10×10 Gb/s or 4×25 Gb/s can be used for the implementation of 100 Gb/s commercial client interfaces [1].

In the case of the direct modulation, the semiconductor laser is biased near the threshold and driven by the electrical sinusoidal signal for analog modulation or electrical signal bit stream for digital modulation [2]. In the case of the external modulation, two types of the external modulators are mainly used: the Mach-Zehnder modulator (MZM) and the electro absorption modulator (EAM) [2, 4]. The channel coupler is a microlens focusing the optical signal onto the optical fiber entrance plane [2]. The bit rate of optical transmitters is limited by electronic components [2].

An optical receiver converts the optical signal into the electrical domain and recovers the transmitted data [2, 4]. The structure of the optical receiver depends on the modulation format. Consider first the on-off keying (OOK) modulation, where an electrical binary stream modulates the optical carrier intensity inside an optical transmitter [2, 4]. In such a case, the directly modulated optical signal after the propagation through the optical fiber is converted in the receiver directly into the original digital signal in the electrical domain [2]. The main component of the receiver is a semiconductor photodetector (PD) converting light into electrical signal through the photoelectric effect [2]. Such a communication system is intensity modulation with direct detection (IM/DD) system [2]. PD should possess high sensitivity, fast response, low noise, low cost, and high reliability [2]. Generally, the transmission quality is characterized by the received signal-to-noise ratio (SNR) given by $SNR = 20log_{10}(A_{max}/A_N)$ dB, where A_{max} and A_N are the maximum amplitude of the given analog signal and the root mean square (RMS) noise amplitude of the analog signal, respectively [2]. The total noise is a stochastic process consisting of a number of components such as mode partition noise (MSN), laser intensity noise, modal noise, shot noise, thermal noise, dark current noise, amplified spontaneous emission (ASE), and crosstalk noise [4]. The shot noise and thermal noise are the two fundamental noise mechanisms responsible for the current fluctuations in optical receivers [2]. The detailed analysis of the different types of noise may be found in Refs. [2–4]. Here, we should note that the thermal noise limits PD performance in practice, whereas the shot noise determines a fundamental limit of PD operation [2]. For instance, in the shot-noise limit, SNR $= 20$ dB can be realized for PD quantum efficiency close to unity and a number of photons $N_p = 100$ in the input optical signal [2].

Unlike the IM/DD system, the performance of the system with a coherent detection technique is limited by the shot-noise alone [2, 4, 5]. Another important advantage of the coherent detection is the possibility of the detection of signals with advanced modulated formats such as frequency-shift keying (FSK), binary phase-shift keying (PSK), quadrature PSK, and 16-quadrature amplitude modulation (QAM) allowing bit rates of 50, 100, and 200 Gb/s [1, 2, 4]. In the case of the coherent detection, the input optical signal with the optical carrier frequency ω_0 is mixed with the continuous wave (CW) optical signal generated at the receiver by a narrow linewidth laser called the local oscillator (LO) with the frequency ω_{LO} [2]. The output contains the signal with the intermediate frequency $\omega_{IF} = \omega_0 - \omega_{LO} \ll \omega_0$ which falls on PD [2]. There are two types of the coherent detection technique: (i) a homodyne detection when $\omega_0 = \omega_{LO}\,;\, \omega_{IF} = 0$; (ii) a heterodyne detection when $\omega_0 \neq \omega_{LO}$ [2, 4]. In the latter case, the detection of the optical signal is carried out in two stages: first, the carrier frequency ω_0 is converted to the intermediate frequency ω_{IF} usually belonging to the radio frequency (RF) range; then, the RF signal is down converted to the modulating signal bit stream [2, 4].

Optical signal processing in optical communication systems is based on the linear and nonlinear optical techniques used for the manipulation and processing of digital, analogue, and quantum information [1]. Ultrafast optical nonlinearities provide an operation rate advantage as compared to the electronic techniques for switching, regeneration, wavelength conversion, performance monitoring, and analog digital conversion (ADC) [1]. In particular, semiconductor optical amplifiers (SOAs) can be used in such applications due to their high operation rate and strong nonlinearity [1–3].

The brief review of the optical communication systems clearly shows that a semiconductor laser is a key component of both the transmitter and receiver. Evidently, for the highly efficient applications in optical communication systems, semiconductor laser performance should be characterized by low-threshold current, high-speed direct modulation, ultrashort optical pulse generation, narrow spectral linewidth, broad modulation bandwidth, comparatively high optical output power, low relative intensity noise (RIN), low cost, and low electrical power consumption [2, 6]. The objective of this chapter is the investigation of the optical injection influence on the performance of the novel semiconductor laser based on quantum dots (QD) in a quantum well (QW) structure. The number of publications concerning semiconductor lasers in general and QD lasers in particular is enormous and hardly observable. For this reason, in Section 2, we briefly discuss the structure, operation principle, and basic characteristics of a semiconductor laser widely used in optical communication systems. In Section 3, the fundamentals of a QD in QW (QDWELL) laser and SOA are discussed. The original theoretical results related to the dynamics and performance of the optically injected QDWELL lasers and SOA are presented in Section 4. The conclusions are presented in Section 5.

2. Structure and operation principle of a semiconductor laser

Semiconductor laser is an electrically pumped p-i-n diode [6]. In the forward-biased diode, electrons in the conduction band and holes in the valence band are injected into the active layer placed between the p-type and n-type cladding layers [2, 6]. The structure of an edge-emitting semiconductor laser is shown schematically in **Figure 2** [2].

The cladding layers are made of the semiconductor material with the bandgap E_{g2} larger than the active layer bandgap E_{g1} [2, 6]. Such a structure consisting of the p-n junctions made of semiconductor materials with different bandgaps is called a heterostructure [2, 7]. As a result, the electrons and holes are confined in the active layer.

The $Al_xGa_{1-x}As/GaAs/Al_xGa_{1-x}As$ heterostructure is shown in **Figure 3** [6]. The concentration of donors N_D and acceptors N_A in the n-type and p-type regions, respectively, should satisfy the condition that the separation $E_{FC} - E_{FV}$ between the electron Fermi energy level E_{FC} and the hole Fermi energy level E_{FV} is larger than the bandgap energy E_g [2]. Recombination of the electron-hole pair in the presence of the optical field results in the photon stimulated emission [2, 6].

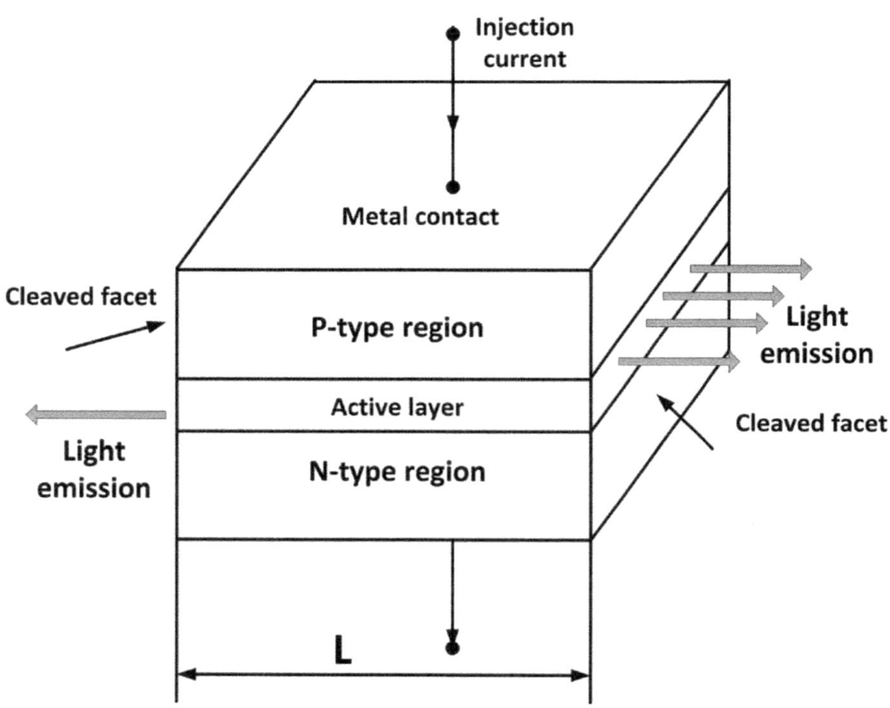

Figure 2. Edge-emitting semiconductor laser.

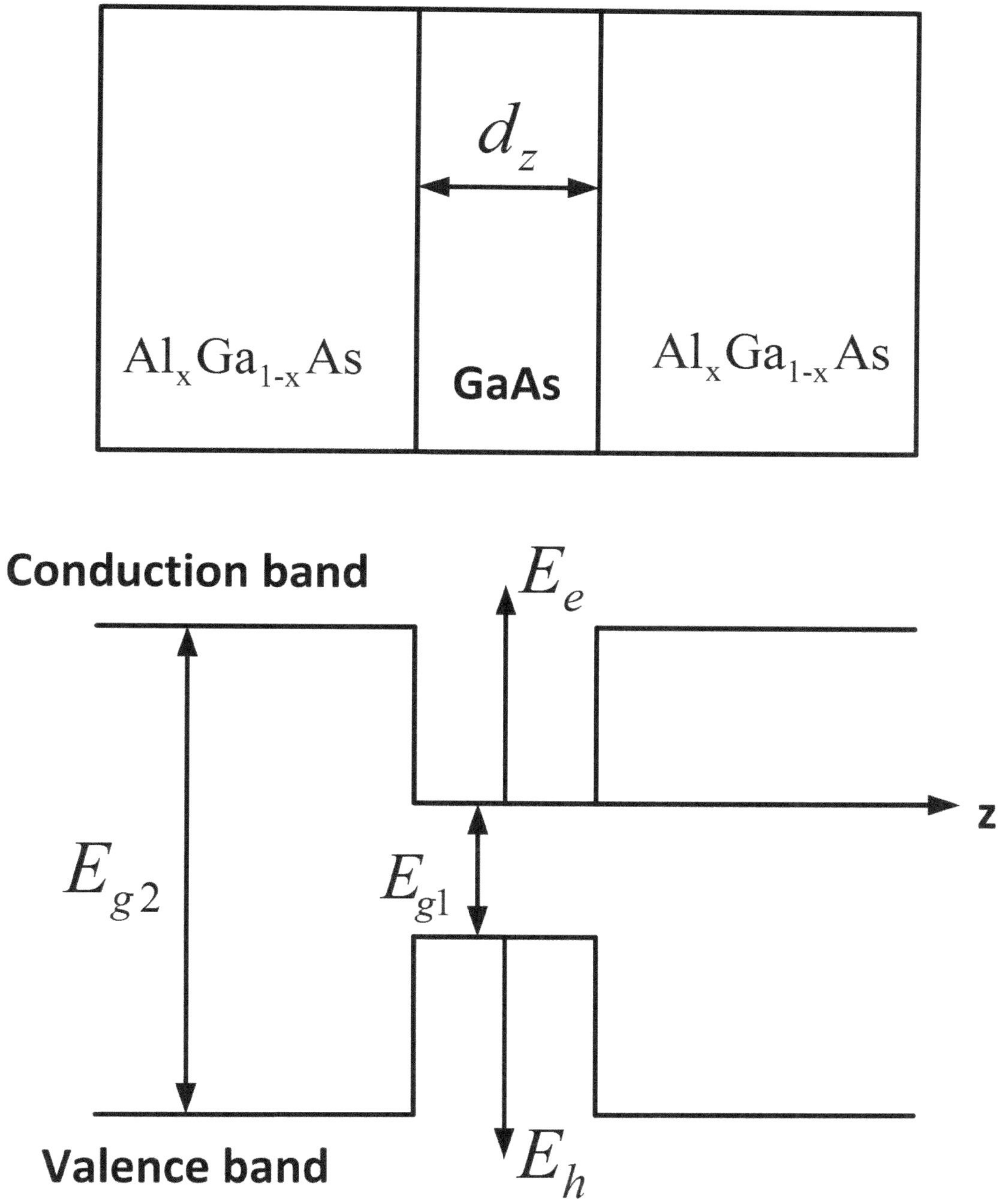

Figure 3. The energy bands of the semiconductor laser based on the $Al_xGa_{1-x}As/GaAs/Al_xGa_{1-x}As$ heterostructure.

Usually, the active layer has a larger refractive index than the cladding layers providing also the confinement of the photons generated in the active layer [2]. When the injected carrier density in the active layer becomes larger than a critical value, the population inversion occurs and the active region manifests an optical gain [2]. Input signal intensity P_{in} in such a case is amplified, and the output signal intensity P_{out} has the form [2]

$$P_{out}(L) = P_{in}(0)\exp(gL) \tag{1}$$

where g is the power gain coefficient and L is the active layer length. Usually, the laser dimensions are less than 1 mm in all three directions which result in compact design [2]. The gain is negative $g < 0$ for the injected carrier density N below the population inversion value. The gain becomes positive and increases with the increase of the injected carrier density N. The peak gain g_p is given by the approximate empirical relationship [2].

$$g_p(N) = \sigma_g(N - N_T) \tag{2}$$

where σ_g is the gain cross section or the differential gain and N_T is the transparency value of the carrier density. For InGaAsP lasers, the typical values of these parameters are following: $\sigma_g \approx (2-3) \times 10^{-16}$ cm^{-2}, $N_T \approx (1-1.5) \times 10^{18}$ cm^{-3} [2]. Eq. (2) can be used in the high-gain region where g_p >100 cm^{-1} [2]. The laser operation also requires an optical feedback provided by placing the gain medium into a Fabry-Perot (FP) cavity formed in our case by the two cleaved laser facets serving as mirrors [2]. In such a case, a critical value of the gain g_{cr} is given by [2]

$$g_{cr} = \alpha_{\text{int}} + \frac{1}{2L}\ln\left(\frac{1}{R_1R_2}\right) \tag{3}$$

where α_{int} is the coefficient of the internal losses caused by free carrier absorption and interface scattering and $R_{1,2}$ are the facet reflectivities. In the laser FP cavity, only the longitudinal modes can exist determined by the condition [2]

$$2\beta L = m\pi;\, \beta = \frac{2\pi\nu n_{\text{mode}}}{c};\, \nu = \nu_m = \frac{mc}{2n_{\text{mode}}L};\, m = 1, 2, \ldots \tag{4}$$

where β, n_{mode}, and c are the mode propagation constant, the mode refractive index, and the free space light velocity, respectively. Eq. (4) shows that the laser frequency ν corresponds to the discrete spectrum of the laser longitudinal modes. A semiconductor laser radiation typically consists of several longitudinal modes with a gain spectrum bandwidth of about 10 THz [2]. The dominant mode is the one with gain closest to the peak gain g_p [2]. The optical modes penetrate into the cladding layers due to the boundary conditions for the optical mode electric and magnetic field while the gain takes place only in the active layer. Hence, the confinement factor of the active layer $\Gamma = g/g_m$ is introduced which is typically less than 0.4 [2]. Here, g_m is the so-called material gain. The modes propagate inside the optical fiber with different speeds because of GVD as it was mentioned above. Consequently, the multimode semiconductor lasers are limiting the optical communication operation rate [2].

There exist single-mode semiconductor lasers emitting light mainly in a single longitudinal mode such as distributed feedback (DFB) lasers, coupled-cavity lasers, tunable lasers, and vertical-cavity surface-emitting lasers (VCSEL) [2]. The detailed analysis of structure and operation principles of these lasers can be found in Ref. [2] and references therein. Here, we only briefly discuss the peculiarities of these lasers following Ref. [2]. DFB laser has a built-in periodic grating with the period d_0 which results in refraction index periodicity. Consequently, the waves propagating in the forward and backward directions in the cavity are coupling, and the Bragg diffraction occurs. The feedback in such a case is distributed throughout the laser cavity length. The coupling occurs only for the modes with the Bragg wavelength λ_B satisfying the Bragg condition $d_0 = (\lambda_B/2n_{\text{mode}})m$, $m = 1, 2, 3, \ldots$ [2]. This condition provides the mode selectivity of the DFB lasers. The strongest coupling between the forward and backward propagating modes occurs for $m = 1$ [2].

In a coupled-cavity semiconductor laser, single-mode operation is realized by coupling the laser cavity to an external cavity. The in-phase feedback occurs only for the laser modes with a wavelength which almost coincides with a certain longitudinal mode of the external cavity [2]. The effective reflectivity of the laser facet close to the external cavity depends on the wavelength, and the losses of certain modes sharply decrease. The longitudinal mode which simultaneously has the highest gain and the lowest cavity loss becomes the dominant mode [2].

A tunable semiconductor laser typically consists of three sections: the active section, the phase-control section, and the Bragg section [2]. Each one of these sections is biased independently by injecting different currents in a following way: the current injected into the Bragg section changes the Bragg wavelength λ_B through carrier induced variations in the refractive index n; the current injected into the phase-control section changes the feedback phase through carrier induced variations in the refractive index of this section [2]. As a result, the laser wavelength can be tuned over the range of 10–15 nm by controlling the injection current in the phase-control and the Bragg sections [2].

VCSEL emits light in a direction perpendicular to the active layer plane and operates in a single longitudinal mode regime due to the small cavity length of about 1 μm in a wide range of wavelengths of about 650–1600 nm [2]. The emitted light has a form of a circular beam which can be inserted into SMF with high efficiency [2].

The dynamics of the single-mode semiconductor laser is described by the following rate equations for the number of electrons N and photons P in the active layer [2].

$$\frac{dP}{dt} = GP + R_{sp} - \frac{P}{\tau_p} \tag{5}$$

$$\frac{dN}{dt} = \frac{I}{q} - \frac{N}{\tau_c} + GP \tag{6}$$

where I is the injection current, q is the electron charge, $G = (\Gamma v_g \sigma_g / V)(N - N_0)$ is net rate of the stimulated emission, V is the active layer volume, $N_0 = N_T V$, v_g is the light group velocity, R_{sp} is the spontaneous emission rate, and τ_p are τ_c are the photon and carrier lifetime, respectively. The threshold current I_{th} for the laser generation regime obtained from the steady-state solution of Eqs. (5) and (6) has the form [2].

$$I_{th} = \frac{q}{\tau_c}\left(N_0 + \frac{V}{G_N \tau_p}\right); \qquad G_N = \frac{\Gamma v_g \sigma_g}{V} \tag{7}$$

The direct modulation of the laser emission can be realized if the injection current includes a time-dependent component. In the case of a sinusoidal modulation, the modulation frequency ω_m is limited by the relaxation oscillation (RO) frequency Ω_R given by [2]

$$\Omega_R = \left[G G_N P_b - \frac{(\Gamma_P - \Gamma_N)^2}{4} \right]^{1/2} \tag{8}$$

where P_b is the steady-state photon number, $\Gamma_P = (R_{sp}/P_b) + \varepsilon_{NL} G P_b; \Gamma_N = \tau_c^{-1} + G_N P_b$,

ε_{NL} is a nonlinear gain parameter, and $\Gamma_R = (\Gamma_N + \Gamma_P)/2$ is a damping rate of the relaxation oscillations. The efficiency of the laser modulation is significantly reduced for $\omega_m > \Omega_R$ [2]. The 3 dB modulation bandwidth f_{3dB} of the semiconductor laser is also determined by the RO frequency Ω_R. For $\Omega_R \gg \Gamma_R$, it has the form $f_{3dB} \approx \left(\sqrt{3}\Omega_R\right)/2\pi$ [2].

Consider now QW semiconductor lasers based on a heterostructure shown in **Figure 3**. If the thickness of the active layer d_z is less than approximately 10 nm, the confined electrons and holes exhibit quantum effects, and such a heterostructure is called a single QW [6, 7]. In such a case, the De Broglie electron wavelength $\lambda = h/p$ in a semiconductor is close to the active layer thickness [7]. Here, h, p are the Planck constant and electron momentum, respectively. The charge carriers in a QW laser are localized in the z direction perpendicular to the active layer plane and represent a two-dimensional (2D) electron and hole gas [6]. The solution of the Schrödinger equation for carriers in QW produces discrete states of energy E_n in the z direction perpendicular to the QW plane and a continuous range of allowed energies in the QW plane [7]. The electron energy E_e in a QW is given by [6, 7]

$$E_e = E_n + \frac{\hbar^2\left(k_x^2 + k_y^2\right)}{2m_e^*} \tag{9}$$

where $\hbar = h/2\pi$, m_e^* is the electron effective mass, and $k_{x,y}$ are the electron continuous momentum components in the QW plane [6, 7]. Similar expression can be written for the hole energy in QW. The threshold current in a 2D structure reduces significantly as compared to the bulk semiconductor laser [6]. The electronic and optical properties of a semiconductor laser strongly depend on the carrier density of states (DOS) $\rho_{e,h}(E_{e,h})$.The electron DOS in a QW with n_e energy levels has the form [6, 7]

$$\rho_e(E_e) = \sum_{n_e} \frac{m_e^*}{\pi\hbar^2} u(E_e - E_n); \, u(x) = \begin{cases} 1, x > 0 \\ 0, x < 0 \end{cases} \tag{10}$$

Similar expression can be written for the hole DOS. It is seen from Eq. (10) that carrier DOS in QW does not depend on the energy. The static and dynamic properties of the QW semiconductor laser can be described by the rate equations for the carrier density in QW N_W and in a separate confinement heterostructure (SCH) layer N_B, and the photon density in the cavity P given by [8]

$$\frac{dN_B}{dt} = \frac{I}{qV_{SCH}} - \frac{N_B}{\tau_r} - \frac{N_B}{\tau_{nb}} + \frac{N_W V_W}{\tau_e V_{SCH}} \tag{11}$$

$$\frac{dN_W}{dt} = \frac{N_B V_{SCH}}{\tau_r V_W} - \frac{N_W}{\tau_n} - \frac{N_W}{\tau_{nr}} - \frac{N_W}{\tau_e} - \frac{v_g G(N_W) P}{1 + \varepsilon P} \tag{12}$$

$$\frac{dP}{dt} = \frac{\Gamma v_g G(N_W) P}{1 + \varepsilon P} - \frac{P}{\tau_p} + \Gamma \beta_{Sp} \frac{N_W}{\tau_n} \tag{13}$$

where τ_n, τ_{nr}, τ_{nb}, τ_r, and τ_e are the radiative recombination lifetime in QW, the nonradiative recombination lifetime in QW, the total recombination lifetime in the SCH region, the carrier transport time across the SCH to QW active region, and the carrier thermionic emission time from QW into the barrier and SCH layers, respectively; β_{Sp} is the spontaneous emission feedback factor, $G(N_W)$ is the carrier density dependent gain, ε is the intrinsic gain compression factor, and V_W and V_{SCH} are the QW and SCH volumes, respectively. The rate equations for QW laser (11)–(13) are different from the bulk laser Eqs. (5) and (6). The additional Eq. (11) includes the effects of the carrier transport in SCH region typically used in QW lasers for the carrier and optical confinement [6, 8].

3. Structure and theoretical model of a quantum dot-in-a-well (QDWELL) laser and SOA

QD is a nanostructure where the electron and hole motion is confined in three dimensions reducing the degrees of freedom to zero [7, 9]. The III-V QDs are grown epitaxially on a semiconductor substrate [9]. The spontaneous formation of 3D islands during strained layer epitaxial growth is known as the Stranski-Krastanov process [9]. A continuous film of a QW thickness beneath the QD is called the wetting layer (WL) [9, 10]. In QD, carriers occupy a finite number of energy levels defined by the solution of the Schrödinger equation for a 3D potential separating the inside of QD from the outside [7]. Typically, the spherical and pyramidal QD models are used [7]. The pyramidal model is more realistic since the QD grown by means of the Stranski-Krastanov technology have approximately pyramidal form [9]. In such a case, the Schrödinger equation for QD can be solved numerically [7, 9]. Carrier DOS in QD is a $\delta-$ function for each energy level [7]. QD has a lateral size of approximately 15–18 nm, a height of about 4–5 nm, and contains several hundreds of thousands of atoms [9, 10]. The QD density per unit area in each layer is about $10^{10} - 10^{11}$ cm^{-2} [9, 10]. QD lasers exhibit extremely low threshold current due to the 3D carrier confinement [9, 10]. The spectral bandwidth of QD lasers is broader than the one of the QW lasers due to the inhomogeneous spectral broadening caused by QD size variations that exhibit a Gaussian size distribution and consequently a nearly Gaussian distribution of the emitted light frequencies [9]. Semiconductor lasers based on self-organized InGaN, InAs, InGaAlP, and In(Ga)As QD are promising candidates for applications in optical communication systems because of their low threshold current, high temperature stability, fast carrier dynamics, and the possibility of the light emission from ultraviolet to the far infrared spectral regime [9–13]. Typically, the active area of a QDWELL laser consists of several QW layers each one of a height of 4 nm with the embedded 3D confined QD having an average size of approximately 4 nm × 18 nm × 18 nm [10]. An electric current is injected into the QW layers which form the carrier reservoir (WL) where the carrier-carrier scattering leads to a filling or depletion of the confined QD levels [10]. The energy band structure of a QDWELL laser is shown in **Figure 4**. Inside the QD, the lower level or the ground state (GS) and several higher levels or excited states (ES) for electrons and holes may exist [11]. We discuss the QDWELL laser model with the GS and one ES where the emission occurs form GS [10]. In QDWELL lasers, the Coulomb interaction is the essential mechanism of the carrier-carrier scattering for high carrier densities in WL [10].

The turn-on dynamics, the small signal and large signal responses of a QDWELL laser have been studied theoretically and experimentally in the fundamental works of Lüdge, Schöll and co-workers [10, 13–18].

It has been shown that the complicated carrier dynamics in a QDWELL laser is determined by the nonradiative carrier-carrier scattering processes between the QD and the QW [17]. The

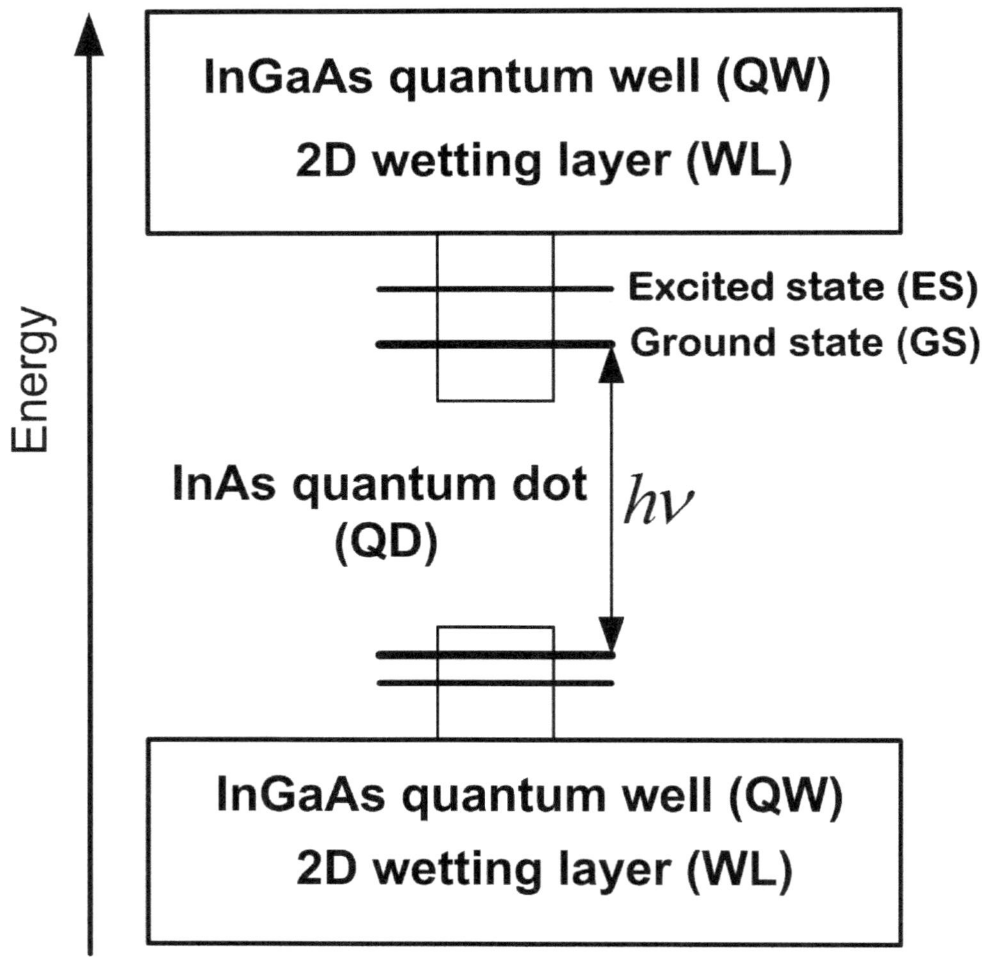

Figure 4. The energy band structure of a QDWELL laser.

corresponding scattering rates S_e^{in}, S_e^{out}, S_h^{in}, S_h^{out} of electrons (e) and holes (h) for scattering in and out of the QD are strongly nonlinear and depend on the electron and hole occupation probabilities in the confined QD levels ρ_e, ρ_h and the electron and hole densities in the QW w_e, w_h [13–18]. The electron and hole scattering times are given by, respectively, $\tau_e = \left(S_e^{in} + S_e^{out}\right)^{-1}$ and $\tau_h = \left(S_h^{in} + S_h^{out}\right)^{-1}$ [14, 16, 18]. The carrier relaxation processes in QD states are much faster than the capture processes from the QW into the QD at high carrier densities in the QW WL, and the influence of the higher QD levels can be adiabatically eliminated [17].

The nonlinearity of the scattering rates S_e^{in}, S_e^{out}, S_h^{in}, S_h^{out} and the separate dynamics of holes and electrons are essential for the QDWELL laser dynamics adequate description [16]. The electron and hole dynamics in the QDWELL laser is desynchronized due to different effective masses and different energy separation between 2D WL band edge and confined QD level [16]. The QD hole density is changing more slowly as compared to the electron density [16]. Desynchronized carrier dynamics results in the deterioration of the laser performance both in the small-signal and in the large-signal regimes [15, 17]. In particular, the QDWELL laser modulation frequency is limited by the RO frequency of 7 GHz [17]. The relatively slow carrier scattering processes from 2D WL carrier reservoir into confined QD levels are responsible for the high damping of the turn-on process and the flat modulation response curve of QDWELL lasers [10]. Usually, the modulation frequency can be increased by increasing the injection current, but in our case, the injection current increase results in the device heating and the further reduction of the RO frequency [17].

The modulation characteristics of semiconductor lasers can be improved by the optical injection locking (OIL) [19]. OIL of semiconductor lasers provides a single-mode regime and near-single-sideband modulation, strongly enhances the RO frequency and bandwidth, reduces nonlinearity, reduces RIN, reduces chirp, and increases link gain [19]. The directly modulated OIL transmission-style system combining the master laser and the slave QDWELL laser in the transmission style is shown in **Figure 5** [19].

In the transmission-style system, the injected light from the master laser enters one slave QDWELL laser facet and output is taken from the other facet [19]. The direct modulation signal is applied to the QDWELL slave laser. An isolator prevents light coupling back to the master laser. A polarization controller (PC) is used to match the overlap of master and QDWELL slave laser polarizations. The master laser light coherently combines with the QDWELL slave laser

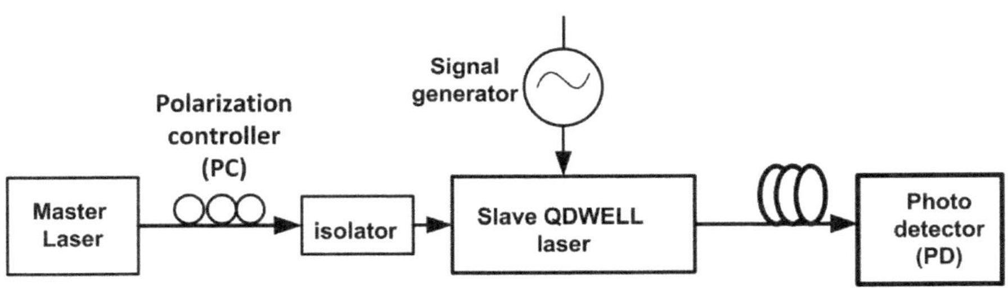

Figure 5. Directly modulated OIL system.

light changing its internal field [19]. The slave laser radiation wavelength tends to the master laser radiation wavelength until the both wavelengths become equal locking the master laser light frequency and phase [19].

The dynamics of the optically injected QDWELL laser has been studied theoretically by using the Lüdge-Schöll (LS) rate equations for the photon density per unit area n_{ph} and the phase Φ of the slave QDWELL laser radiation, the QD electron and hole occupation probabilities ρ_e, ρ_h, and the electron and hole densities in the QW w_e, w_h [20].

$$\frac{dn_{ph}}{dt} = n_{ph}\left[2\tilde{W}Z_a^{QD}\left(\rho_e+\rho_h-1\right)-2\kappa\right]+\frac{\beta_{Sp}}{A}2Z_a^{QD}R_{sp}\left(\rho_e,\rho_h\right)+\frac{2K}{\tau_{in}}\sqrt{n_{ph}n_{ph}^0}\cos\left(\Phi-2\pi\Delta\nu_{inj}t\right) \tag{14}$$

$$\frac{d\Phi}{dt}=\frac{\alpha}{2}\left[2\tilde{W}\left(\rho_e+\rho_h-1\right)-2\kappa\right]+\frac{2K}{\tau_{in}}\sqrt{\frac{n_{ph}^0}{n_{ph}}}\sin\left(\Phi-2\pi\Delta\nu_{inj}t\right) \tag{15}$$

$$\frac{d\rho_{e,h}}{dt}=-\tilde{W}A\left(\rho_e+\rho_h-1\right)n_{ph}-R_{sp}\left(\rho_e,\rho_h\right)+S_{e,h}^{in}(w_e,w_h)\left(1-\rho_{e,h}\right)-S_{e,h}^{out}(w_e,w_h)\rho_{e,h} \tag{16}$$

$$\frac{dw_{e,h}}{dt}=\frac{j}{q}-2N^{QD}\left[S_{e,h}^{in}(w_e,w_h)\left(1-\rho_{e,h}\right)-S_{e,h}^{out}(w_e,w_h)\rho_{e,h}\right]-\tilde{R}_{sp} \tag{17}$$

Here, A is the in-plane area of the QW, ν_L and ν_{inj} are the close frequencies of the QDWELL laser and the injected light, respectively, $K=\sqrt{T_{inj}n_{inj}/n_{ph}^0}$ is the injection strength, T_{inj} is the transmission coefficient of the cavity mirror, n_{inj} is the injected photon density in the QDWELL laser active region, and n_{ph}^0 is the steady-state photon density without injection; $\Delta\nu_{inj}=\nu_{inj}-\nu_L$ is the input detuning, $Z_a^{QD}=a_LAN_a^{QD}$ is the number of active QDs inside the waveguide, a_L is a number of self-organized QD layers, N_a^{QD} is the density per unit area of the active QD of a lasing subgroup, N^{QD} is the density per unit area of all QDs; 2κ are the optical intensity losses; $\tilde{W}$, and W are the Einstein coefficients for the coherent and incoherent interaction of the QD with all resonator modes, respectively; α is the linewidth enhancement factor (LEF), β_{Sp} is the spontaneous emission factor, $R_{sp}\left(\rho_e,\rho_h\right)$, and $\tilde{R}_{sp}$ are the spontaneous emission rates in QD and QW, respectively; τ_{in} is the time of one round trip of the light in the cavity of length L. We do not present here the complicated explicit expressions of the scattering rates S_e^{in}, S_e^{out}, S_h^{in}, S_h^{out} which can be found in Ref. [20]. It has been shown that the QDWELL laser's dynamic properties under optical injection are strongly influenced by the Coulomb scattering processes which can be taken into account by using the microscopically calculated carrier lifetimes $\tau_{e,h}$ [20]. The nonlinear dynamics of the optically injected QDWELL laser is determined by the injection strength and the input detuning [20]. The frequency-locking range depends on the pump current. The photon density as well as the electron and hole densities in QD and QW increase with increasing the pump current which results in the change of scattering rates and the QDWELL turn-on behavior [20]. The QDWELL laser under the optical injection manifests complex oscillatory dynamics for a comparatively large detuning $\Delta\nu_{inj}$ [20]. The large regions of stable CW operation occur for the short carrier lifetimes of about 10 ps typical for strongly damped relaxation oscillations [20].

Under certain conditions, QDWELL laser can operate as a SOA for high-speed applications [9]. Generally, SOA is the LED amplifying the input optical signal. SOAs are divided into two groups depending on their structure: (i) traveling wave amplifiers (TWA) and (ii) FP cavity amplifiers [9]. The block diagram of the TWA QDWELL SOA is shown in **Figure 6**.

SOA attracted a wide interest for applications in optical communication systems due to their small size, high gain, strong nonlinearity, and the possibility of on-chip optoelectronic integration [2, 9]. QD SOAs are particularly promising candidates for the amplification of ultrafast pulses in the relatively broad spectral range due to their ultrafast carrier dynamics and inhomogeneous spectral broadening [9]. The SOA dynamics is described by the carrier rate equations, equation for the time-dependent SOA gain and the equations for the slowly varying envelopes (SVE) of the optical pulses [2, 9].

The system of rate equations (14)–(16) should be modified for the case of a TWA QDWELL SOA. Eq. (17) for the electron and hole densities in the QW w_e, w_h does not change. The modified equations for the pumping and signal photon densities per unit area $n_{ph,p,s}$, phases $\Phi_{p,s}$, and the QD electron and hole occupation probabilities ρ_e, ρ_h in QDWELL SOA have the form [20–23].

$$\frac{\partial n_{ph,p,s}(z,\tau)}{\partial z} = \left(g_{\text{mod }p,s} - 2\kappa\right) n_{ph,p,s}(z,\tau) \tag{18a}$$

$$\frac{\partial \Phi_{p,s}}{\partial z} = -\frac{\alpha}{2} g_{\text{mod }p,s} \tag{18b}$$

$$\frac{d\rho_{e,h}}{dt} = -\tilde{W}A(\rho_e + \rho_h - 1)(n_{ph,p} + n_{ph,s}) - R_{sp}(\rho_e, \rho_h) + S_{e,h}^{in}(w_e, w_h)\left(1 - \rho_{e,h}\right) - S_{e,h}^{out}(w_e, w_h)\rho_{e,h} \tag{19}$$

where $\tau = t - z/v_g$, and $g_{modp,s}$ is the modal gain [22].

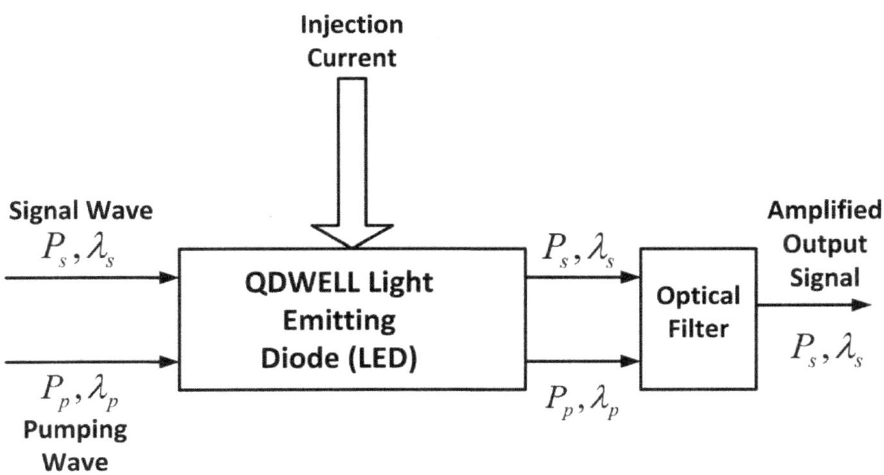

Figure 6. The block diagram of TWA QDWELL SOA.

$$g_{\mathrm{mod}} = \frac{2\Gamma}{r_{QD}} N^{QD} \int d\omega_i F(\omega_i)\sigma_i(\omega_0)\left(\rho_e + \rho_h - 1\right) \tag{20}$$

Here, r_{QD} is the mean size of QD, $\sigma_i(\omega_0)$ is the cross section of interaction of photons of frequency ω_0 with carriers in QD at the transition frequency ω_i:

$$\sigma_i(\omega_0) = \frac{\sigma_{res}}{1 + (\omega_i - \omega_0)^2 T_2^2} \tag{21}$$

where σ_{res} is the resonant cross section, $T_2 = 2\gamma_{hom}^{-1}$ is the dephasing time, and γ_{hom} is the homogeneous full width of half maximum (FWHM). The distribution function $F(\omega_i)$ of the transition frequency in the QD ensemble describes the inhomogeneous broadening caused by the QD size variation. It is assumed to be Gaussian and has the form [22]

$$F(\omega_i) = \frac{1}{\Delta\omega\sqrt{\pi}} \exp\left[-\frac{(\omega_i - \overline{\omega})^2}{(\Delta\omega)^2}\right] \tag{22}$$

where $\overline{\omega}$ is the average transition frequency, $\Delta\omega = \gamma_{\mathrm{inhom}}\left(2\sqrt{\ln 2}\right)^{-2}$, and γ_{inhom} is the inhomogeneous FWHM.

QD SOA can be used for the ultrafast and distortion-free amplification at pulse repetition rates of 20, 40, and 80 GHz with pulse durations of 710, 1.9, and 2.2 ps, respectively [9]. It should be noted that the amplification without patterning effect in QD SOA can be achieved at much lower pump current densities than in QW SOA [22].

4. Optical synchronization of a QDWELL laser dynamics and its influence on QDWELL laser and SOA performance

We investigated theoretically the optical synchronization of the carrier dynamics in the optically injected QDWELL laser and SOA and its influence on the QDWELL laser and SOA performance [23–27]. We solved numerically the LS equations (14)–(17) for the optically injected QDWELL laser [24–26] and Eqs. (17)–(22) for the QDWELL SOA [23, 27] using the typical values of QDWELL parameters and explicit expressions for the carrier scattering rates S_e^{in}, S_e^{out}, S_h^{in}, S_h^{out} [17, 20]. We used the MATLAB environment for the numerical simulations. We obtained a large number of the simulation results for the QW and QD carrier dynamics, QDWELL laser and SOA output signal eye diagrams, QD electron and hole phase trajectories, constellations for the QDWELL laser output 4-quadrature amplitude (4-QAM) modulation signal, QDWELL SOA gain wavelength and power dependence, QDWELL SOA chirp and extinction ratio (ER) dependence on the detuning between the pump and signal waves. They can be found in Refs. [23–27]. In this section, we summarize the main results and present the figures that demonstrate the influence of the optical injection on the QDWELL laser and QDWELL SOA performance.

It has been shown that the sufficiently strong optical injection leads to the synchronization of the electron and hole dynamics in QD [24]. In the transient turn-on regime, the bias current and the optical output pulses are strongly synchronized in the case of a comparatively high optical injection power of several milliwatts and the moderate bias current density $j = 3j_{th}$, where $j_{th} \approx 6.7A/m^2$ is the threshold current density of the QDWELL laser [24]. In such a case, $\rho_e + \rho_h < 1$, and the term describing the inversion of the two-level system with electrons and holes is negative: $(\rho_e + \rho_h - 1) < 0$. It is seen from Eqs. (14)–(16) that the optically induced transitions in QD are dominant, and the QD electron and hole occupation probabilities ρ_e, ρ_h increase. Their increase is limited by the condition $\rho_e + \rho_h = 1$. Then, the first term in Eqs. (14)–(16) vanishes, and the QD carrier dynamics is determined by the spontaneous transition rate $R_{sp}(\rho_e, \rho_h)$ and essentially different carrier scattering rates S_e^{in}, S_e^{out}, S_h^{in}, S_h^{out} in and out of QD. The photons are mainly absorbed as it is shown in Eq. (14). In the case of a low level of optical injection less than 1 mW, large bias current density $j = 6j_{th}$ and a significant detuning $\Delta\nu_{inj} = 5GHz$ close to the RO frequency of 7 GHz electron occupation probability ρ_e strongly oscillates while the QD hole and QW carrier dynamics remains stable.

The large-signal response of the QDWELL laser is important for digital communication systems. In the case without the optical injection, the QDWELL laser large-signal dynamics is determined by the QW carrier densities w_e, w_h and their lifetime [17]. The sufficiently strong optical injection of the master laser is inserted instead of a large bias current provides a large enough QD carrier density [24]. In such a case, a device heating by the bias current does not occur. The radiative transition rates in QD are determined by the optical injection power and the frequency detuning $\Delta\nu_{inj}$ between the master and slave lasers [24]. The fast stimulated transitions in QD are dominant, and the increase of the optical injection power results in the stronger electron and hole dynamics synchronization. The QW carrier density is varying weakly. In the case of the electrical pseudorandom binary sequence (PRBS) input signal, the modulation frequency may be increased up to 30 GHz due to the enhanced RO frequency [24]. The boundaries of the stable OIL regime are determined by the following condition [28]

$$-K\frac{E_{inj}}{E_0}\sqrt{1+\alpha^2} \leq 2\pi\Delta\nu_{inj} \leq K\frac{E_{inj}}{E_0} \tag{23}$$

where E_0 and E_{inj} are the slave and master laser field amplitudes, respectively. The instability occurs and the QDWELL output signal manifests strong oscillations in the case of the large detuning, low bias current density $j = 2j_{th}$, and low optical injection level [25]. The eye diagrams of the QDWELL under strong optical injection power $P_{inj} > 1$mW start to deteriorate at the repetition frequency of 30 GHz [24, 25].

The directly modulated QDWELL laser can be introduced into an analogous optical link (AOL) in the framework of ultra wideband (UWB) radio-over-fiber (UROOF) technology [26, 29, 30]. The UWB high-speed AOL shown in **Figure 7** consists of electrical/optical (E/O) converter, a standard SMF (SSMF) optical fiber, and optical/electrical (O/E) converter [30]. We solved numerically Eqs. (14)–(19) for the AOL based on the optically injected QDWELL laser and standard SMF (SSMF) [26]. We considered the small UWB signal with the modulation frequency of 60 GHz, the zero detuning $\Delta\nu_{inj} = 0$ between the master laser and the QDWELL slave laser radiation, the 4-quadrature amplitude modulation (4-QAM) format, and

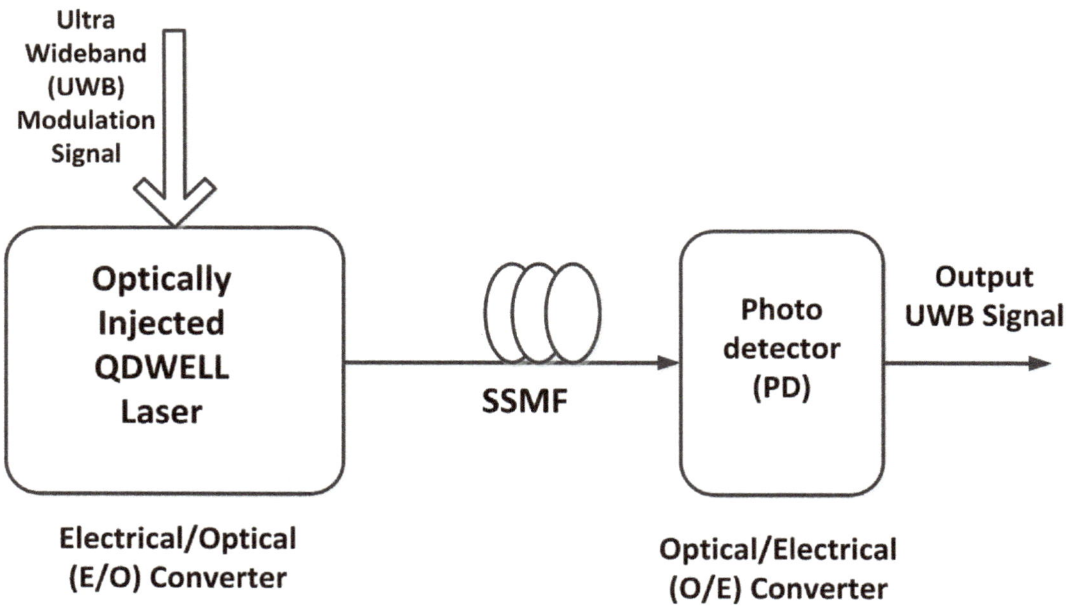

Figure 7. UWB analogous optical link (AOL) with the optically injected QDWELL laser (see **Figure 5**).

128 subcarriers with a 500 MHz bandwidth, and the direct bias current density $j = 2j_{th}$. The dynamics of QD and QW carriers under the strong optical injection of about 1 mW is synchronized, which results in the improvement of the AOL performance.

The calculated constellation of the output 4-QAM modulation signal under the strong optical injection of about 1 mW is shown in **Figure 8**.

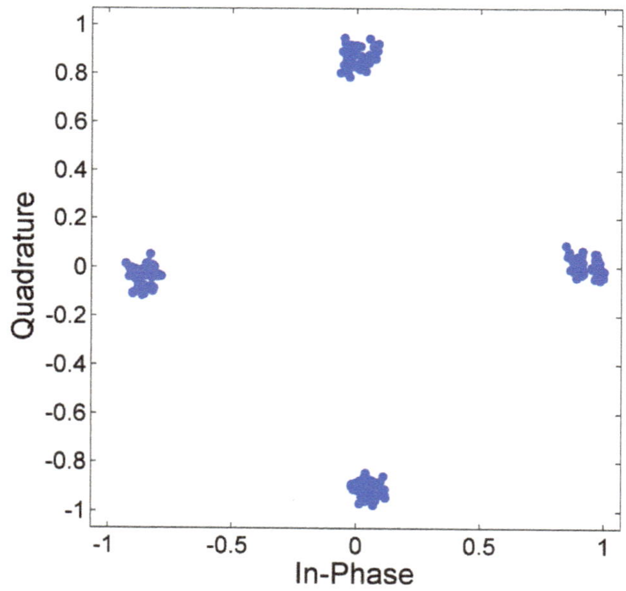

Figure 8. The calculated constellation of the output 4-QAM modulation signal for the modulation frequency 60 GHz, zero detuning, and the optical injection power of 1 mW.

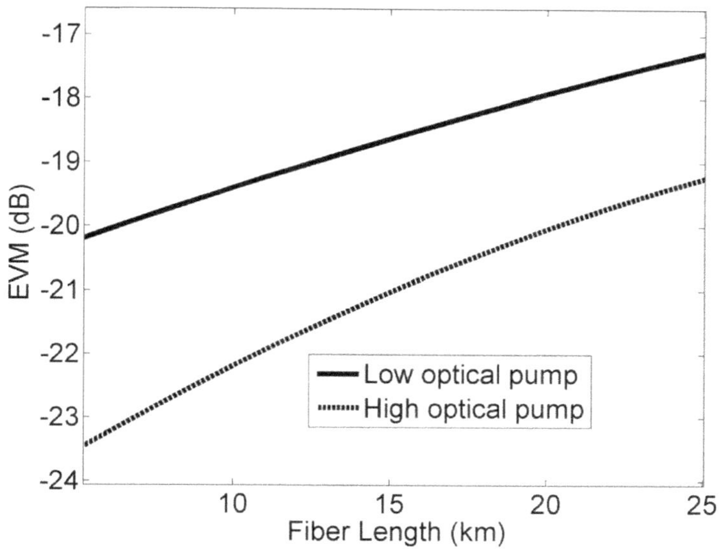

Figure 9. The calculated AOL EVM dependence on the distance for the low optical injection power of 10^{-3} mW (the upper curve) and for the high optical injection power of 1 mW (the lower curve).

The constellation shown in **Figure 8** is practically identical with the constellation of the input signal unlike the case without the optical injection [26]. The AOL error vector magnitude (EVM) dependence on the distance for the optical injection power of 10^{-3} and 1 mW is presented in **Figure 9**.

It is shown in **Figure 9** that in the case of the strong optical injection, the AOL EVM decreases significantly due to the enhancement of the QDWELL laser modulation frequency [26].

The spectrum of the detected UWB signal for the modulation frequency of 60 GHz, zero detuning between the master and the slave lasers, high optical injection power of 1 mW, and the propagation distance of 50 km is shown in **Figure 10** [26].

Consider now the influence of the optical pumping on the QDWELL SOA performance [23, 27]. We investigated the copropagating pumping and signal optical waves characterized by the optical power $P_{p,s}$ and the wavelengths $\lambda_{p,s}$, respectively [23]. Eqs. (17)–(22) have been solved numerically for the pulse regime and the large PRBS signal response [23]. The central pumping and signal wavelengths are $\lambda_p = 1.25$ μm and $\lambda_s = 1.35$ μm, respectively, and the optical pumping power is $P_p = 2,\ \ 5,\ \ 10$ mW [23]. The interaction of the pumping and signal wave in a QDWELL SOA results in the nonlinear optical phenomenon of the cross-gain modulation (XGM) [31]. During the XGM process, the strong modulated pumping wave imposes the modulation on the weak CW signal wave [31]. In such a case, SOA is acting as a wavelength converter (WC) transposing information at one wavelength to the signal at another wavelength [31]. In the case of a QDWELL SOA, the strong pumping wave at the same time synchronizes the QD and QW carrier dynamics. Indeed, when the pumping power $P_p = 1$ μW is small compared to the signal power $P_s = 1$ mW, the electron and hole dynamics in QD is desynchronized, the gain recovery time is determined by the carrier scattering rates

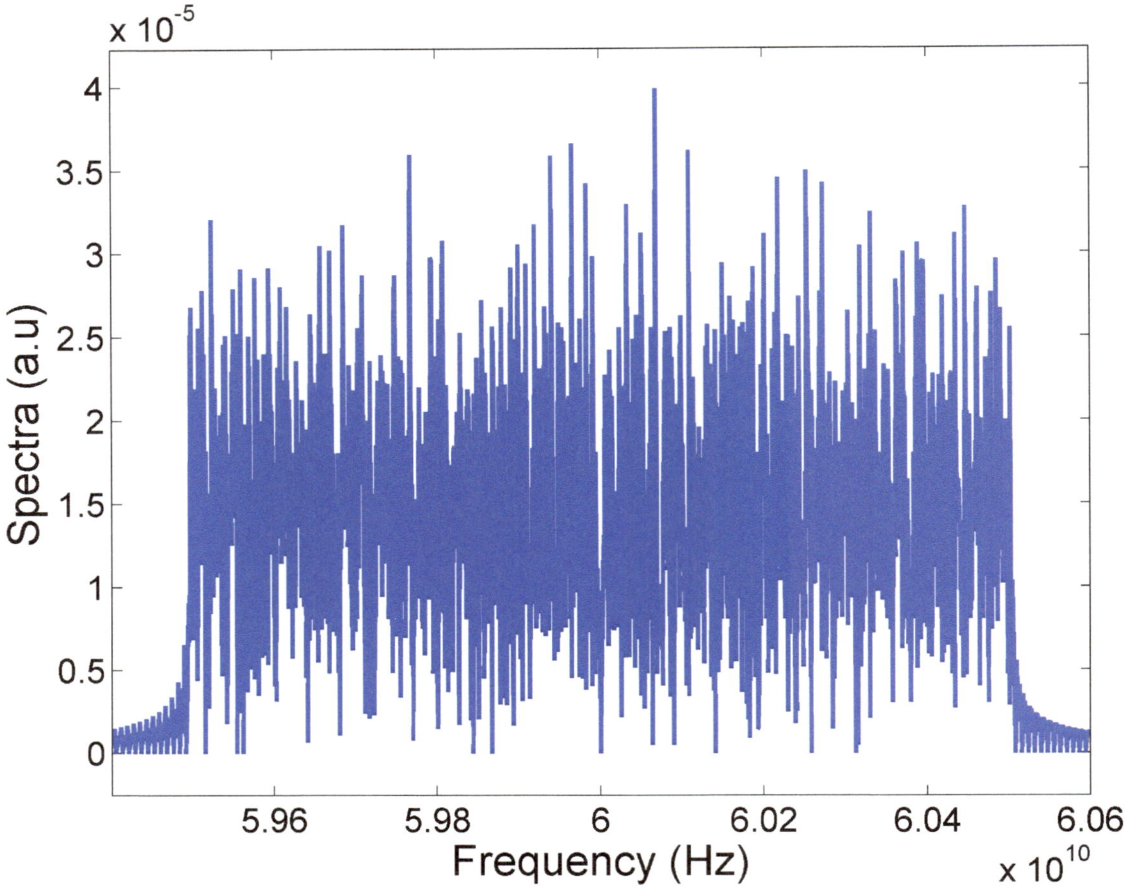

Figure 10. The calculated spectrum of the detected UWB signal after the propagation over the distance of 50 km for the modulation frequency of 60 GHz, zero detuning, and high optical injection power of 1 mW.

S_e^{in}, S_e^{out}, S_h^{in}, S_h^{out}, and the gain recovery process is slow [23]. In the opposite case of the strong pumping power of about 1 mW, the QD electron and hole dynamics is synchronized, the QD carrier occupation probabilities remain practically constant, and the gain recovery time decreases. The oscillation period of the QD carrier occupation probabilities ρ_e, ρ_h is determined by the optically enhanced gain recovery time [23]. The signal wave gain decreases with the increase of the pumping wave power due to XGM and the gain saturation at a large optical power. The bias current density influence on the gain is weak because of the weak connection between QW carrier reservoir and QD during the XGM process [23]. Under the strong optical of 1 mW and the low bias current density of $2j_{th}$, the QDWELL SOA performance improves. The eye diagram of the QDWELL SOA in such a regime is shown in **Figure 11**.

Figure 11 shows that the pattern effect in the eye diagram vanishes up to the repetition frequency of 140 Gb/s. The output signal optical power and chirp time dependence are shown in **Figure 12**. It is shown in **Figure 12** that due to the carrier dynamics synchronization provides the XGM process without the pattern effect, and the chirp reduces to about 5GHz and becomes symmetric [23, 27].

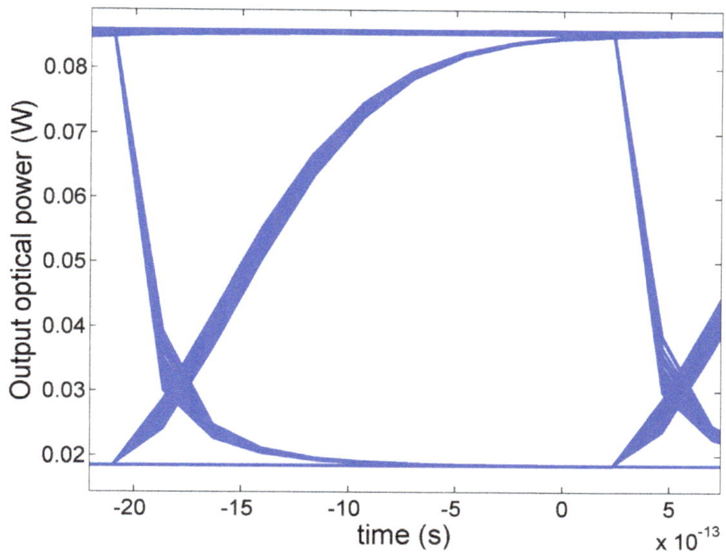

Figure 11. The output signal eye diagram for the signal power of 1 mW, optical pumping power of 2 mW, the bias current density of $2j_{th}$, and modulation bit rate of 140 GB/s.

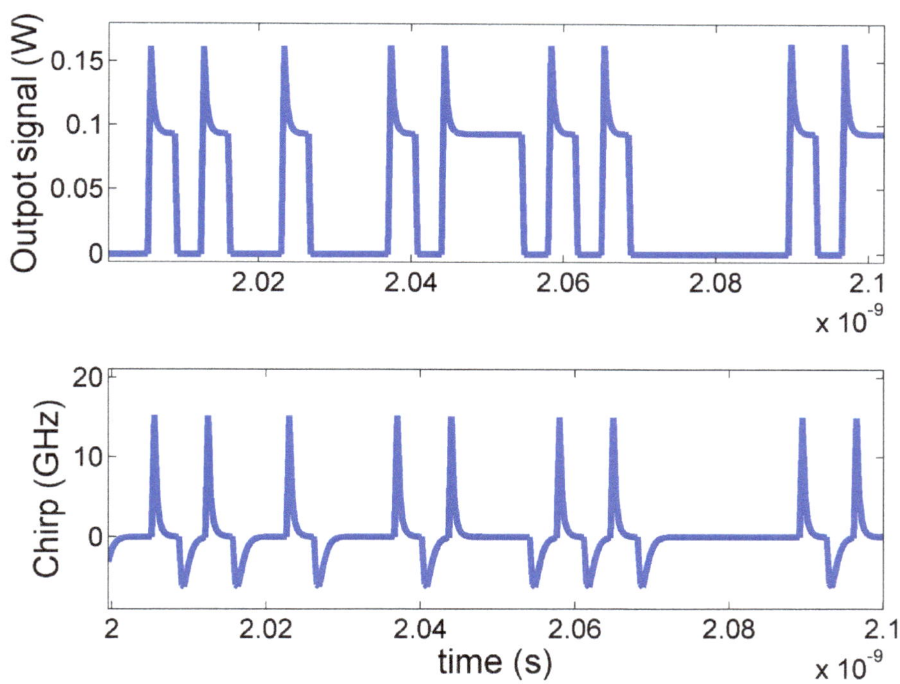

Figure 12. The optical power (the upper box) and the chirp (the lower box) of the output signal.

The strong optical pumping wave of several milliwatts during the XGM process may enhance the QDWELL SOA bandwidth up to 100 nm for a central wavelength of 1350 nm [27]. We investigated the ER dependence on the XGM detuning for the signal wave wavelength

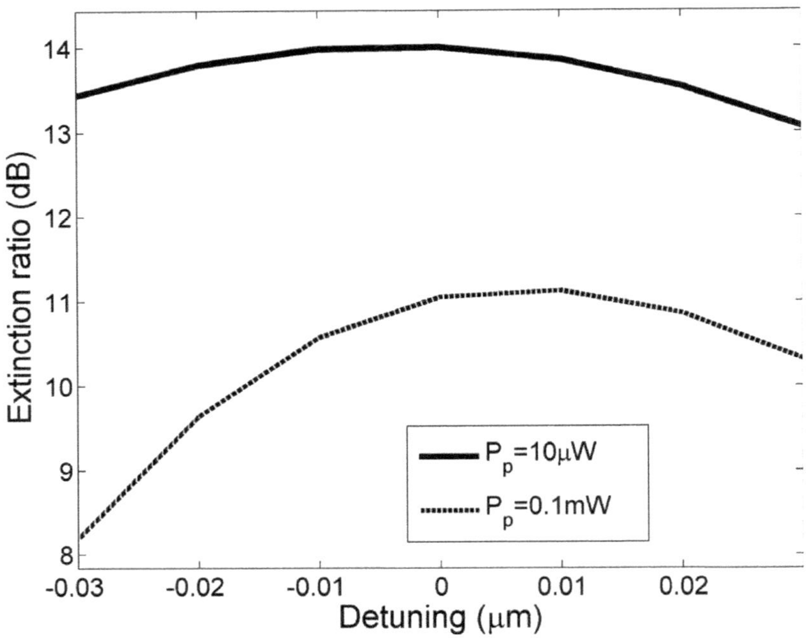

Figure 13. ER dependence on the XGM detuning for the signal wave wavelength $\lambda_s = 1.35$ μm, the signal wave optical power $P_s = 1$ mW, the bias current density of $2j_{th}$, and the CW pumping wave power P_p of 0.1 mW (the lower curve), 0.01 mW (the upper curve).

$\lambda_s = 1.35$μm and optical power $P_s = 1$ mW, and the CW pumping wave power P_p of 0.1 and 0.01 mW [27]. The simulation results are shown in **Figure 13**.

ER reduces with the increase of the CW pumping wave power P_p as it is shown in **Figure 13**.

5. Conclusions

QW and QD semiconductor lasers and SOA are promising candidates for applications in modern optical communication systems due to their comparatively low threshold current, high operation rate, high modulation bandwidth, low chirp, patterning-free operation, and stability with respect to temperature variations. These QW and QD advantages are mainly caused by the one-dimensional (1D) carrier localization in QW and three-dimensional carrier localization in QD. QDWELL lasers are based on the number of QW layers with the embedded QD. In such a case, 2D electron and hole gas also exists in a QW carrier reservoir, i.e., WL. QDWELL lasers are characterized by the lowest possible threshold current and fast QD carrier dynamics due to the 3D localization of electrons and holes in QD. However, the carrier dynamics in QW and QD is desynchronized due to the strong nonlinear carrier scattering rates for the electron and hole transitions in and out of QD. As a result, the modulation frequency of QDWELL lasers is limited by a comparatively low RO frequency of about 7 GHz. The modulation frequency can be enhanced by an increase of the injection current. But the increase of the injection current leads to the QDWELL laser heating and significant performance deterioration.

The modulation frequency and operation rate of semiconductor lasers and SOA can be improved by OIL using the master laser for the optical injection of the slave laser. We solved numerically the system of LS rate equations for QDWELL laser including the optical injection, the explicit expressions for the nonlinear carrier scattering rates, and the inhomogeneous spectral broadening specific for QD. The simulation results show that the carrier dynamics in QW and QD is synchronized by the sufficiently strong optical injection power $P_{inj} > 1$ mW since the fast stimulated transitions become dominant. As a result, the modulation frequency increases up to 30 GHz when the QDWELL output signal eye diagrams start to deteriorate. The high level of the carrier density in QD is provided by the optical injection, and the influence of the bias current is negligible. The strong bias current is not necessary, and the device heating and degradation are avoided. The optical injection process strongly depends on the frequency detuning between the master laser and the QDWELL slave laser. For the low optical injection power and the large detuning of about 10 GHz, the instability occurs accompanied by the strong oscillations of the QDWELL laser output signal. We solved the LS rate equations for AOL based on the optically injected QDWELL laser in the direct modulation regime for the optical injection power of 1 mW, zero detuning, 4-QAM modulation format, and 128 subcarriers with a 500 MHz bandwidth. In such a case, AOL manifests high performance up to the modulation frequency of 60 GHz.

TWA QDWELL LED with an input optical signal operates as a QDWELL SOA. QDWELL SOA is described by the modified system of rate equations including the equations for the pumping and signal wave photon densities. In such a case, the sufficiently strong pumping wave simultaneously plays a role of the optical injection and synchronizes the carrier dynamics in QW and QD. The nonlinear optical processes of XGM and XPM occur in a QDWELL SOA. QDWELL SOA performance is improved, the fast gain recovery takes place, the gain bandwidth is enhanced, ER and chirp decreases.

Author details

Yossef Ben Ezra[1,2] and Boris I. Lembrikov[1]*

*Address all correspondence to: borisle@hit.ac.il

1 Department of Electrical Engineering and Electronics, Holon Institute of Technology (HIT), Holon, Israel

2 MER Cellos, Holon, Israel

References

[1] Agrell E, Karlsson M, Chraplyvy AR, et al. Roadmap of optical communications. Journal of Optics. 2016;**18**(6):1-40. DOI: 10.1088/2040-8978/18/6/063002

[2] Agrawal GP. Fiber-optic Communication Systems. 4th ed. New York: Wiley; 2010. ISBN: 978-0-470-50511-3

[3] Agrawal GP. Nonlinear Fiber Optics. 5th ed. New York: Academic Press; 2013. ISBN: 978-0-12397-023-7

[4] Shieh W, Djordjevic I. Orthogonal Frequency Division Multiplexing for Optical Communications. London: Academic Press Elsevier; 2010. ISBN: 978-0-12-374879-9

[5] Ip E, Pak Tao Lau A, Barros DJF, Kahn JM. Coherent detection in optical fiber systems. Optics Express. 2008;**16**:753-791. ISSN: 1094-4087

[6] Zhao B, Yariv A. Quantum well semiconductor lasers. In: Kapon E, editor. Semiconductor Lasers I. Fundamentals. New York: Academic Press; 1999. pp. 1-121. ISBN: 0-12-397630-8

[7] Harrison P. Quantum Wells, Wires and Dots. Chichester, England: Wiley; 2005. p. 482. ISBN-13: 978-0-470-01079-2

[8] Nagarajan R, Bowers JE. High-Speed lasers. In: Kapon E, editor. Semiconductor Lasers I Fundamentals. New York: Academic Press; 1999. pp. 177-290. ISBN: 0-12-397630-8

[9] Rafailov EU, Cataluna MA, Avrutin EA. Ultrafast Lasers Based on Quantum Dot Structures. Weinheim: Wiley-VCH; 2011. p. 250. ISBN: 978-3-527-40928-0

[10] Lüdge K. Modeling Quantum-Dot-Based devices. In: Lüdge K, editor. Nonlinear Laser Dynamics. Weinheim, Germany: Wiley-VCH; 2012. pp. 1-33. ISBN: 978-0-470-85619-2

[11] Ustinov VM, Zhukov AE, Egorov AY, Maleev NA. Quantum Dot Lasers. Oxford: Oxford University Press; 2003. ISBN: 0 19 852679 2

[12] Ledentsov NN, Bimberg D, Alferov ZI. Progress in epitaxial growth and performance of quantum dot and quantum wire lasers. Journal of Lightwave Technology. 2008;**26**:1540-1555. DOI: 10.1109/JLT.2008.923645

[13] Coleman JJ, Young JD, Garg A. Semiconductor quantum dot lasers: A tutorial. Journal of Lightwave Technology. 2011;**29**:499-5102011. DOI: 10.1109/JLT.2010.2098849

[14] Malić E, Bormann MJP, Hövel P, Kuntz M, Bimberg D, Knorr A, Schöll E. Coulomb damped relaxation oscillations in semiconductor quantum dot lasers. IEEE Journal of Selected Topics in Quantum Electronics. 2007;**13**:1242-1247. DOI: 10.1109/JSTQE.2007.905148

[15] Lüdge K, Bormann MJP, Malić E, Hövel P, Kuntz M, Bimberg D, Knorr A, Schöll E. Turn-on dynamics and modulation response in semiconductor quantum dot lasers. Physical Review B. 2008;**78:**035316-1-035316-11. DOI: 10.103/PhysRevB.78.035316

[16] Lüdge K, Schöll E. Quantum-dot lasers—Desynchronized nonlinear dynamics of electrons and holes. IEEE Journal of Quantum Electronics. 2009;**45**:1396-1403. DOI: 10.1109/JQE.2009.2028159

[17] Lüdge K, Aust R, Fiol G, Stubenrauch M, Arsenijević D, Bimberg D, Schöll E. Large-signal response of semiconductor quantum-dot lasers. IEEE Journal of Quantum Electronics. 2010;**46**:1755-1762. DOI: 10.1109/JQE.2010.2066959

[18] Lüdge K, Schöll E. Nonlinear dynamics of doped semiconductor quantum dot lasers. The European Physics Journal D. 2010;**58**:167-174. DOI: 10.1140/epjd/e2010-00041-8

[19] Lau EK, Wong LJ, Wu MC. Enhanced modulation characteristics of injection-locked lasers: A tutorial. IEEE Journal of Selected Topics in Quantum Electronics. 2009;**15**:618-633. DOI: 10.1109/JSTQE.2009.2014779

[20] Pausch J, Otto C, Tylaite E, Majer N, Schöll E, Lüdge K. Optically injected quantum dot lasers: Impact of nonlinear carrier lifetimes on frequency-locking dynamics. New Journal of Physics. 2012;**14**:1-20. DOI: 10.1088/1367-2630/14/053018

[21] Agrawal GP, Olsson NA. Self-phase modulation and spectral broadening of optical pulses in semiconductor optical amplifiers. IEEE Journal of Quantum Electronics. 1989;**25**:2297-2306. DOI: 0018-9197/89/1100-2297

[22] Uskov AV, Berg TW, Mork J. Theory of pulse-train amplification without patterning effects in quantum-dot semiconductor optical amplifiers. IEEE Journal of Quantum Electronics. 2004;**40:**306-320. DOI: 10.1109/JQE.2003.82032

[23] Ben Ezra Y, Lembrikov BI. Semiconductor optical amplifier based on a quantum Dot-in-a-Well (QDWELL) structure under optical pumping. IEEE Journal of Quantum Electronics. 2014;**50**:340-347. DOI: 10.1109/JQE.20142308393

[24] Ben Ezra Y, Lembrikov BI. Synchronized carrier dynamics in quantum Dot-in-a-Well (QDWELL) laser under an optical injection. IEEE Journal of Selected Topics in Quantum Electronics. 2013;**19**:1-8. DOI: 10.1109/JSTQE.2013.2246770

[25] Ben Ezra Y, Lembrikov BI. Quantum Dot-in-a-Well (QDWELL) laser dynamics under optical injection. Optical and Quantum Electronics. 2014;**46**:1239-1245. DOI: 10.1007/s1 1082-013-9829-3

[26] Ben Ezra Y, Lembrikov BI. UWB system optical link based on a Quantum-Dot-in-a-Well (QDWELL) laser. Optical and Quantum Electronics. 2015;**47**:1527-1533. DOI: 10.1007/s1 1082-015-0154-x

[27] Ben Ezra Y, Lembrikov BI. Investigation of a cross-gain modulation (XGM) in a semiconductor optical amplifier (SOA) based on a quantum Dot-in-a-Well (QDWELL) structure. IET Optoelectronics. 2015;**9**:43-51. DOI: 10.1049/iet-opt.2014.0061

[28] Chang-Hasnain CJ, Zhao X. Ultrahigh-speed laser modulation by injection locking. In: Kaminov IP, Li T, Willner AE, editors. Optical Fiber Telecommunications. V A. Components and Subsystems. London: Elsevier Academic Press; 2008. pp. 145-182. ISBN: 978-0-12-374171-4

[29] Ran M., Ben Ezra Y., Lembrikov B. I. Ultra-wideband radio-over-optical fibre technologies. In: Kraemer R, Katz M., editors. Short-Range Wireless Communications: Emerging Technologies and Applications. Harvard: Wiley; 2009. pp. 271-327. ISBN: 978-0-470-69995-9

[30] Ran M, Ben Ezra Y, Lembrikov BI. High performance analog optical links based on quantum dot devices for UWB signal transmission. In: Lembrikov BI, editor. Ultra Wide-band. Croatia: InTechOpen; 2010. pp. 75-96. ISBN: 978-953-307-139-8

[31] Connely MJ. Semiconductor Optical Amplifiers. London: Kluwer; 2002. p. 169. ISBN: 0-7923-7657-9

3

Physical-Layer Encryption Using Digital Chaos for Secure OFDM Transmission

Xuelin Yang, Adnan A.E. Hajomer and Weisheng Hu

Additional information is available at the end of the chapter

Abstract

Due to the broadcasting nature of passive optical network (PON), data security is challenging. For the transmission of orthogonal frequency division multiplexing (OFDM) signals, the high peak-to-average power ratio (PAPR) is considered as one of the major drawbacks. This chapter reviews the digital chaos-based secure OFDM data encryption schemes, where the transmission performance is improved via PAPR reduction. The digital chaos is incorporated into the signal scrambling approaches: selective mapping (SLM), partial transmit sequence (PTS); and precoding approaches: discrete Fourier transform (DFT) and Walsh-Hadamard transform (WHT) for PAPR reduction. Multi-fold data encryption is achieved with a huge key space provided by digital chaos, to enhance the physical-layer security for OFDM-PON, while the pseudo-random properties of digital chaos are applied for PAPR reduction, which consequently improves the transmission performance. The evidences of these encryption approaches are presented in terms of theories, simulations, as well as experimental demonstrations. The chaotic data encryption schemes could be promising candidates for next-generation OFDM-PON.

Keywords: orthogonal frequency division multiplexing (OFDM), peak-to-average power ratio (PAPR), digital chaos, passive optical network (PON)

1. Introduction

Over the last decades, passive optical network (PON) has been playing a vital role in data traffic explosion driven by broadband services such as high-definition television (HDTV), cloud computing, 3D television, video on demand (VoD) [1, 2], and so on, because it offers several potential benefits such as high capacity, low cost, and energy efficiency. In fact, PON is a broadcasting structure that extends for ~20–100 km from optical line terminal (OLT) to optical network units (ONUs), in which no active component (such as Erbium-doped fiber amplifier, EDFA) is employed. The broadcasting nature in the downstream data traffic and the huge number of subscribers in PON make the data more susceptible to be eavesdropped or attacked by illegal ONUs. For instance, during the ranging process, the OLT has to broadcast the serial number and ID information of the ranged ONUs; a malicious user could make use of this information for spoofing.

Comparing with the encryption in higher layers, for instance, media access control (MAC) layer, physical-layer encryption can protect the data as well as the control and header information. If the physical-layer data encryption is implemented, this type of spoofing can be avoided since the header and ID information are all encrypted within the physical-layer; thus, it becomes desirable for security enhancement in PON. For security reasons, churning procedure of scrambling the data for downstream connection has been defined in ITU-T G.983.1 standard (Section 8.3.5.6 [3]), which is based on a key sent from ONU to OLT through a secure channel with a defined protocol; however, the key is vulnerable to be broken due to its limited short key length.

Chaos communication has been proposed to provide a huge key space for data security enhancement attributed to its unpredictable nature of randomness, noise-like nature, and broadband. However, the implementation of chaotic optical communication (i.e., fast changing of chaotic optical carriers) requires identical devices with identical parameters for transmitter and receiver, which is quite restricted from real implementation. On the other hand, digital chaos has attracted notable attention recently as a flexible alternative to avoid the implementation difficulty for device-based optical chaos [4–10]. Because it offers very appealing properties from the perspective of data encryption such as ergodicity, pseudo-randomness, and high sensitivity to the initial values, digital chaos provides a huge key space for security applications. Moreover, due to flexible digital signal processing (DSP) in electric domain, digital chaos is easier to be applied.

Optical orthogonal frequency division multiplexing (OFDM) is regarded as a promising modulation technique for next-generation PON, owing to the advantages in high spectrum efficiency, cost-effectiveness, and tolerance to fiber dispersion. High-speed data rate of OFDM signals is achieved by parallel transmission of partially overlapped spectra, lower rate frequency domain tributaries [11]. Moreover, the generation, modulation, and demodulation of OFDM signals have to be performed using DSP in electric domain, therefore it provides a natural physical-layer environment, where digital chaos can be incorporated into OFDM data encryption during transmission. However, OFDM modulation often leads to high peak-to-average power ratio (PAPR), which is one of the most detrimental factors in OFDM signal transmission, as it causes power saturation and nonlinear distortion at the optical receiver while degrading the transmission performance. The pseudo-random properties of digital chaos are helpful for PAPR reduction of OFDM signals, which consequently improves the transmission performance.

In this chapter, OFDM data encryption schemes are reviewed in detail for physical-layer security enhancement based on digital chaos during transmission, while jointly the transmission performances are significantly improved because of the effective reduction in PAPR of OFDM signals. The rest of the chapter is organized as follows. In Section 2, the fundamental theory of PAPR reduction of OFDM signals is shown. The properties of digital chaos are presented in Section 3. In Section 4, the encryption schemes are illustrated in details. Conclusions are given in Section 5.

2. PAPR of OFDM signals

OFDM modulation is the superposition of many independent signals modulated onto individual subcarriers with equal-spaced bandwidth. **Figure 1** shows the overlapping of the subcarriers in frequency domain. High PAPR could be inevitable especially when a large number of subcarriers are in phase. As a result, the optical receiver with a wide linear range is required to accommodate a large dynamic range of PAPR [12].

If a block of N symbols is denoted as the vector $X = [X_0, X_1, \ldots, X_{N-1}]$ for OFDM signals, the vector X is oversampling by g (i.e., $g(N-1)$ zero-padding, ZP), where g is an integer greater than or equal one. Therefore, the complex envelop of OFDM signals is

$$x[n] = \frac{1}{\sqrt{N}} \sum_{i=0}^{Ng-1} X_i e^{\frac{2\pi i n}{Ng}}, 0 \leq n \leq Ng-1 \quad (1)$$

By definition, the PAPR of OFDM signals is

$$PAPR = 10\log(\max(|x[n]|^2)/E(|x[n]|^2)) \quad (2)$$

From Eqs. (1) and (2), the oversampling factor must be greater than one for sufficient accuracy of PAPR calculation [13]. To evaluate the PAPR performance, the complementary cumulative distribution function (CCDF) is commonly simulated, which is defined as the PAPR probability exceeding a given threshold for a certain OFDM data block. Based on the central limit theorem, the real and imaginary parts of the complex OFDM signals after inverse fast Fourier transformation (IFFT) have Gaussian distribution in the case of sufficient large number of subcarriers, thus the amplitudes follow the Rayleigh distribution. For instance, if the CDF of the amplitude of a signal sample is given by

$$F(z) = 1 - e^{-z} \quad (3)$$

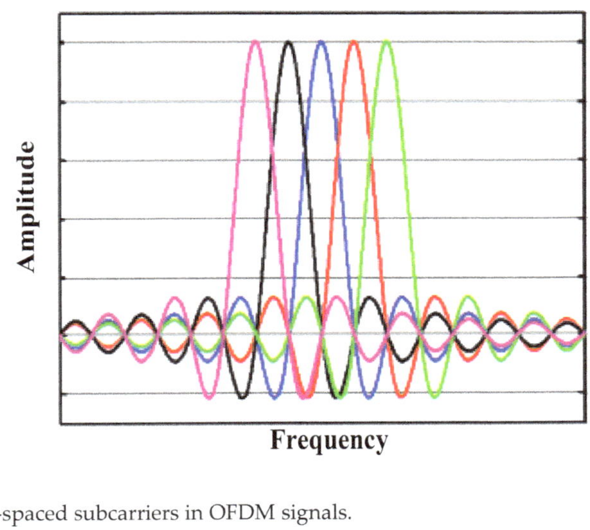

Figure 1. Spectrum of equal-spaced subcarriers in OFDM signals.

and N is a large number of samples, the CCDF of PAPR of the signal is [14]

$$P(PAPR > z) = 1 - P(PAPR \leq z) = 1 - F(z)^N = 1 - (1 - e^{-z})^N \quad (4)$$

3. Characteristics of digital chaos

Digital chaos has recently attracted numerous applications in OFDM-PONs [4–9], especially for data encryption, which is mainly due to the chaotic characteristics including pseudo-randomness, ergodicity, and high sensitivity to the initial values. Secure optical OFDM transmission is achieved by digital chaos, in which a huge key space is generated and predetermined by chaotic equations. Since the initial values and the other control parameters are utilized as the secure keys between OLT and ONUs, it provides a huge key space, which guarantees the physical-layer confidentiality. At ONUs, the same chaotic sequences are generated using the same keys for data recovering after reception.

The fundamental properties of digital chaos can be described, for example, via a 4-dimensional (4D) hyper chaos [15]

$$\begin{cases} \dot{x} = a(-x + y) + yzu \\ \dot{y} = b(x + y) - xzu \\ \dot{z} = cy - u + dxyu \\ \dot{u} = -eu + xyz \end{cases} \quad (5)$$

where a, b, c, d, and e are constant parameters. When $a = 35$, $b = 10$, $c = 80$, $d = 0.5$, and $e = 10$, these appropriate initial values bring the system into chaotic zones. Eq. (5) can be solved by Runge-Kutta method with a time step of $h = 0.001$. The solutions of Eq. (5) output the attractor diagrams of the 4D chaos, as plotted in **Figure 2**, where excellent chaotic behaviors in terms of pseudo-randomness and phase dynamics are illustrated.

In digital chaos, the chaotic state is very sensitive with respect to the initial values, thus even a tiny change or modification of the original initial values will let it enter into another different chaotic state. In **Figure 3(a)**, the variation curve is illustrated for the digital chaotic sequence $\{y_i\}$, under a slight change (1×10^{-15}) in the initial value of y_0. The auto- and cross-correlation functions $R_{ac}(\tau)$ and $R_{cc}(\tau)$ are plotted in **Figure 3(b) and (c)** respectively, which reveal the high sensitivity associated with the chaotic initial values, where τ is the time lag. The good quality of randomness observed in **Figure 3** is essential to guarantee high-level security reliability for data encryption.

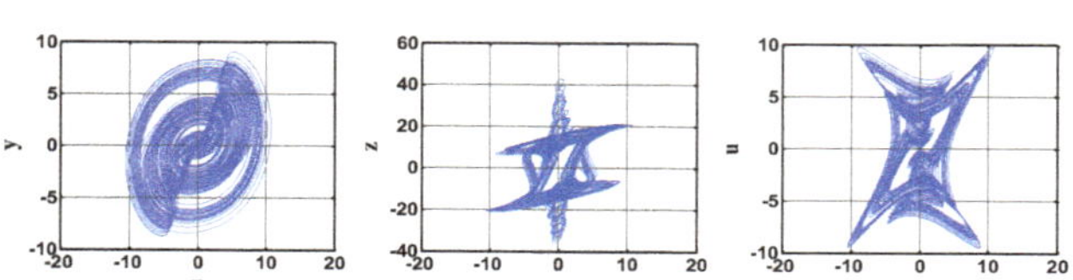

Figure 2. Chaotic attractor diagrams in phase planes of (x, y), (x, z), and (x, u).

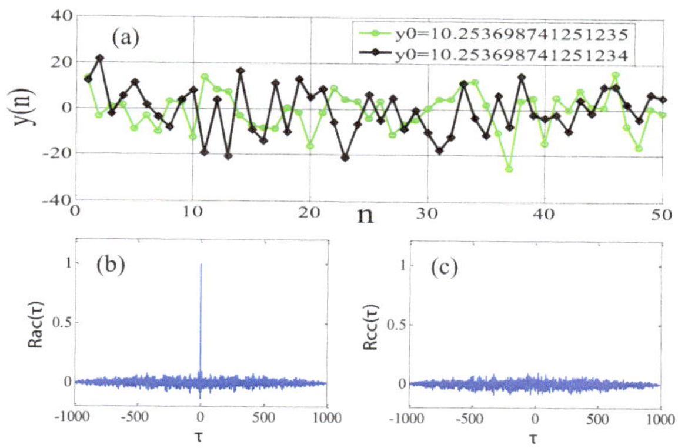

Figure 3. (a) Chaotic sequence of $\{y_i\}$ under tiny change of the initial values; (b) autocorrelation for y_0 =1.428121243912452; (c) cross-correlation for y_0 =1.428121243912452 and y_0 =1.428121243912453.

For a conservative estimate, a tiny change (~1×10^{-15}) of the initial values leads to a totally different chaotic state, as shown in **Figure 3**, therefore, the key space of 4D digital hyper chaos is ~10^{60} ($10^{15}\times10^{15}\times10^{15}\times10^{15}$).

Furthermore, the iteration times of chaotic differential equations and even the equations themselves can be served as the additional secure keys as well, so the actual key space will be >>10^{60}. Currently, the fastest computing speed is about $2.5\times10^{13}s^{-1}$, thus it will take ~1.3×10^{39} years to work out the possible initial keys of the 4D chaos via brutal-force trials [16]. Consequently, the chaotic data encryption provides a huge key space, which is large enough to resist exhaustive attacks.

4. Chaotic OFDM encryption with PAPR reduction

A variety of chaotic encryption schemes has been proposed for the physical-layer security in OFDM-PON [4–9], in which digital chaos has been incorporated into OFDM signal generation or modulation to enhance data security during transmission. However, in the existing chaotic secure schemes, the encryption is generally achieved without considering the improvement of transmission performance through the effective PAPR reduction in OFDM signals.

This section presents some systematic novel approaches to enhance the physical-layer security and jointly improve the OFDM transmission performances. The key idea behind these schemes is the combination of chaos-based encryption along with PAPR reduction; these approaches mainly belong to two categories of PAPR reduction: signal scrambling and matrix precoding, such as

- Chaotic signal scrambling: Chaotic partial transmit sequence (PTS) and chaotic selective mapping (SLM).
- Chaotic precoding: Discrete Fourier transform (DFT) and Walsh-Hadamard transform (WHT).

4.1. Chaotic signal scrambling

Signal scrambling is one of the traditional approaches in PAPR reduction. Chaotic sequences can be employed to scramble an input data block of OFDM symbols randomly, and then one of the scrambled OFDM signals with minimum PAPR is selected for final transmission. With the use of chaotic sequences in OFDM symbol scrambling, the physical-layer security is enhanced since the initial keys are required for correct recovery of the original OFDM data via correct counter-scrambling. At the same time, the transmission performance is improved simultaneously, owing to the effective reduction in PAPR via signal scrambling.

4.1.1. Chaotic partial transmit sequence (PTS)

The block diagram of OFDM data encryption using chaotic PTS is plotted in **Figure 4** [17]. After serial-to-parallel (S/P) conversion, a pseudo-random binary sequence (PRBS) is mapped onto QAM subcarriers, and then is sent for data encryption, which is applied via the following threefold encryption: chaotic training sequence (TS) insertion, chaotic random partition generation, and chaotic phase weighing factors generation.

Figure 5(a)–(c) shows the different partition examples in PTS scheme. Compared with the adjacent and interleaved partitions, the random partition provides the weakest cross-correlation among the sub-blocks, which leads to the largest PAPR reduction after selecting the optimum phase weighing factors. The PAPR can be minimized after searching for the optimized OFDM symbols for chaotic (random) signal scrambling.

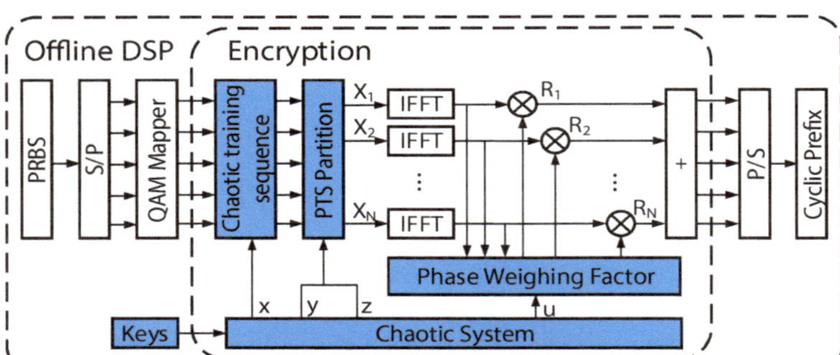

Figure 4. Schematic diagram of OFDM signal encryption using chaotic PTS.

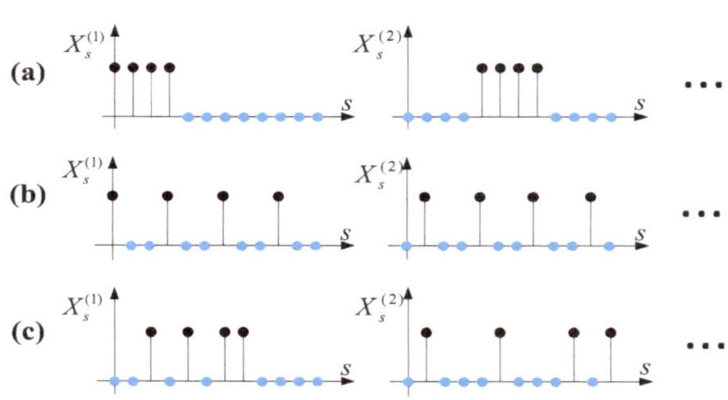

Figure 5. Partition design in PTS, for (a) adjacent; (b) interleaved; (c) random partitions.

To generate digitized chaotic sequences for OFDM signal encryption, the chaotic sequences obtained in Eq. (5) are processed using [18]

$$D_{x,i} = \mathrm{mod}(Extract(x_i, m, n, p), M) \tag{6}$$

where $Extract(x_i, m, n, p)$ outputs an integer, which is obtained by the mth, nth, and pth digits in the decimal part of x_i, mod(a, b) outputs the remainder of a divided by b, M is the maximum digital value in the digital sequence, which is 256 in all of our schemes [4]. Similarly, the other sequences $\{y_i\}$, $\{z_i\}$, and $\{u_i\}$ can be digitalized into $\{D_{y,i}\}$, $\{D_{z,i}\}$, and $\{D_{u,i}\}$. The details of chaotic PTS are given as follows.

The first sequence $\{D_{x,i}\}$ is applied to generate the chaotic TS for time synchronization of OFDM symbols. Since the illegal ONU does not have the information of chaotic TS, it has to try $(N_0+N)/N_0$ times on average to realize correct synchronization and demodulation, where N_0 is the length of cyclic prefix (CP). Moreover, due to the pseudo-random characteristic of digital chaos, it produces an autocorrelation similar to the δ function, which is advantageous for accurate OFDM symbol synchronization, as shown in **Figure 6(a)**. A legitimate ONU can correctly demodulate the original data; however, with a slight change in the initial keys, the peak does not appear, so that the data cannot be fully recovered, as illustrated in **Figure 6(b)**.

In chaotic (random) PTS, the chaotic sequences $\{D_{y,i}\}$, $\{D_{z,i}\}$, and $\{D_{u,i}\}$ can be employed to generate the partition information and the phase weighing factors. First, the digitized chaotic sequence $\{D_{y,i}\}$ is applied to generate the binary sequence $\{A_1\}$ using $\{\mathrm{mod}(D_{y,i}, 2)\}$. Similarly, the binary sequence can be generated by $\{D_{z,i}\}$. After bitwise logical operation of these two binary sequences, as shown in **Figure 7**, the chaotic random partition information $\{P^{(l)}\}$ is formed. The sub-blocks are then generated using $X^{(l)} = X \cdot P^{(l)}$, $l = 1, 2, \cdots, L$. where L denotes the total number of sub-blocks.

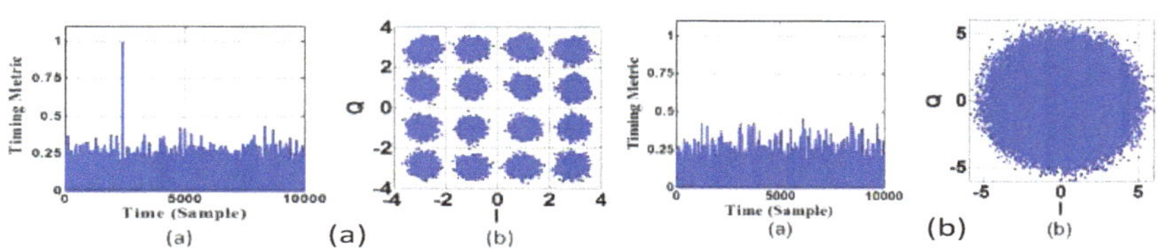

Figure 6. Effect of OFDM symbol time synchronization with (a) the right key; (b) a wrong key.

	N															
A_1	1	0	1	0	1	0	0	0	1	0	0	1	⋯	1	1	1	0
A_2	1	1	0	1	0	0	1	1	1	0	1	0	⋯	0	0	0	1
$P^{(1)} = A_1 A_2$	1	0	0	0	0	0	0	0	1	0	0	0	⋯	0	0	0	0
$P^{(2)} = A_1 \bar{A}_2$	0	0	1	0	1	0	0	0	0	0	0	1	⋯	1	1	1	0
$P^{(3)} = \bar{A}_1 A_2$	0	1	0	1	0	0	1	1	0	0	1	0	⋯	0	0	0	1
$P^{(4)} = \bar{A}_1 \bar{A}_2$	0	0	0	0	0	1	0	0	0	1	0	0	⋯	0	0	0	0

Figure 7. Generation of partition information (four sub-blocks) in chaotic PTS.

Second, the digitalized chaotic sequence $\{D_{u,i}\}$ is applied to generate the phase weighing factors R_l in PTS,

$$R_l = \exp\left(j \cdot 2\pi \cdot D_{u,i}/M\right) \quad (i = 1, 2, \ldots, K) \tag{7}$$

where K is the total number of phase weighing factors. After multiplying the phase weighing factors and transforming from frequency into time-domain, the corresponding time-domain vector becomes

$$x = \sum_{l=1}^{L} R_i x^{(l)} \tag{8}$$

$$x_n^{(l)} = \frac{1}{\sqrt{N}} \sum_{s=0}^{N-1} X_s^{(l)} \exp\left(j \frac{2\pi sn}{N}\right) \tag{9}$$

where $x^{(l)}$ is the IFFT of $X^{(l)}$, s and n denote the indices of OFDM subcarrier and symbol, respectively.

In PTS, the combinations of sub-blocks and phase weighing factors are calculated to find the minimum PAPR, thus the final PAPR is

$$PAPR_{\mathrm{PTS}} = \min\left(PAPR\left(\sum_{l=1}^{L} R_i x^{(l)}\right)\right) \tag{10}$$

The final OFDM vector becomes

$$x_{pts} = \sum_{l=1}^{L} R_{l,\,\mathrm{opt}} x^{(l)} \tag{11}$$

where $R_{l,opt}$ is the optimized phase weighing factors. At ONUs, the decryption and demodulation procedures are the inverse operations that are implemented at OLT.

Since the initial values in digital chaos serve as the secure keys to recover the original OFDM signals, only legal ONUs can generate the same chaotic scrambling information with the correct secure keys that applied at OLT. By introducing digital chaos into signal scrambling, not only the security confidentiality of OFDM signal transmission is greatly enhanced, but also a performance-improved transmission can be expected with a lower PAPR.

As for the computational complexity, it is the same for the chaotic PTS and conventional random PTS. If compared with the case of OFDM transmission without PTS, the computational complexity of chaotic PTS scheme will be L^k times higher, which is inevitable for PAPR optimization. In our case, assuming four sub-blocks and two phase weighing factors, the computational complexity is 16 times higher for PAPR optimization; however, it guarantees both higher security and lower PAPR. Since the number of possible PTS partition states is 4^N ($4^{256} \approx 10^{154}$ in our case), it is hardly to be broken through brutal-force attacks. However, chaotic PTS still requires sideband information to transmit the partition and the phase factor information, which reduces the overall spectral efficiency in transmission.

4.1.2. Chaotic selected mapping (SLM)

As shown in **Figure 8**, chaotic SLM is another alternative data encryption scheme based on signal scrambling using digital chaotic sequences. In chaotic SLM, the digitized chaotic sequences can be obtained by processing of the chaotic sequences from 2D Henon chaos [19],

$$\begin{cases} x_{i+1} = 1 - \eta x_i^2 + y_i \\ y_{i+1} = \psi x_i \end{cases} \tag{12}$$

The input OFDM sequence is denoted as $X = [X_0, X_1, \ldots, X_{N-1}]$ in frequency domain, where X_k represents the complex data of the *k*th subcarrier, and N represents the total number of subcarriers. First, to implement the digital chaos into SLM scheme, the total number of the chaotic phase sequences is set to be V, and all of the phase sequences $P_v = [P_{v,0}, P_{v,1}, \ldots, P_{v,N-1}]$ $(1 \leq v \leq V)$ are obtained from the chaotic sequence $\{D_{y,i}\}$, where $P_{v,k} = exp(j \cdot 2\pi \cdot D_{y,k}/M)$, $(0 \leq k \leq N-1)$. By component-wise vector multiplication of the input OFDM sequence X with the phase sequence $\{P_v\}$, the input OFDM signals are encrypted fully via the chaotic sequences obtained in Eq. (12). The encrypted OFDM sequence $\{Y_v\}$ is

$$Y_v = X \otimes P_v = [X_0 P_{v,0}, X_1 P_{v,1}, \cdots, X_{N-1} P_{v,N-1}] \tag{13}$$

where $\otimes$ denotes the component-wise vector multiplication. Second, the IFFT is performed for all of the encrypted OFDM sequences $\{Y_v\}$

$$x_v = \text{IFFT}(Y_v) \tag{14}$$

Finally, the OFDM signal sequence $\{x_v\}$ with minimum PAPR is chosen for OFDM transmission. It should be mentioned that in terms of computational complexity, better PAPR reduction can be expected with a larger V; however, the complexity will also increase due to the increased times of IFFT.

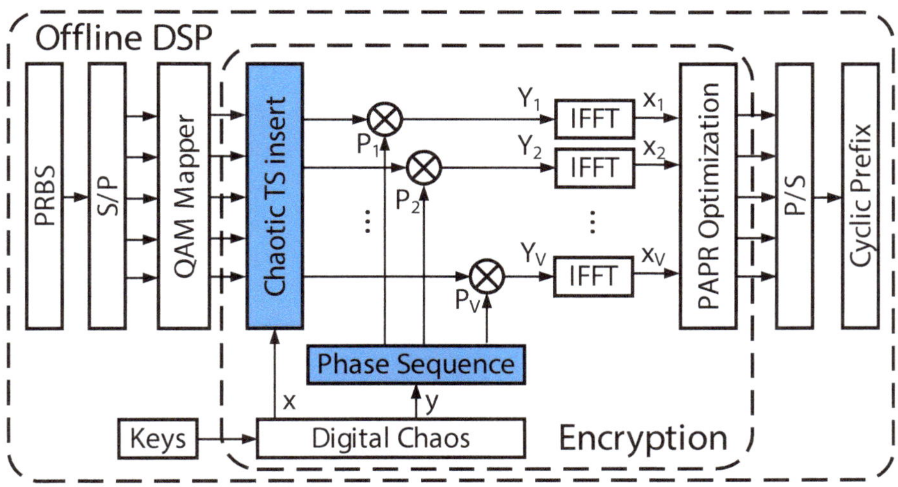

Figure 8. Schematic diagram of OFDM signal encryption using chaotic SLM.

4.1.3. Experiment results of chaotic PTS and SLM

The experimental setup of encrypted OFDM-PON transmission using chaotic PTS/SLM is shown in **Figure 9**. The input data stream was mapped onto 129 OFDM subcarriers at OLT, where 64 subcarriers were in the format of 16-QAM, and one subcarrier was unfilled with DC. The other 64 subcarriers had to be set as the complex conjugate of the aforementioned 64 subcarriers, where Hermitian symmetry had to be satisfied to realize intensity modulation and direct detection (IM/DD) after IFFT. After TS insertion for OFDM symbol time synchronization, chaotic PTS/SLM optimization and parallel-to-serial (P/S) conversion, a CP of 1/8 length of OFDM symbol was added into each OFDM symbol.

The above processing steps were performed offline using Matlab programs. The encrypted signal was then loaded into an arbitrary waveform generator (AWG), which had a sample rate of 10 Gs/s of electrical OFDM signal generation. A continuous-wave (CW) laser ($\lambda = 1550.3$ nm) was applied as the optical carrier. After that, the electrical signal in AWG was changed into optical by an optical Mach-Zehnder modulator (MZM). Thus, the net data rate applied was actually at 8.9 Gb/s (10 Gs/s $\times 4 \times 64/256 \times 8/9$), with the electrical bandwidth of 2.5 GHz (10 Gs/s $\times 65/256$). At ONU, the received data were recorded with a 20 Gs/s real-time oscilloscope after direct detection via a 10-GHz photodiode (PD). The waveform, electric and optical spectra of OFDM signals are measured and plotted in **Figure 10**.

The CCDFs of encrypted OFDM signals are plotted in **Figure 11**, where 10,000 OFDM symbols with 16-QAM modulation format were employed. From **Figure 11** it can be noted that, by employing chaotic signal scrambling, significant PAPR reduction is achieved, with respect to the conventional OFDM transmission without any signal scrambling. The total number of sub-blocks and phase factors applied in PTS was $L = 4$ and $K = 2$ respectively, and the phase factor in SLM was $V = 4$.

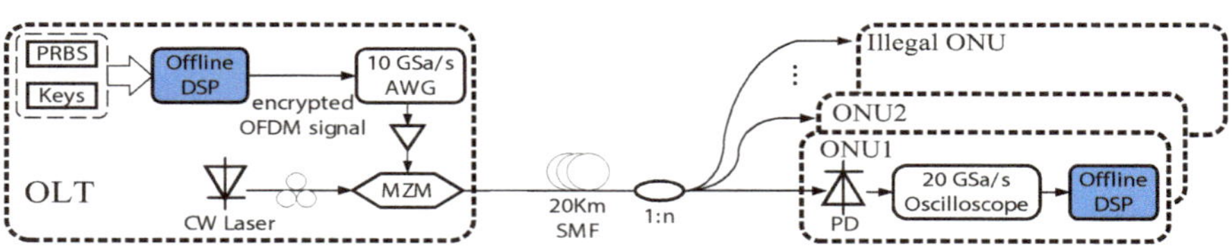

Figure 9. Experimental setup of encrypted OFDM transmission using chaotic PTS/SLM.

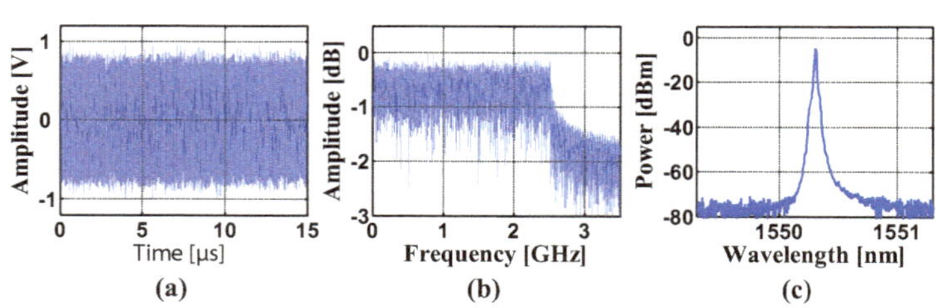

Figure 10. Encrypted OFDM signals. (a) Waveform; (b) electric spectrum; (c) optical spectrum.

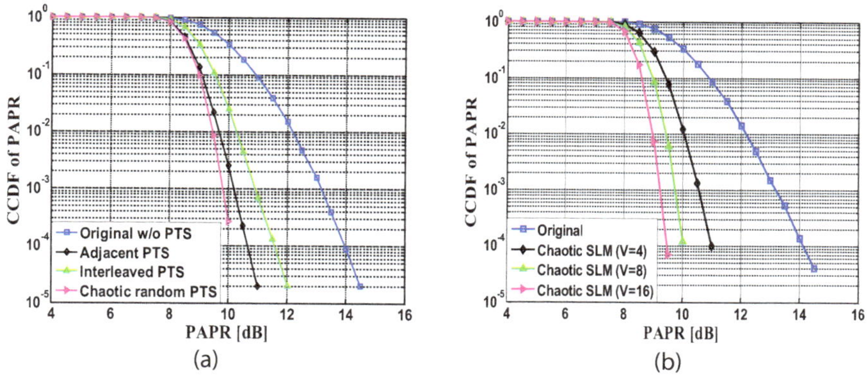

Figure 11. CCDFs of PAPR in chaotic (a) PTS; (b) SLM.

The bit error ratio (BER) measurements of secure OFDM transmission using chaotic PTS/SLM are plotted in **Figure 12**. The OFDM signals can be correctly decrypted after transmission over 20 km standard single-mode fiber (SSMF) for legitimate ONUs with the correct keys, while any illegal ONU cannot decrypt OFDM signals with a wrong key, which is even a tiny discrepancy of 1×10^{-15} away from the correct key. The corresponding constellations are plotted as the insets in **Figure 12**. If compared with the conventional OFDM transmission, the BER performance was improved ~1 dB (BER@10^{-3}) for encrypted OFDM signals, if compared with back-to-back (B2B) signals, which can mainly be attributed to the effective reduction of PAPR through the signal scrambling schemes, either chaotic PTS or SLM.

4.2. Chaotic precoding of OFDM signals

Matrix precoding is a set of typical approaches for PAPR reduction, which provides unique properties such as lower computational complexity and higher spectral efficiency (without requirement of sideband), if compared with the signal scrambling schemes. Actually, PAPR

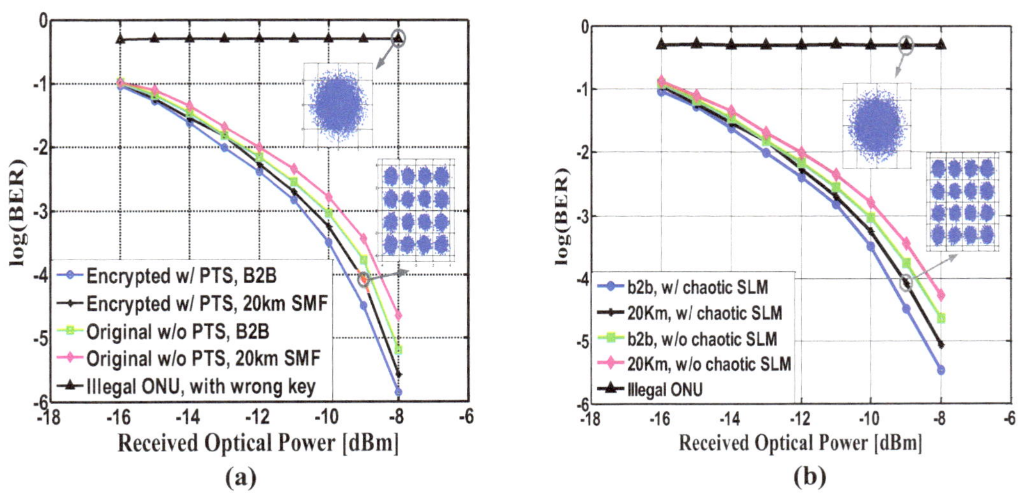

Figure 12. BERs of encrypted OFDM transmission using chaotic (a) PTS; (b) SLM.

reduction is achieved through the proper selection and distribution of power between OFDM blocks. In this section, the digital chaos will be employed into the precoding schemes, aimed to construct or reconfigure the standard precoding matrices into alternative chaotic ones, so that the security can be enhanced during transmission via the introduction of digital chaos into the procedure of OFDM signal precoding, and jointly the PAPR can be reduced.

4.2.1. Chaotic discrete Fourier transform spread OFDM (DFT-S-OFDM)

As depicted in **Figure 13**, the DFT-S-OFDM data encryption is implemented multi-fold, such as chaotic TS insertion, chaotic DFT matrix generation, and chaotic subcarrier allocation, all of which are predetermined by digital chaotic sequences [20]. The details of encryption are given as follows.

In the first fold of encryption, one chaotic sequence is used to generate TS, the same way as described in the previous chaotic signal scrambling schemes. The second fold of encryption is to generate the chaotic reconfigurable DFT precoding matrix. Assuming $\mathbf{F}$ is the standard $M \times M$ DFT matrix, the matrix element is given by

$$\mathbf{F}_{\alpha,\beta} = \frac{1}{\sqrt{M}} e^{-j2\pi(\alpha-1)(\beta-1)/M},\ 0 \le \alpha, \beta \le M-1 \tag{15}$$

where α, β are the indexes of row and column, respectively. From Eq. (15), a reconfigurable matrix can be generalized

$$\mathbf{F}'_{\alpha,\beta} = \frac{1}{\sqrt{M}} e^{-j2\pi(\alpha-m)(\beta-n)/M},\ 0 \le \alpha, \beta \le M-1 \tag{16}$$

where m, n are the real constants, which can be predetermined using chaotic sequences obtained in Eq. (5) as

$$m_{y,i} = \cdots + 10n_1 + n_2 + 0.1n_3 + 0.01n_4 + \cdots \tag{17}$$

where n_j donates the n_jth digit in the decimal part of y. Similarly, $\{n_{z,i}\}$ can be generated from $\{z_i\}$ in the same way as $\{m_{y,i}\}$.

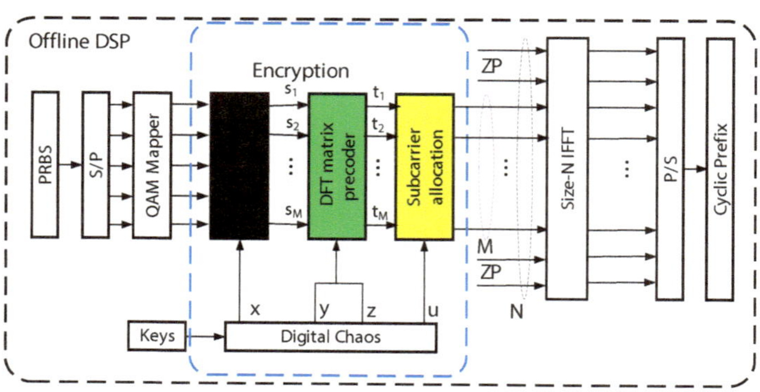

Figure 13. Schematic diagram of chaotic encrypted DFT-S-OFDM signal transmission.

Mathematically, the same basic features in the original standard **F** are kept for the reconfigurable DFT matrix **F**′. For example, the inverse matrix of **F**′ is equal to its conjugate matrix. Moreover, all of the elements in **F**′ are in geometric sequence with the same common ratio of $e^{-j2\pi/M}$ for the same row or column indexes.

To verify the capability of PAPR reduction of OFDM signals by the new DFT matrix **F**′, the upper bound of the peak factor should be considered, which is the square root of PAPR and depends on the aperiodic autocorrelation functions of the input symbols.

$$\gamma \leq 1 + \frac{2}{N}\sum_{n=1}^{N}|\rho_n| \tag{18}$$

$$\rho_n = \sum_{l=1}^{N-l} a_l a_{l+n}, \quad n = 1,\ 2, \ldots N-1 \tag{19}$$

where ρ_n are the aperiodic autocorrelation coefficients, a_n are the input symbols (QAM, QPSK, etc.), and N is the total number of subcarriers. Eq. (18) shows that, if the autocorrelation coefficients in the corresponding input symbols are small, the peak factor of OFDM signals will be small.

Assuming the kth OFDM symbol is $s^{(k)} \triangleq [s_1^{(k)}, s_2^{(k)}, \cdots, s_M^{(k)}]^T$, where T denotes the transpose of matrix, the DFT precoded OFDM signal becomes

$$\boldsymbol{t}^{(k)} \triangleq [t_1^{(k)}, t_2^{(k)}, \cdots, t_M^{(k)}]^T = \mathbf{F}'\boldsymbol{s}^{(k)} \tag{20}$$

As plotted in **Figure 14(a)**, the same non-periodic autocorrelation functions are obtained for the OFDM symbol $\boldsymbol{t}^{(k)}$, either precoded with DFT matrix **F** or **F**′. From **Figure 14(a)**, it should be noted that different shapes of sidelobe appear in the autocorrelation functions, while the sidelobe shapes of OFDM signals with DFT precoding are lower than those without DFT precoding, when the same modulation format is applied. This verifies that, effective PAPR reduction of OFDM signals is achieved by DFT precoding with the reconfigurable matrix **F**′, the same effect as the standard DFT matrix **F**.

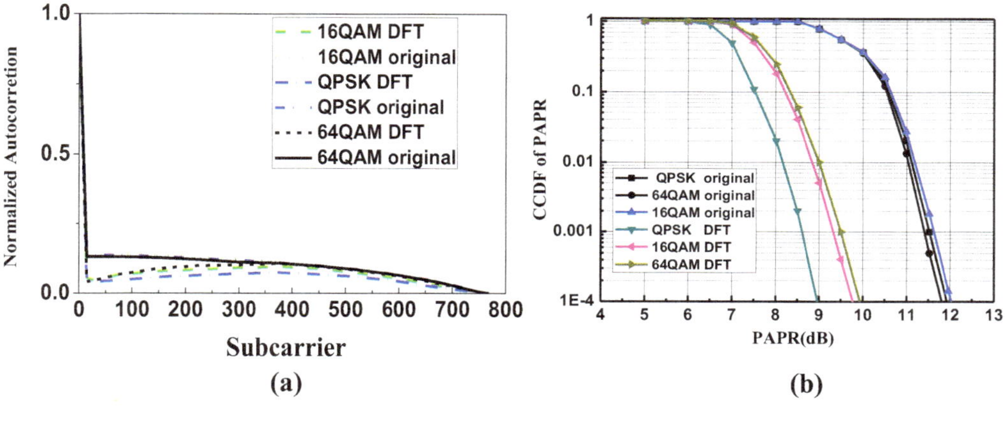

Figure 14. PAPR reduction using chaotic DFT, (a) normalized aperiodic autocorrelation function; (b) CCDFs of PAPR.

To directly evaluate the PAPR reduction, the CCDF curves is plotted in **Figure 14(b)**, for the cases of OFDM signals with and without the reconfigurable DFT precoding, where 100,000 OFDM symbols are applied in the formats of QPSK, 16-QAM, and 64-QAM, while Nyquist sampling rate is adopted. It can be observed that, if compared with the cases without DFT, the OFDM signals precoding with the reconfigurable matrix DFT can reduce the PAPR, regardless of the modulation formats, while the differences at the probability of 10^{-4} are about 3, 2.3, and 1.9 dB respectively for QPSK, 16-QAM, and 64-QAM. The CCDF curves in **Figure 14(a)** prove that, the new reconfigurable DFT does significantly reduce the PAPR of OFDM signals.

Finally, the fourth chaotic sequence $\{u_i\}$ is used to generate the scrambling criterion for subcarrier allocation before ZP. After an array $\{w_i\}$ $(i=1,\ldots,M)$ is obtained from $\{u_i\}$, where M is the total number of subcarriers (as shown in **Figure 13**), the elements in $\{w_i\}$ are swapped into $\{w_i'\}$ in the new order with respect to their values, for instance, from the lowest value to the highest. Then the new index order of subcarriers is rearranged as that of $\{w_i'\}$, details as shown in **Figure 15**. For example, assuming an original chaotic array of $\{0.8, 0.3, 0.5, 1.2, 1.1\}$, it is swapped from the small value to the big one, and then the new array $\{0.3, 0.5, 0.8, 1.1, 1.2\}$ can be obtained. In the original OFDM symbol, the first subcarrier is changed into the third position of the new OFDM symbol; similarly, the second is assigned to the first position, and so on. Following this scrambling criterion, a key space of $M!$ is created for OFDM data encryption.

The setup of IM/DD OFDM transmission experiment was the same as applied in the previous signal scrambling schemes; however, with the FFT points of 1024 and the number of subcarriers of 384, consequently the data rate was 13.3 Gb/s (10Gs/s$\times 4 \times$384/1024$\times$8/9). The BER measurements were shown in **Figure 16**, for the encrypted 16-QAM OFDM signals. The encrypted signals with DFT-S-OFDM precoding were fully recovered by the legal ONUs. Due to the effective PAPR reduction via DFT precoding, the BER performance was improved ~2 dB (BER@10^{-3}) for the encrypted DFT-S-OFDM signals, if compared with the cases of original OFDM transmission without DFT precoding. The power penalty was only 0.9 dB (at BER@10^{-3}) between the case of B2B and transmission after 20 km SSMF. Moreover, for any illegal ONU with a wrong key, even a very small deviation of 1×10^{-15}, it was not possible to decrypt the original OFDM signals. The corresponding constellations for correct and incorrect chaotic keys are plotted as insets in **Figure 16**, which verified the security feasibility of chaotic DFT-S-OFDM signal transmission.

To quantitatively evaluate the performance dependence on the mismatches of row or column parameters Δm, Δn in the reconfigurable DFT matrix $\mathbf{F}'$, the BERs are shown with respect to the parameters Δm, Δn in **Figure 17**, where the modulation format is 16-QAM, and the received optical power is -9 dBm. In **Figure 17(a)**, when $\Delta m > 0.01$, the BER is significantly deteriorated

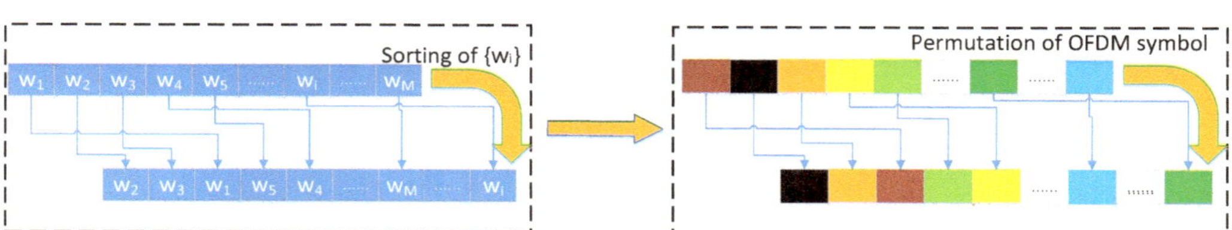

Figure 15. Subcarrier allocation in OFDM symbol using digital chaotic sequences.

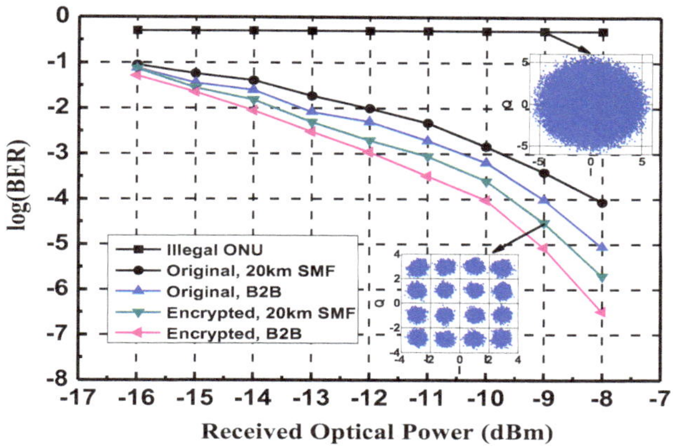

Figure 16. BERs of chaotic encrypted DFT-S-OFDM transmission.

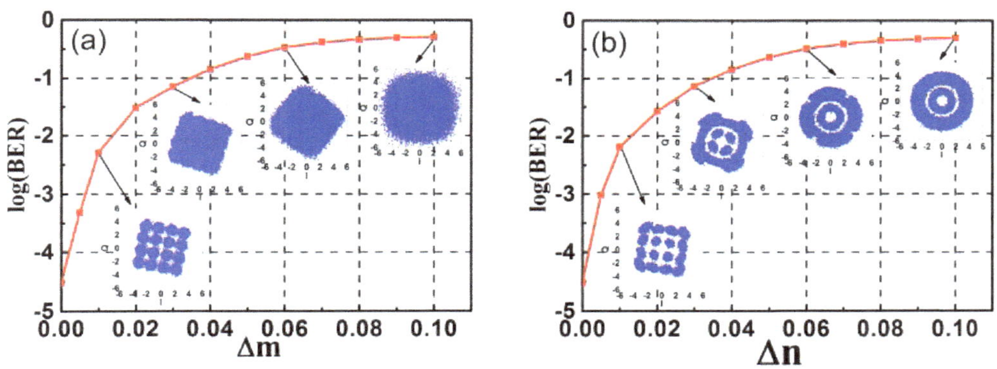

Figure 17. BERs versus the index mismatches in (a) Δm (b) Δn of chaotic DFT matrix.

(BER>10^{-3}), thus the data cannot be correctly received, even after using forward error correction approaches. It can also be noted that, if Δm is increased further, deterioration becomes more serious, as shown in the corresponding BERs and constellations. Similarly, the BERs variation with respect to the column parameter mismatch in the reconfigurable DFT matrix in $\mathbf{F}'$ is plotted in **Figure 17(b)**. From **Figure 17(b)**, the original data can be only recovered when $\Delta n < 0.01$. The results also verify that, the rotation angles in the corresponding constellations become larger with the increase of Δn. The difference of BERs and constellations with respect to Δm, Δn again confirms that it is helpful to set two independent values of m, n in the new reconfigurable DFT matrix in $\mathbf{F}'$, for the purpose of key space expansion in encryption scheme using chaotic DFT precoding.

For the proposed multi-fold DFT-S-OFDM data encryption, the total key space can be given as follows. First, the parameters in DFT matrix (both of row and column) can be varied from 0 to 384 with the sensitivity of 0.01, which generates a key space of ~$1.5\times10^9(38,400\times38,400)$ for each OFDM symbol. Second, the subcarrier allocation generates a key space of 384!. Finally, the chaotic TS for OFDM symbol time synchronization enlarges the key space by a factor of ~10.

4.2.2. Chaotic Walsh-Hadamard transform (WHT)

Chaotic WHT is an alternative encryption technique using signal precoding, which provides lower computational complexity if compared with DFT precoding. The encryption scheme based on a reconfigurable chaotic WHT precoding matrix is shown in **Figure 18** [21], where the modified version of reconfigurable WHT matrix is also determined by digital chaos.

Theoretically, it can be proven that the permutation on the standard WHT matrix does not degrade the precoding properties. Since the WHT matrix is an orthogonal matrix whose elements are constrained from the set $\{-1, 1\}$,

$$\mathbf{H}\mathbf{H}^{\mathrm{T}}=\mathbf{H}^{\mathrm{T}}\mathbf{H}=\mathbf{nI} \tag{21}$$

Assuming $\mathbf{Q}_{n\times n}$ and $\mathbf{R}_{n\times n}$ are the row and column permutation matrices on WHT matrix $\mathbf{H}_{n\times n}$ respectively, then the chaotic row/column permutated WHT matrix becomes $\mathbf{P}_{n\times n}$.

$$\mathbf{P}=\mathbf{QHR},\quad \mathbf{P}^{\mathrm{T}}=\mathbf{R}^{\mathrm{T}}\mathbf{H}^{\mathrm{T}}\mathbf{Q}^{\mathrm{T}} \tag{22}$$

$$\mathbf{P}\mathbf{P}^{\mathrm{T}}=\mathbf{QH}(\mathbf{R}\mathbf{R}^{\mathrm{T}})\mathbf{H}^{\mathrm{T}}\mathbf{Q}^{\mathrm{T}}=\mathbf{QHIH}^{\mathrm{T}}\mathbf{Q}^{\mathrm{T}}=\mathbf{QnIQ}^{\mathrm{T}}=\mathbf{nI} \tag{23}$$

To specify the PAPR reduction capability of chaotic WHT matrix, the same analysis based on the autocorrelation coefficient can be applied as for chaotic DFT.

As plotted in **Figure 19(a)**, the original 16-QAM OFDM symbols are applied and compared with the standard and chaotic WHT. The autocorrelation function is reduced in sidelobe if either chaotic or standard WHT is employed. An enlarged sidelobe for a specific zone is inserted in **Figure 19(a)** for clear view. Moreover, the sidelobe shape is exactly the same for both the standard and chaotic WHT precoders.

Furthermore, the CCDF curves are calculated for the cases of chaotic WHT, standard WHT, and un-precoded original OFDM signals, as shown in **Figure 19(b)**. The PAPR is effectively reduced by ~1 dB by either the standard or chaotic WHT precoder, compared to the un-precoded OFDM signals. In addition, the CCDFs are exactly the same for the standard and chaotic WHT, which again verifies the above theoretical analysis.

Therefore, row/column permutated WHT can be still applied as a precoder not only to reduce the PAPR but also to enhance the physical-layer security of OFDM data, since the correct

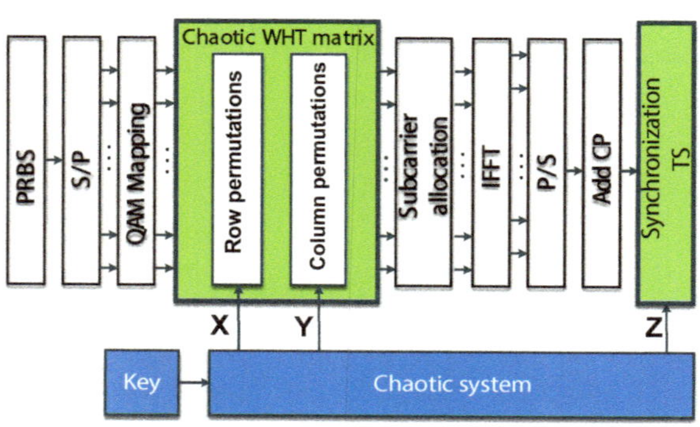

Figure 18. Schematic diagram of OFDM encryption precoded with chaotic WHT.

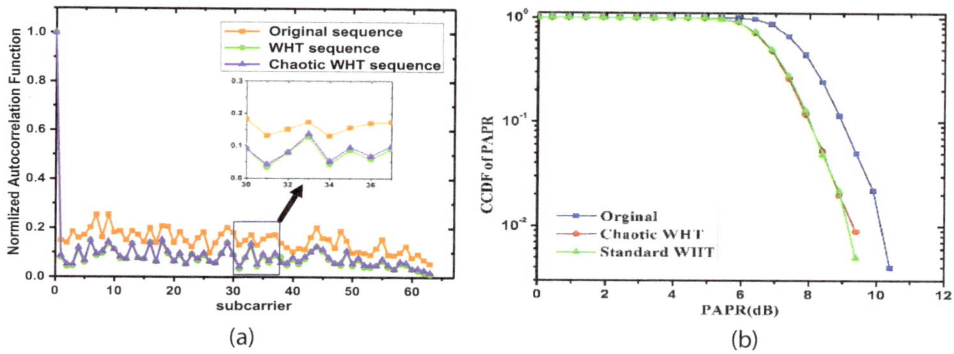

Figure 19. PAPR reduction precoded by the standard/chaotic WHT, (a) normalized aperiodic autocorrelation function; (b) CCDF of PAPR.

original OFDM data can be recovered only after obtaining the correct row/column sequences to reconstruct the chaotic WHT matrix through legal ONUs.

In the proposed chaotic WHT scheme, a 3D hyper Chen's chaos is implemented to generate the digital chaotic sequences for data encryption [22]:

$$\begin{cases} \dot{x} = -ax + y \\ \dot{y} = (c-a)x + cy + xz \\ \dot{z} = bz + xy \end{cases} \tag{24}$$

where a, b, and c are constants. Generally, with a=35, b=3, $c\in[20, 28.4]$, it enters into chaotic zones. To generate the permutation vectors D_x, D_y, and D_z, each of the chaotic sequences $\{x\}$, $\{y\}$, and $\{z\}$ is post-processed as follows.

First, we assume that the length of chaotic sequences $\{x\}$ and $\{y\}$ is M, where M is the order of WHT matrix, and then

$$D_x = sort\{x\}, \quad D_y = sort\{y\}, \quad D_z = sort\{z\} \tag{25}$$

where *sort* function returns an index vector, according to the ascending order of the values in the chaotic sequences.

The procedure of the proposed WHT chaotic encryption is given as follows. The two permutation vectors D_x and D_y, are used to permute the row and column indexes of the standard WHT matrix, respectively. Assuming the standard WHT matrix $\mathbf{H}^{(0)}_{M\times M}$ is

$$\mathbf{H}^{(0)}_{M\times M} = [\boldsymbol{r}_1^T, \boldsymbol{r}_2^T, \cdots, \boldsymbol{r}_M^T] \tag{26}$$

where $r_1, r_{2,\ldots}, r_M$ are the row vectors. After row permutation, it becomes

$$\mathbf{H}^{(1)}_{M\times M} = [\boldsymbol{r}_{\boldsymbol{D}x(1)}^T, \boldsymbol{r}_{\boldsymbol{D}x}{}^T{}_{(2)}, \cdots \boldsymbol{r}_{\boldsymbol{D}x}{}^T{}_{(M)}] = [\boldsymbol{c}_1, \boldsymbol{c}_2 \cdots \boldsymbol{c}_M] \tag{27}$$

Similarly, the columns are permutated via vector D_y, and the corresponding chaotic WHT matrix after both row and column permutations becomes

$$\mathbf{H}^{(2)}_{M\times M} = \left[c_{D_y(1)}, c_{D_y(2)}, \cdots, c_{D_y(M)}\right] = \left[v_1^T, v_2^T, \cdots, v_M^T\right] \tag{28}$$

The chaotic sequence D_z is used to generate the chaotic TS for OFDM symbol synchronization, which is composed of random $\{-1, 1\}$ and defined as [23]

$$\text{TS} = \{(\text{mod}(D_{z,i},\ 2) - 0.5)\times 2\} \tag{29}$$

Since the hyper-chaotic sequences have high sensitivity to the initial values (~1×10^{-15}) and present exceptional random behavior, the initial values are served as the security key between OLT and ONUs. A secure channel has to be employed to transfer the keys between OLT and ONUs, for instance, using quantum key distribution (QKD), which is out of the scope of this chapter and will not be given in detail here. **Figure 20** shows the sensitivity of the row/column permutation indexes of chaotic WHT matrix with respect to a tiny change in the initial conditions.

Figure 21 shows the experimental setup of the encrypted transmission of OFDM signals using chaotic WHT precoding. The IM/DD was also applied, and the details were described in Section 4.1. At OLT, there were 256 points in IFFT, among which 64 subcarriers were used for OFDM data. The effective data rate was 8.9 Gb/s (10 Gs/s×4×64/256/(1+1/8)). The signal waveform, electric and optical spectra are plotted as insets in **Figure 21**.

The transmission performance was evaluated for OFDM signals precoded by chaotic WHT via BER measurements of 128 OFDM symbols, for the cases of B2B and transmission after 20 km

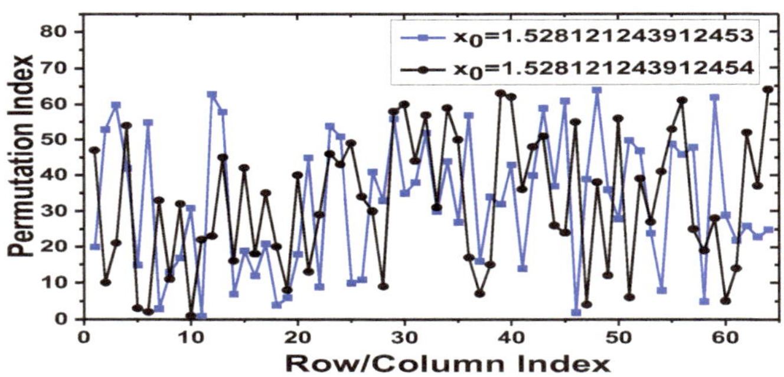

Figure 20. Chaotic permutation index versus a slight change in the initial values.

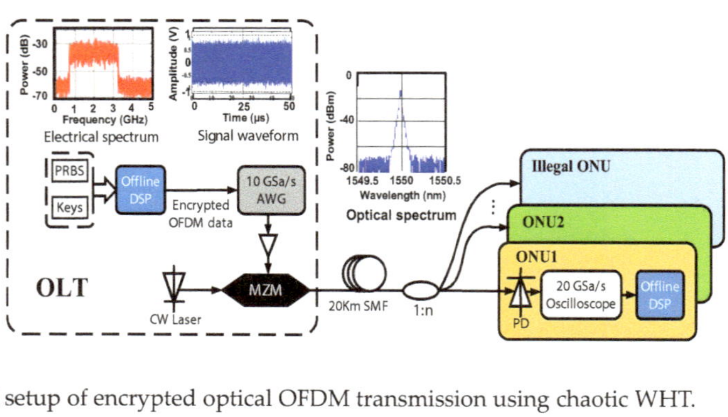

Figure 21. Experimental setup of encrypted optical OFDM transmission using chaotic WHT.

SSMF, as shown in **Figure 22**. The BER was improved by ~1 dB (BER@10^{-3}) in receiver sensitivity for chaotic WHT, compared with the un-precoded OFDM signals. This improvement can be mainly attributed to the precoding gain from chaotic WHT, since BER was improved also for the case of B2B, which verified the effective PAPR reduction in OFDM signals through chaotic WHT precoding. For transmission after 20 km SSMF, the effect of fiber dispersion is negligible due to the narrow spectrum of OFDM signals.

The robustness of the multi-fold data encryption is evaluated via the total key space created by chaotic WHT, since only legal ONUs can generate the correct chaotic WHT matrix to recover the original OFDM data. Assuming that the dimension of chaotic WHT matrix is 64×64, it creates a total key space of $64! \times 64!$ (~10^{178}). In addition, the chaotic TS for OFDM symbol synchronization further increases the key space by a factor of ~10. As a result, a total key space of ~10^{179} is achieved in chaotic WHT encryption scheme.

The computational complexities the chaotic encryption schemes with PAPR reduction are listed in **Table 1**. The computational complexity is reduced significantly in WHT, since it does not require complex multiplication if compared with DFT, SLM, and PTS, because the elements in WHT matrix are constrained to the value set of {1,−1}. Meanwhile, WHT has the same

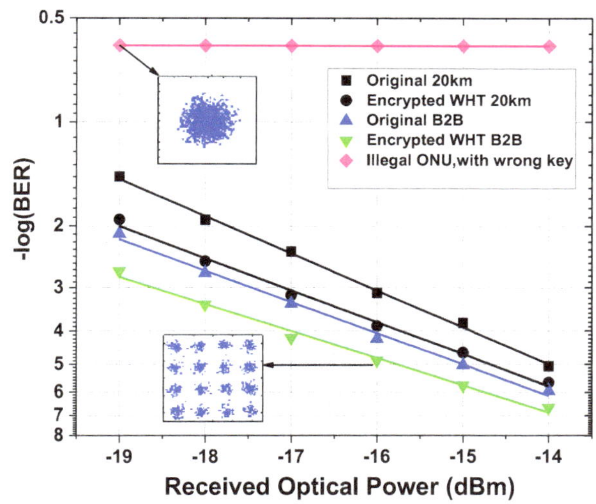

Figure 22. BER measurements of the OFDM transmission with and without chaotic WHT.

	WHT	DFT	SLM	PTS
Multiplication	None	N^2	$B\frac{gN}{2}\log_2 gN$	$V\frac{gN}{2}\log_2 gN$
Addition	$N(N-1)$	$N(N-1)$	$BgN \log_2 gN$	$VgN\ log_2\ gN + (V-1)gNW^V$
Sideband	None	None	Needed	Needed

B is the number of candidate signals in SLM, *V* is the number of sub-blocks, and *W* is the number of phase factors.

Table 1. Computation complexity and sideband information of chaotic encryption schemes with PAPR reduction.

addition complexity as DFT. In addition, if compared with PTL or SLM schemes, simpler optical OFDM transmission structure and higher spectral efficiency can be expected for the WHT and DFT precoding schemes, since no additional sideband information needs to be transmitted.

5. Conclusions

We have reviewed a serial of novel data encryption schemes for physical-layer security enhancement for optical OFDM transmission. By incorporating multi-fold digital chaos-based data encryption in various DSP procedures in OFDM signal generation, modulation as well as OFDM symbol synchronization, data security is greatly enhanced with a huge key space. Moreover, the OFDM transmission performance is significantly improved because the pseudo-random properties of digital chaos are employed for effective PAPR reduction. Moreover, for the chaotic precoding schemes, the reconfigurable chaotic precoding matrices are predetermined by digital chaos; therefore, the sideband information is no longer necessary, which also improves the spectral efficiency in transmission. The proposed chaotic data encryption schemes have been successfully demonstrated by ~10 Gb/s OFDM transmission experiments, which verifies that these schemes could be promising candidates for next-generation secure OFDM-PON.

Acknowledgements

This work was supported in part by International S&T Cooperation Program of China, 2016YFE0104500, Natural Science Fund of China under Grants 61571291, 61431009, and 61221001.

Author details

Xuelin Yang*, Adnan A.E. Hajomer and Weisheng Hu

*Address all correspondence to: x.yang@sjtu.edu.cn

State Key Laboratory of Advanced Optical Communication Systems and Networks, Shanghai Jiao Tong University, Shanghai, China

References

[1] Fang Y, Yu J, Zhang J, Chi N, Xiao J, Chang G. Ultrahigh capacity access network architecture for mobile data backhaul using integrated W-band wireless and free-space optical links with OAM multiplexing. Optics Letters. 2014;**39**(14):4168–4171. DOI: 10.1364/OL.39.004168

[2] Reis J, et al. Terabit+ (192 × 10Gb/s) Nyquist shaped UDWDM coherent PON with upstream and downstream over a 12.8 nm band. Journal of Lightwave Technology. 2014;**32**(4):729–735. DOI: 10.1109/JLT.2013.2283017

[3] ITU-T Recommendation G.983.1, Broadband Optical Access Systems Based on Passive Optical Networks, International Telecommunication Union-Telecommunication Sector (ITU-T); 2005

[4] Zhang L, Xin X, Liu B, Yin X. Physical secure enhancement in optical OFDMA-PON based on two-dimensional scrambling. Optics Express. 2012;**20**(26):B32–B37. DOI: 10.1364/OE.20.000B32

[5] Liu B, Zhang L, Xin X, Wang Y. Physical-layer security in OFDM-PON based on dimension-transformed chaotic permutation. IEEE Photonics Technology Letters. 2014;**26**(2):127–130. DOI: 10.1109/LPT.2013.2290041

[6] Zhang L, Xin X, Liu B, Wang Y. Secure OFDM-PON based on chaos scrambling. IEEE Photonics Technology Letters. 2011;**23**(14):998–1000. DOI: 10.1109/LPT.2011.2149512

[7] Liu B, et al. Piecewise chaotic permutation method for physical-layer security in OFDM-PON. IEEE Photonics Technology Letters. 2016;**28**(21):2359–2362. DOI: 10.1109/LPT.2016.2594042

[8] Zhang W, et al. Joint PAPR reduction and physical-layer security enhancement in OFDMA-PON. IEEE Photonics Technology Letters. 2016;**28**(9):998–1001. DOI: 10.1109/LPT.2016.2522965

[9] Zhang W, Zhang CF, Jin W, Chen C, Jiang N, Qiu K. Chaos coding-based QAM IQ-encryption for improved security in OFDMA-PON. IEEE Photonics Technology Letters. 2014;**26**(19):1964–1967. DOI: 10.1109/LPT.2014.2343616

[10] Yang X, Hu X, Shen Z, He H, Hu W, Bai C. Physical layer signal encryption using digital chaos in OFDM-PON. In: 10th International Conference on Information Communications and Signal Processing (ICICS); November 2015. DOI: 10.1109/ICICS. 2015.7459872

[11] Cvijetic N. OFDM for next generation optical access networks. Journal of Lightwave Technology. 2012;**30**(4):384–398. DOI: 10.1109/JLT.2011.2166375

[12] Popoola WO, et al. Pilot-assisted PAPR reduction technique for optical OFDM communication systems. Journal of Lightwave Technology. 2014;**32**(7):1374–1382. DOI: 10.1109/JLT.2014.2304493

[13] Jayalath ADS, Tellambura C. Reducing the peak-to-average power ratio of orthogonal frequency division multiplexing signal through bit or symbol interleaving. Electronics Letters. 2000;**36**(13):1161–1163. DOI: 10.1049/el:20000822

[14] Sharif M, Gharavi-Alkhansari M, Khalaj BH. On the peak-to-average power of OFDM signals based on oversampling. IEEE Transactions on Communications. 2003;**51**(1):72–78. DOI: 10.1109/TCOMM.2002.807619

[15] Tang L, Zuo Q, Cui W. Synchronization scheme using four-dimensional chaotic system for OFDM. Journal of Communication. 2010;**31**(1):73–84 (in Chinese).

[16] Alvarez G, Li S. Some basic cryptographic requirements for chaos-based cryptosystems. International Journal of Bifurcation and Chaos. 2006;**16**(8):2129–2151. DOI: 10.1142/S0218127406015970

[17] Hu X, Yang X, Shen Z, He H, Hu W. Chaos-based partial transmit sequence technique for physical-layer security in OFDM-PON. IEEE Photonics Technology Letters. 2015;**27**(23):2429–2432. DOI: 10.1109/LPT.2015.2466092

[18] Cheng M, et al. Security-enhanced OFDM-PON using hybrid chaotic system. IEEE Photonics Technology Letters. 2015;**27**(3):326–329. DOI: 10.1109/LPT. 2014.2370757

[19] Hu X, Yang X, Hu W. Chaos-based selected mapping scheme for physical-layer security in OFDM-PON. Electronics Letters. 2015;**51**(18):1429–1431. DOI: 10.1049/el.2015.1261

[20] Shen Z, Yang X, He H, Hu W. Secure transmission of optical DFT-S-OFDM data encrypted by digital chaos. IEEE Photonics Journal. 2016;**8**(3). DOI: 10.1109/JPHOT.2016.2564438

[21] Hajomer A A E, Yang X, Hu W. Chaotic Walsh–Hadamard Transform for Physical Layer Security in OFDM-PON. IEEE Photonics Technology Letters. 2017;**29**(6): 527–530. DOI: 10.1109/ LPT. 2017. 2663400

[22] Chen G, Ueta T. Yet another chaotic attractor. International Journal of Bifurcation and Chaos. 1999;**9**(7):1465–1466. DOI: 10.1142/S0218127499001024

[23] Wang X, et al. SSBI mitigation at 60GHz OFDM-ROF system based on optimization of training sequence. Optics Express. 2011;**19**(9):8839–8846. DOI: 10.1364/OE.19.008839

4

Adaptive High Linearity Intensity Modulator for Advanced Microwave Photonic Links

Benjamin Dingel, Nicholas Madamopoulos and Andru Prescod

Additional information is available at the end of the chapter

Abstract

This chapter, first, presents the motivation behind the need for adaptive, highly linear electro-optic modulators and an overview of the different optical linearization approaches of electro-optic modulators. Then, the figures of merits in terms of linearity performance are described and analyzed. Next, the chapter focuses on one excellent linearization approach called interferometric modulator with phase-modulating and cavity-modulating components (IMPACC). Here, we model IMPACC by simulating each of the key building blocks separately before putting them together as IMPACC modulator. This adaptive IMPACC design is compared to typical Mach-Zehnder interferometer (MZI) based modulators, and ring-assisted Mach-Zenhder interferometer (RAMZI) modulators. Theoretical analysis and results show that the IMPACC provides both superior linearity performance and unique adaptive feature that can be used to compensate for manufacturing tolerances, thus, providing extra flexibility in terms of device manufacturability as well as system integration.

Keywords: linear optical intensity modulator, spurious free dynamic range, bandwidth, Mach Zehnder interferometer (MZI) modulator, resonator-assisted Mach Zehnder interferometer (RAMZI), interferometric modulator with phase-modulating and cavity-modulating components (IMPACC), broadband communication

1. Introduction

1.1. Technology trends in microwave photonic links (MPLs)

The current state of microwave photonic links (MPLs) is in the midst of transformation toward more dynamic and flexible MPLs, in response to three major on-going technical shifts, as depicted in **Figure 1**. The first technical shift is an outgrowth of the concept of *software-defined*

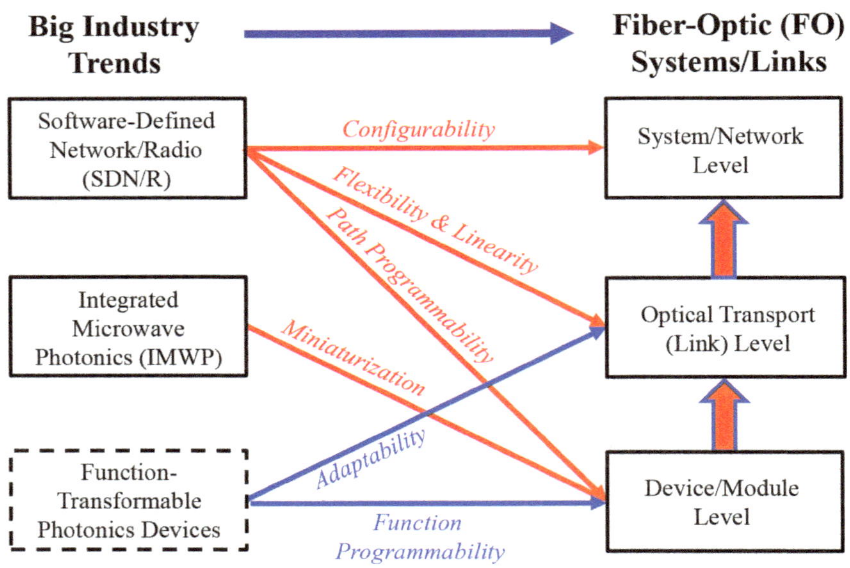

Figure 1. Technology shifts with SDN, integrated microwave photonic (IMWP) and function-transformable photonic devices are generating changes in the architecture, transport and module levels in the traditional fiber-optic links and systems.

networks (SDN) or software-defined radio (SDR) [1–6]. SDN/SDR is going to be investigated, demonstrated, and applied by the microwave photonics community from many different fronts. These include (i) architecture level for future networking needs, (ii) link level to support flexible optical transport layer needs, and even (iii) module level that requires programmable functions that are accessible by direct software control. This shift is driven further by ever expanding applications of MPLs in various "bandwidth-hungry" communication fields like cellular [7] and optical wireless [8], besides the traditional applications of cable television (CATV) [9], subcarrier multiplexing (SCM) optical communication systems [10–12], defense and military [13–16]. The overall goal is toward some form of *software-defined MPLs*.

Figure 2 shows one potential MPL in the context of a bigger configuration defined by SDN—a generic analog optical link, which we will refer to as SD-MPL. It consists of (i) a reconfigurable intensity-modulated direct-detection (IM-DD) link architecture, (ii) control plane, and (iii) network controller. The reconfigurable IM-DD comprises of three modules, namely (1) tunable transmitter, (2) fiber, and (3) receiver. The transmitter module consists of a tunable CW laser source, intensity modulator, RF driver, and RF amplifier, whereas the receiver module could be made up of fiber amplifier and photodetectors for a simple case, and might include DSP module with analog-to-digital converter (ADC) for a more sophisticated/advanced case.

Within the context of SDN, the control plane assembles, enables, and coordinates all pertinent control signals coming from and through the transmitter and receiver modules. Then, the network controller accesses the control plane through OpenFlow interface [5, 17–20] to configure the transmitter and receiver modules. For this generic architecture to function back to traditional, standalone MPLs, the network controller and control plane can be removed and

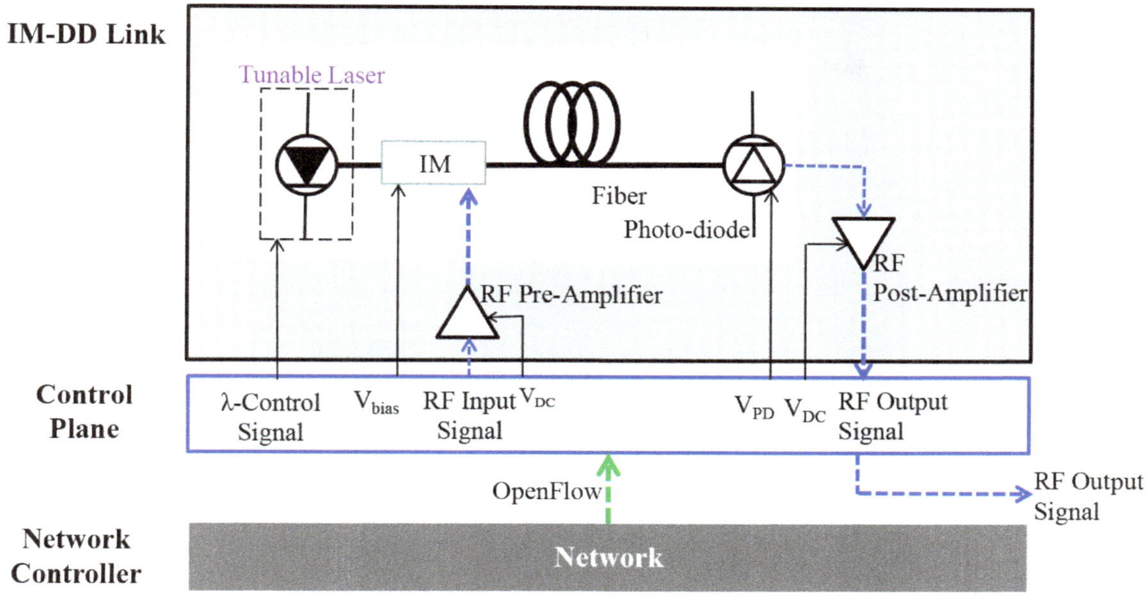

Figure 2. A generic software defined IM-DD-based MPLs in context of a bigger configuration defined by SDN.

the individual control signals are now transferred to the transmitter and receiver modules of the MPLs.

The second technical shift is the rapid growth of the field *integrated microwave photonics (IMWP)* [21–23], which used to be a niche research area some three decades ago. Today, *IMWP* aims to *miniaturize key photonic components, which are essential in the MPLs,* by using the emerging photonic integrated circuits (PICs) [23–26] technology to the level that would be comparable to the performance of state-of-the-art microwave components and systems [21]. Clearly, the rapid developments and advances [22, 23–26] in fabrication technologies re-enforce more advanced IMWP devices and support the prospect of future *software-defined* MPLs.

1.2. Device programmability

The third technical shift is happening at the module level with the development of optical programmable devices and components. These programmable devices and components can have the traditional features of reconfigurability, tunability, or selectability. There are growing number of reports and demonstrations of so-called *microwave photonics processors* [27–33].

Another implementation of these devices is based on the concept of a photonic field-programmable gate array (**FPGA**) or "Photonics FPGA". It is the analogous version of the electronic **FPGA**, which is an integrated circuit designed to be configured by a customer or a designer after manufacturing. "Photonic FPGA" is geared toward arbitrary function generation or function programmability in miniaturized form, which is essential in the full implementation of SDN. This technology is still in the early stages of development, but the growing numbers of demonstrations of on-chip optical processing have been reported and they look promising [27–40].

1.3. Linearity

However, to really support the march toward future IMWP-assisted, software-defined MPLs, it is imperative that these new photonic components not only have inherent *programmable features* that can be accessed by software control, but also *superior linearity characteristics*. Note that microwave photonic links, in its basic form, is essentially a linearization engineering challenge. Regardless of whether we are talking about current MPLs or the future SD-MPLs, the core requirement of the links is still the same—high linearity links! *Superior device's linearity* translates to *high* overall link *performance*. The combination of superior linearity and programmability features in key photonics devices would lead to adaptive, highly linearity SD-MPLs.

1.4. Intensity-modulated direct detection (IM-DD) link and figure of merits

One of the most dominant optical transport schemes in MPLs is the intensity modulated direct detection (IM-DD) because of its simplicity and low cost. IM-DD scheme is shown in the upper portion of **Figure 2**. In its basic form, it consists of a RF data signal, laser source (either in the form of (i) directly modulated laser or (ii) externally modulated laser, using a combination of CW laser transmitter and optical modulator), optical fibers, optical and electrical amplifiers, and photodetector.

For most low-bandwidth (<1 GHz) applications, the directly modulated laser is used as a laser source for economic reason. However, for applications that require higher linearity, larger link gain and wider bandwidth (e.g., 1–60 GHz), the use of an externally modulated laser is the most viable and common approach. It offers the best approach toward achieving higher link linearity compared with directly modulated lasers since CW lasers with very low RIN noise are commercially available. Furthermore, the high output power of a laser can act "as an indirect RF power booster" for the RF input signal to provide a positive gain rather than a negative gain (or equivalent loss) [41]. This important feature adds advantage to the link using an external modulator compared with a directly modulated laser. *From this point forward of this book chapter, when we talk about the links or MPLs, we will specifically mean only externally modulated links.*

The quality of the overall links performance is largely measured by four important figures-of-merit [42], namely: (1) its dynamic range in terms of the spurious-free-dynamic-range (SFDR), (2) link gain (G), (3) noise figure (NF), and (4) bandwidth. The link gain (G) describes the relationship of the RF input signal power to RF output signal power of the IM-DD link, whereas the noise figure (NF) defines the signal-to-noise ratio (SNR) degradation in the link. The SFDR depicts the RF signal power range that can be accommodated by the link, taking into account the effects of noise and nonlinear distortions. It represents the highest signal to noise ratio when the intermodulation signal power is equal to the noise floor. Thus, it combines noise and linearity performance of the link. More detailed technical discussions on these figures-of-merit are found in references [14–16, 43].

The nonlinearities in any part of the links (such as laser, modulator, fiber, photodetector, amplifier, etc.) translate directly to signal noise and distortion. Nonlinear performance at the fiber occurs when the optical powers of the signals are high enough; hence, this is usually mitigated by proper engineering of the optical signal powers to avoid fiber-based nonlinearity. Similarly, detector nonlinearity is mitigated by proper optical power engineering so that the optical power impinging on the photodetector is within the linear response of the photodetector. The nonlinearity of the modulator affects the distortion and dynamic range of fiber optic links. From among these factors, the distortion produced by the external optical modulators has the most negative impact on the overall performance of the link. More specifically, the inherent nonlinear transfer function of the external intensity modulator is the dominant limitation in the performance of MPLs.

1.5. Adaptive linearized intensity modulator

As mentioned earlier, the major source of distortion in IMDD links is the nonlinearity of the intensity modulator. The quest to develop high linearity external modulator with wide bandwidth at low cost has been an on-going challenge for decades.

Besides linearity, other key and important attributes of an external intensity modulator include, (1) low half-wave (V_π) voltage, (2) low insertion loss (*IL*), (3) high bandwidth (*BW*) capability, (4) low-cost, and (5) low complexity design. The low V_π voltage is related to the slope of the transfer function of the modulator. In general, the steeper the slope of the transfer function of the modulator, the smaller the V_π voltage. The immediate impact is an increase of the link gain (G). A modulator with low V_π voltage, as well as low insertion loss, is very attractive since it lowers the necessary input power level of the link and increases gain. Thus, low V_π and low *IL* lead to higher dynamic range, and low noise figure (*NF*), while large bandwidth capability makes it better for high-end applications. Minimal complexity and low cost are interlinked and can open the use of high bandwidth **MPLs** applicable to more commercial applications.

Lastly, any linearized modulator must also have a high tolerance to changes in the RF power level, bias drifts, optical wavelength drifts, etc. This requirement calls for modulator with adaptive capability to provide better maintenance and stable operation of the overall MPLs.

Thus, the focus of this book chapter is two-fold. First, it provides a comprehensive overview of the different linearization approaches that have been proposed and implemented to obtain a highly linear external intensity modulator. Second, it reviews a special type of intensity modulator called interferometric modulator with phase-modulating and cavity-modulating components (IMPACC) that is not only highly linear but also adaptive. These combined features, as far as we know, are not available with any previously reported intensity modulators.

1.6. Objectives and organization

This chapter of the book is organized as follows. First, we give a brief outline of the different linearization technologies used for intensity modulator in Section 2. After a broad overview, we introduce the different optical approaches in linearizing the modulators. Then, we focus on one particular subgrouping known as MZI-based linearized modulator. In Section 3, we present a special type of MZI-based linearized modulator called IMPACC. IMPACC provides both superior linearity and unique adaptive feature. Here, we discuss its basic principle, configuration, and modeling. The comparison and discussion of the original IMPACC and ring-assisted Mach-Zenhder interferometer (RAMZI) modulator are also presented. In Section 4, we provide and analyze IMPACC and RAMZI's respective performances before giving our conclusion in Section 5.

2. Survey of linear intensity modulators

In this section, we discuss the different linear intensity modulators. Before presenting these modulators, we first discuss an important figure-of-merit called spurious free dynamic range (SFDR) which quantifies the linearity of a given link. In Section 2.1, we discuss its derivation using a 2-tone test and graphically describe its significance. In Section 2.2, we provide an overview of the different general techniques to linearize intensity modulators before focusing on one general approach called optical linearization in Section 2.3. Optical linearization has many implementation flavors depending on the particular optical structure used for the modulator. In Section 2.4, we narrow down to a subgroup known as MZI-based modulator because of its popularity and mature technology. Within this subgroup, there are different configurations that have been proposed and implemented, and we spotlight one configuration called IMPACC.

2.1. Measure of linearity: SFDR

A 2-tone frequency test is typically used to evaluate SFDR of the link. **Figure 3(a)** shows a typical IM-DD based MPL consisting of a laser, external modulator RF signal amplifier, optical fiber, photodetector, and RF output amplifier. Note that when the focus of the link is to establish the performance of the external modulator, the common practice is to consider only the intrinsic MPLs. Here, the calculation of dynamic range does not contain the effects of the optical fiber, electrical amplifiers, and optical amplifier, if present.

In **Figure 3(a)**, RF input signal with two closely spaced RF frequencies, f_1 and f_2, of equal-power levels, are simultaneously injected into a nonlinear external optical modulator. The nonlinear interaction between these two signals will create new output frequencies, as shown in the bottom portion of **Figure 3(a)**. Thus, the detected RF output signal will contain two fundamental frequencies, (e.g., f_1 and f_2) together with the second-order IMD output frequencies, (e.g., $f_1 + f_2$ and $f_2 - f_1$), second-order harmonics, (e.g., $2f_1$ and $2f_2$), and third-order IMD outputs (e.g., $2f_1 - f_2$ and $2f_2 - f_1$). Other higher order harmonics and IMD effects (e.g., fourth, fifth, etc.) can also be present, but their effects are generally small and can be filtered out. The second-order SFDR is only important for systems whose bandwidth is more than one octave because the second-order intermodulation products will fall outside the passband of a suboctave system. However, the most important and often most difficult to filter out are the third-order effects which occur very close to the fundamental frequencies within the system bandwidth of interest as shown in **Figure 3(a)**.

Instead of giving the mathematical derivation of SFDR and other relevant parameters or merits, we present a graphical treatment in the discussion of SFDR [16]. For detailed mathematical derivation, we direct readers to relevant references [14–16]. **Figure 3(b)** depicts the output RF

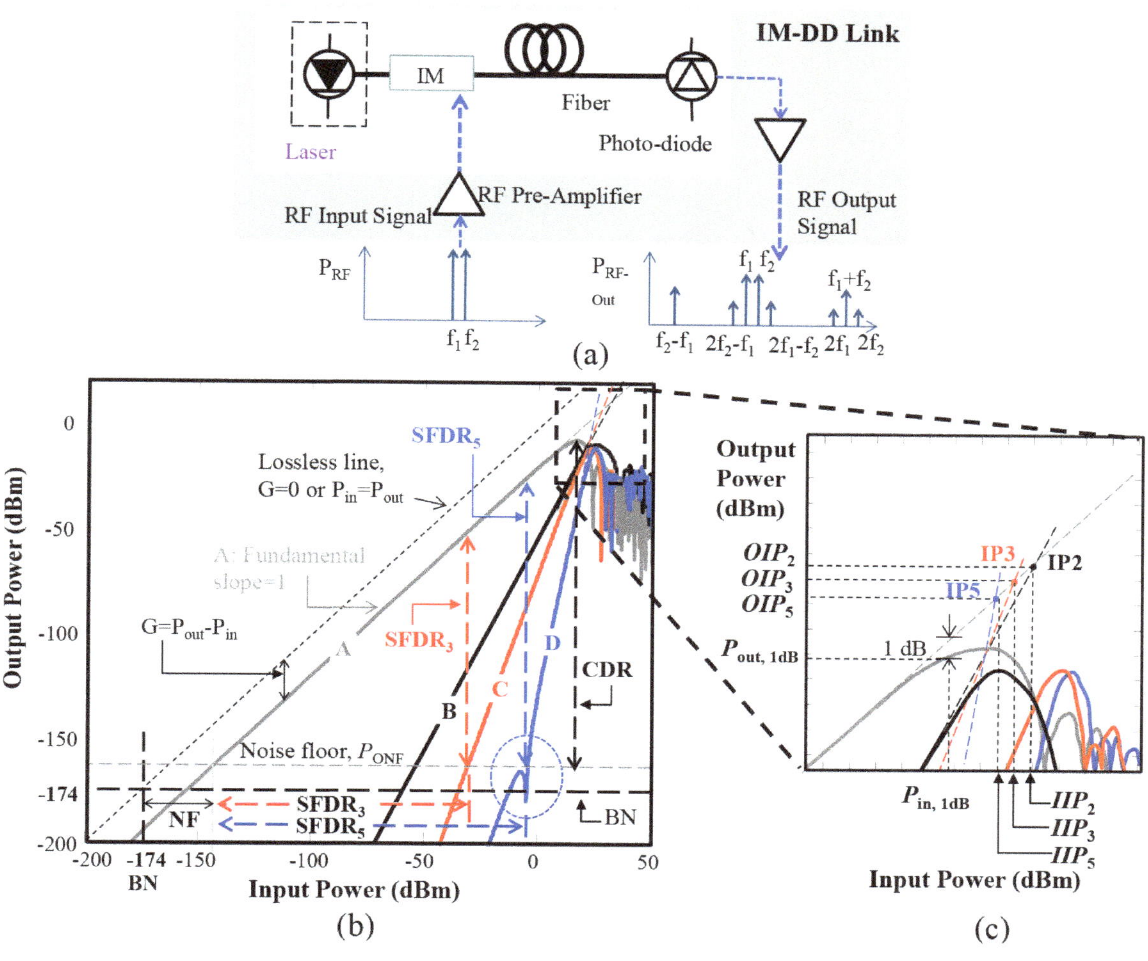

Figure 3. (a) Typical IMDD fiber optic link with input and output RF spectrum; (b) output RF power versus input RF power, showing the important MPL RF parameters/figures of merit, (A) fundamental curve with slope of 1, (B) second-order harmonics and intermodulation terms with slope of 2, (C) third-order intermodulation terms with slope of 3, (D) fifth-order intermodulation terms, with nonlinear slope and not the same slope in the entire frequency range, typical slope at the linear portion is 5 and corresponding SFDR calculations; (c) details of inset of (b) with the definition of the 1-dB compression point, and the *IPn*, with $n = 2, 3, 5$.

signal power versus the RF input power of the different frequency signals generated by the nonlinear interaction within the nonlinear external modulator. This single plot also demonstrates the relationships of the different parameters which we summarize below:

- **P_{ONF}, *BN*, P_{ON},** and **P_{IN}**: P_{ONF} refers to the output noise floor, which is the minimum discernable signal, whereas ***BN*** is the background noise or thermal noise limit and equal to −174 dBm/Hz at a temperature of 25°C. P_{in} is the RF input signal power and P_{out} is the RF output signal power.

- **Link Gain, G:** The MPL gain, in dB, is defined as the difference of the power at the output of the MPL, P_{out}, to the available power of the input, P_{in} [16]:

$$G = P_{out} - P_{in} \tag{1}$$

We should note here that G is typically referred to as small signal gain, because only in the small-signal regime, the gain is independent of input signal power. That is, for any input power, the associated output power is G dB units higher than the input power. Hence, a straight line is obtained. However, as we continue to increase the modulation power (P_{in}), the straight gain line begins to saturate, until a maximum output power is reached and then starts curving downwards. Beyond this, any increase in the input power leads to lower output power. We should note that G is not necessarily positive. We use it here in its general form, where positive G means increase in the output power, whereas, negative G means reduction in the optical power or loss.

- **Different Signal Slope Lines:** The fundamental curve, third-order IMD curve, and fifth-order IMD curve have a slope of 1, slope of 3, and slope of 5 in the linear region, respectively. Note that oftentimes at the fifth-order IMD, the slope is not necessarily the same throughout the entire frequency band, but it can change from slope 3 to slope 5 [16, 44]. **Figure 3(b)** shows the associated slope lines marked as A (fundamental curve with slope of 1), B (second-order harmonics and intermodulation terms with slope of 2), C (third-order intermodulation terms with slope of 3), and D (fifth-order intermodulation terms, with nonlinear slope and not the same slope in the entire frequency range, typical slope at the linear portion is 5.

- **Noise figure (NF):**
 is defined as the 10 log(F), where F is the noise factor of the MPL and defined as the quotient of the signal-to-noise ratio (SNR) at the input to that at the output, taken for a thermal-noise-limited input at T_s or –174 dBm. Typically this is described by the following equation [16, 43]:

$$NF[dB] \equiv 10\, log(F) = 174 + N_{out}[dBm/Hz] - G[dB], \tag{2}$$

- **Spurious free dynamic range (SFDR):** Graphically, the SFDR is the difference measured from the corresponding point on the fundamental curve to the associated point in the intermodulation distortion (IMD) curve, where IMD intersects or equals the P_{ONF}. Physically, this means that this is the area for which any IMD terms are just below the noise floor and cannot be "seen" by the system. This approach determines the SFDR value, as discussed by Bridges et al. [45]. Mathematically, SFDR is defined as [45]:

$$SFDR = \frac{2}{3}\left(OIP3 - P_{ON} - 10\log_{10}(B)\right) \tag{3}$$

for a system limited by third-order IMD, where OIP3 is the output power at the third-order intercept point and B is the system bandwidth (with units Hz). It is simply two-thirds the difference between the largest distortion-less signal that can be input into the system and the smallest detectable signal in the system. The lower end of this range (the minimum discernable signal) is the sum of the output noise floor (P_{ONF}) and the system bandwidth, B. The measurement bandwidth is assumed to be 1 Hz. In this case, the units of SFDR are dB/Hz$^{2/3}$. However, in the case of more linear systems that are limited by fifth-order IMD, SFDR is mathematically defined as [45]:

$$SFDR = \frac{4}{5}\left(OIP5 - P_{ON} - 10\log_{10}(B)\right) \tag{4}$$

where OIP5 is the output power at the fifth-order intercept point. And, the units of SFDR for a system limited by fifth-order IMD are dB/Hz$^{4/5}$. These respective SFDRs for the third-order and fifth-order are labeled as $SFDR_3$ and $SFDR_5$, respectively in **Figure 3(b)**.

- **1-dB Compression point:** This is the point at which the small signal gain (e.g., straight extended line in **Figure 3(b)**) is 1-dB higher than the actual gain curve. This point signifies the end of the linear regime. A closer look of **1-dB Compression point** is depicted in **Figure 3(c)**.
- **Compression dynamic range (CDR)**: is the difference between the output power at the 1-dB compression point (expressed in dBm) to the smallest such input power (e.g., noise floor).
- ***IP_n*–*n*th-order intercept point**: If we extend the linear slope of the fundamental and the *n*th-IMD curve, then these lines intersect at a point. We called this point the *n*th-order intercept point (e.g., IP3 if the IMD is the third-order IMD curve, IP5 if the IMD is the fifth-order IMD curve). The point is defined by the input intercept power (IIP_n) and the output intercept power (OIP_n). These points are depicted in **Figure 3(c)**.

Note that in some new modulator designs, the fifth-order IMD curves are not always straight lines with a constant slope, but have a slight deviation from a straight line just above the noise floor. This is clearly seen in **Figure 3(b)**, in the circled region just below the output system noise level (P_{ONF}—output noise floor), where the slope of an IMD curve increases beyond 5. Since the nonlinearity in the IMD curve is not taken into account in Eqs. (3) and (4), the graphical approach is best suited to be used in calculating the SFDR for the new modulator. In this approach, the SFDR is simply defined as the difference between the fundamental and inter-modulation output powers at the input power location where the IMD curve intersects the output noise floor (P_{ONF}).

2.2. Different linearization approaches

Figure 4 summarizes the three general approaches used to linearize a generic external intensity modulator based on the physical mechanism used, namely: (1) electrical, (2) optical, and (3) digital (post detection). The electrical and digital linearization approaches can further be subdivided into smaller categories (as in the case of optical linearization) but we will not review this in much detail here, however more detailed information can be found in [13, 21–22, 41, 46–55].

Briefly speaking, the electrical linearization approach employs different techniques [46–49] such as: (i) RF predistortion, (ii) RF Feed-forward, (iii) combined RF Feed-forward and predistortion, and (iv) others. These techniques generate circuits with either (a) inverse function

(ArcSin) of the modulator to predistort the RF signal before it is injected into the modulator, or (b) split a portion of the RF signal, amplify, reverse its sign, and add it to delayed original RF signal. The overall goal is to cancel or eliminate the original distorted components in the RF signal. These techniques are generally simple and inexpensive but they are limited by the frequency bandwidth of the electrical circuits [51, 53].

On the other hand, a digital linearization approach is based on the application of digital signal processing, DSP [50, 52, 54] techniques to post-detected output signal. The digital approach is very promising because of its flexibility and future growth of DSP. However, its practical usefulness will depend on the progress and development in analog-to-digital converter (ADC) and digital-to-analog converter (DAC) technologies. Unfortunately, ADC is limited by the sampling rate of <10 G/sample at this time [55]. Although there are ADC/DAC technology with >10 G/sample rates, they are usually for instrument purposes [55] and not for field deployment.

In contrast to electrical and digital linearization approaches, optical linearization is free from circuit frequency modulation limitation and does not utilize ADC/DAC technologies. For these reasons, we focus on the optical linearization approach in this chapter.

2.3. Optical linearization

As shown in **Figure 4**, the optical linearization approach can be subdivided further into 5 subgroupings, depending on the key optical structure of the modulator namely; (1) electroabsorption (EA) modulator [56–58], (2) directional coupler (DC) modulator [59–62], (3) all-optical-based approach [63–65], (4) ring resonator modulator [66–72], and (5) Mach-Zenhder interferometer (MZI)-based modulator [73–109]. These different modulators have been implemented and produced different varying degrees of linearity performance. From among these general configurations, the MZI-based modulator is the most well understood, most developed and popular optical structure in the technical community. Hence, we concentrate on the optical linearization approach with a focus on MZI-based linearized modulator.

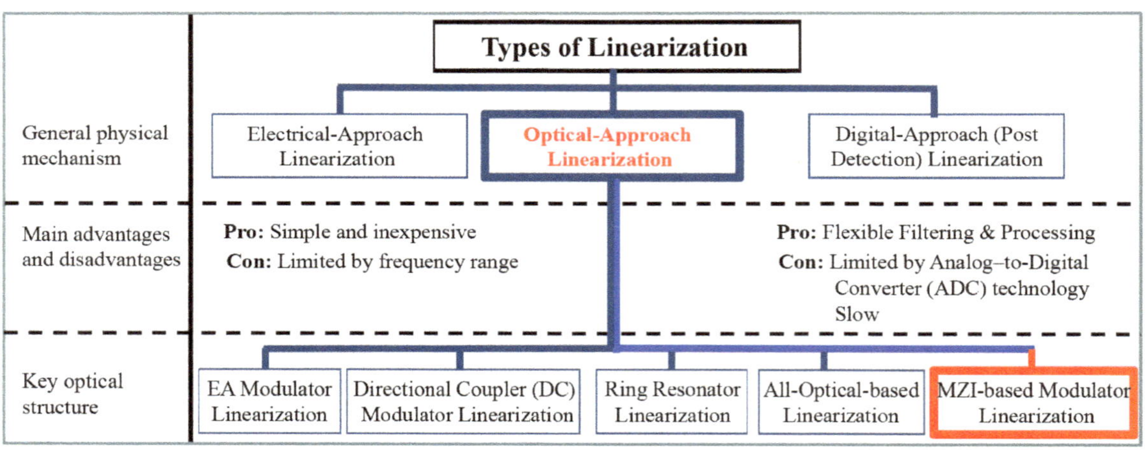

Figure 4. General classification of the different linearity approaches proposed/implemented in fiber-optics links.

2.4. MZI-based linearized modulator

Technically speaking, the MZI-based modulator is the most well-studied type of modulator. A typical nonlinearized MZI modulator has a SFDR value of around 90–112 dB-Hz$^{2/3}$. In the last decade, a wide range of linearization approaches has been proposed and implemented to improve the standard MZI modulator to have 120–130 dB-Hz$^{2/3}$ SFDR range using different configurations [74–76, 78–79, 81–84]. Recently, the goal is to push the SFDR further toward >130 dB-Hz$^{4/5}$. Nevertheless, most of them include complex designs and low modulation bandwidth.

Figure 5 shows these different types of MZI-based linearized electro-optic modulators, which can be divided into four groups depending on the actual implementations. The first group can be called dual-signal MZI-based modulator design [74–75, 83–84]. It comprises of one modulator but with two optical signals injected into the device, as depicted in **Figure 5(a)**. The two input signal amplitudes and phases must be properly matched with predetermined values of the RF amplitude phase signal. These two injected optical signals could be implemented using (i) two polarizations [74–75], (ii) two wavelengths [84], or (iii) bidirectional signals [83]. The major advantage of this family of modulators is the relative simplicity and low cost. Its disadvantages are design inflexibility, nonoptimum performance, and tight tolerance requirements.

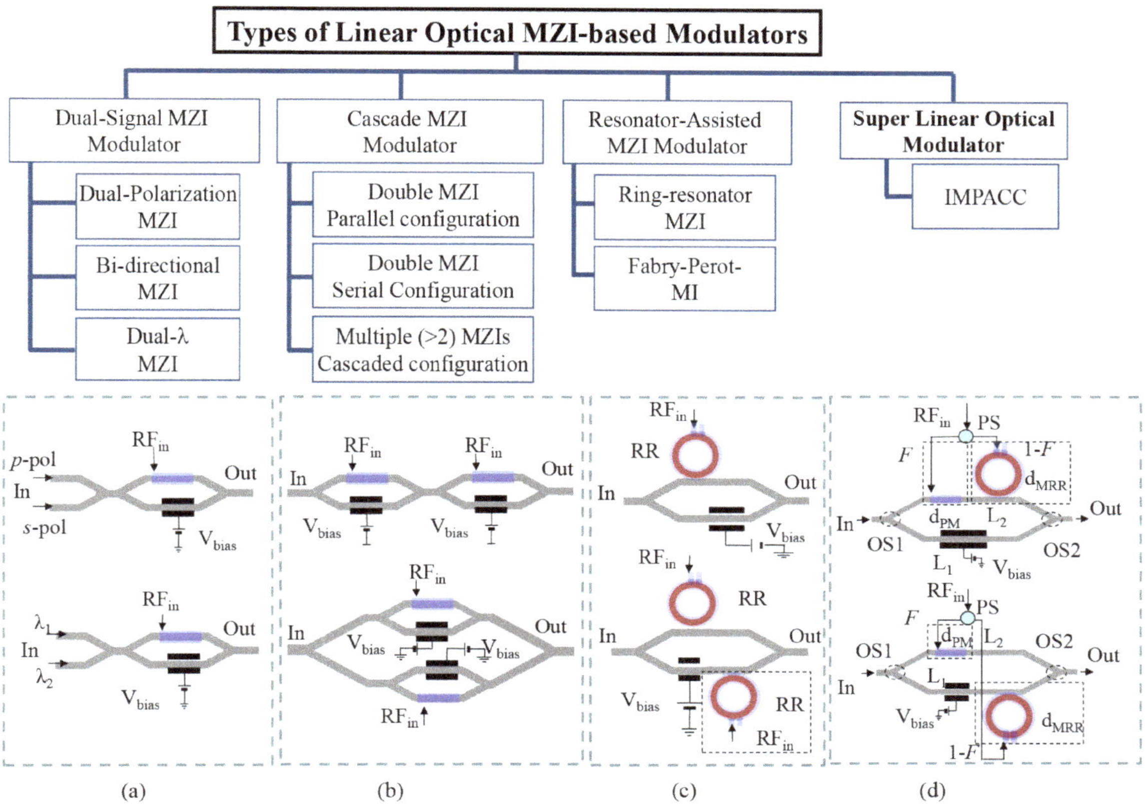

Figure 5. Different types of linear optical modulators based on MZI together with their respective implementation configurations for (a) dual-signal MZI modulator, (b) cascaded modulator, (c) RAMZI, and (d) IMPACC.

The second group can be referred to as cascaded modulator design which consists of two or more standard MZI modulators connected in series [76, 78–79, 80–82] or in parallel [100, 109] arrangements as illustrated in **Figure 5(b)**. It is a generalization of the principle used in the first group. Its major drawbacks are the tight manufacturing tolerance requirements, higher optical loss, higher cost, due to use of multiple modulators, and complicated compensation arrangement.

The third group is generally designated as RAMZI modulators [87–88, 97–99], which has received increased interest in recent years. Some implementations are shown in **Figure 5(c)**. The RAMZI uses ring resonator(s) (RRs) instead of the standard phase modulator (PM) which is coupled in the arm(s) of the MZI. This design gives higher SFDR performance but often at higher manufacturing complexity, limited RF bandwidth range, and stricter transmission coefficient control requirement.

The fourth and last group of optically linearized MZI-based modulators was introduced by our group [89, 93–94, 96, 101–105, 107–108]. This modulator design is a family of modulators which we referred to as **IMPACC**. There are numerous configurations and variations within the family of IMPACC designs [93]. Two of these implementations are shown in **Figure 5(d)**. One of these specific structures of IMPACC is discussed in more details in Section 3 and its performance in Section 4.

3. Design and modeling of IMPACC

This section discusses the basic principle, configuration, and modeling of IMPACC. In Section 3.1, we present an overview of IMPACC configuration. For both educational as well as clarity purposes, we identify the different elements of IMPACC and discuss them separately in Section 3.2. Lastly, we bring all these elements together to come out with IMPACC modulator. While this section gives the basic principle of IMPACC, Section 4 presents its performance.

3.1. Configuration of IMPACC

The configuration of our modulator, known as IMPACC, is shown in **Figure 6**. IMPACC comprises of a MZI configuration, in which one of its arms (e.g., lower) has an active microring resonator (MRR) modulator, whereas the other arm (upper) has an active phase modulator (PM). It uses the standard PM as the phase modulating component and the MRR as the cavity modulating component within a MZI structure [89, 93–94, 96, 101–108]. The exact positions of the PM and MRR relative to one another offer some variations of the implementation, as shown in **Figure 5(d)**.

Interestingly, the IMPACC configuration can be seen as a generalization of both MZI modulator and RAMZI. First, the normal combination of MZI structure and active PM in one of the arm of MZI leads to a typical *single-electrode MZ modulator*. Second, the typical combination of a MZI structure and an active MRR modulator in one of its arms results in a *single-electrode RAMZI modulator* [88, 91–92, 98]. When MRR and PM are present in MZI's arms, then this configuration becomes IMPACC.

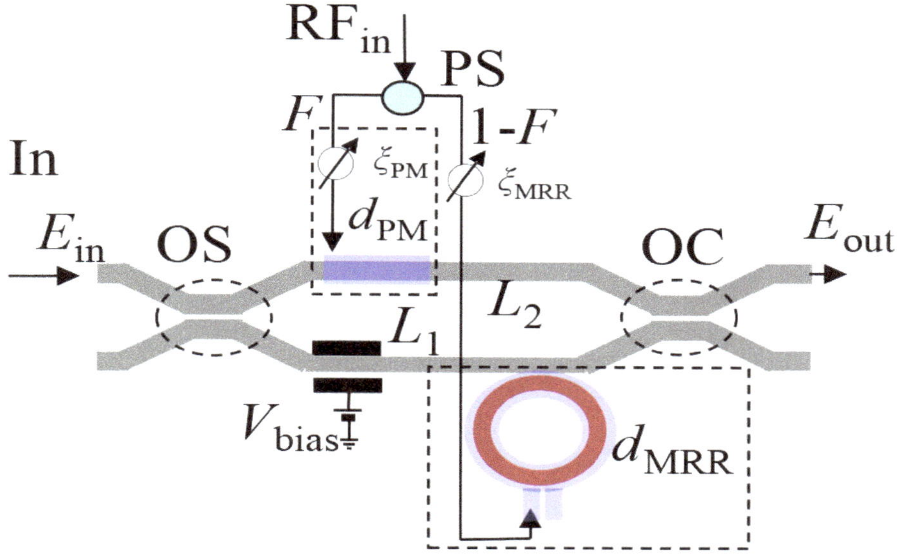

Figure 6. Basic configuration of IMPACC.

Another important aspect of IMPACC is that the RF power signal is split between the MRR and PM by the RF power splitter (PS) with a power split ratio of F:(1$-F$). This is a critical aspect in the operation of the IMPACC where PM and MRR are driven by the same frequency RF signal, but with different RF power, controlled by the power split ratio (F) using a variable or fixed RF power splitter (PS). Furthermore, the portions of the RF signals directed to MRR and PM electrodes are also delayed by an amount ξ_{MRR} and ξ_{PM}, respectively. These external control parameters from PS are not only instrumental in providing superior linear performance but also have key roles in delivering the unique adaptive mode of IMPACC. This adaptive feature is not available with any known modulators, as far as we know.

The input and output optical splitters (OS and OC) in the MZI configurations are generally assumed to have a (50:50) split so that the optical signal is divided equally into two portions. However, in this chapter of the book, we generalize this condition by taking OS and OC to have arbitrary optical power split ratios namely, P_{OS}:1$-P_{OS}$ and P_{OC}:1$-P_{OC}$. This condition allows us to investigate the effect of unbalanced power split ratio of OS and OC on the performance of IMPACC.

3.2. Elements of IMPACC

The details of the design of IMPACC have been previously described [77, 89, 93–94, 101–108]. Structurally speaking, IMPACC comprises of four important building blocks namely; (1) Mach Zehnder interferometer (MZI), (2) phase modulator (PM), (3) microring resonator (MRR) modulator, and (4) RF power splitter. In order to appreciate its rich properties, we will first briefly discuss the fundamentals of each of the four important constituents namely; (i) PM, (ii) MZM, (iii) RR, and (iv) RAMZI, before they are combined and operated under certain conditions to create IMPACC.

3.2.1. Phase modulator (PM)

The basic principle of phase modulators is based on the Pockels or linear electro-optic effect [73]. Depending on the material used, the specific axes orientations and the relative alignment of the incoming polarization to the index ellipsoid of the material used, the net effect is the refractive index modulation based on the external applied field. There are two typical configurations, the longitudinal electro-optic modulation and the transverse electro-optic modulation. In the first case, the electric field of the modulating field is parallel to the direction of the optical beam propagation (except near the edges of the electrodes). In the second case, the electric field of the modulating field is perpendicular to the direction of the optical beam propagation. This second case offers refractive index changes induced by the external electric field E that are proportional to the field E. Hence, the electrically induced phase change (or retardation) for the optical beam through the material is proportional to the product EL, where L is the length of the optical beam and electric field interaction and $E=V/d$ is the applied field, with V the applied voltage at the electrodes, which have a separation of d. The net effect is that the ratio L/d can be engineered large enough to provide devices which require low driving voltages.

Referring to **Figure 7**, an incident beam propagating along the y-direction, with polarization along the z-axis (typical case for $LiNbO_3$ based electro-optic devices), will experience an electrically induced phase change, φ_{pm} of [73]:

$$\phi_{PM} = \frac{2\pi}{\lambda} n_e L - \frac{\pi}{\lambda} n_e^3 r_{33} \frac{L}{d} V = \phi_0 - \Delta\phi, \tag{5}$$

where r_{33} is the electro-optic coefficients implicated in this crystal case and orientation [73], λ is the wavelength of the optical beam, n_e is the extraordinary index of refraction seen by the optical beam along this polarization orientation, L is the length of the crystal (or the interaction length of the optical and electric fields), and d is the separation of the electrodes. Note that the total phase change consists of two terms. The first one (e.g., ϕ_0) is the natural propagation phase change, whereas, the second term ($\Delta\phi$) is the electrically induced phase change. Because of the natural birefringence term, a phase compensator is required to adjust the phase until the total phase retardation in the absence of the electric field is an odd multiple of $\pi/2$. The half wave voltage for this case is [73]:

$$V_\pi = \frac{d}{L} \frac{\lambda}{n_e^3 r_{33}} \tag{6}$$

If the applied voltage V is sinusoidal $V = V_m \sin(\omega_m t)$, where ω_m is the modulation frequency and V_m is the applied magnitude of the voltage, then the transmitted beam will be phase modulated which can be expressed as [73]:

$$E_{(y,t)} = A \exp\left[j(\omega t - ky - \phi_0 + \delta \sin \omega_m t)\right] \tag{7}$$

where A is the constant amplitude, and the modulation index δ is given as [73]

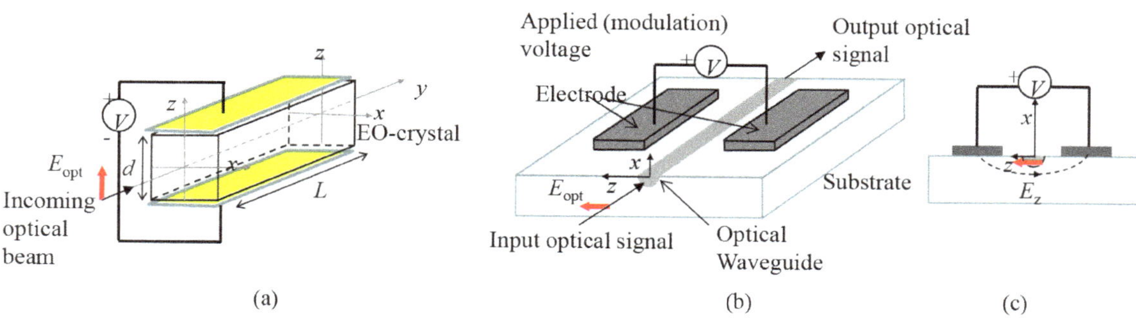

Figure 7. (a) Bulk electro-optic phase modulator, (b) photonic integrated waveguide electro-optic phase modulator in perspective view, (c) waveguide electro-optic phase modulator in cross-section view.

$$\delta = \pi \frac{L}{d} \frac{n_e^3 r_{33}}{\lambda} V_m = \pi \frac{V_m}{V_\pi} \tag{8}$$

We should note here that the phase modulator in a photonic integrated waveguide circuit is slightly different as it has to be modified to accommodate a substrate structure and the electrodes. A typical, but not the only, configuration is shown in **Figure 7(b)** and **(c)**. The optical waveguide is placed at the center of the gap between the two electrodes. In the optical waveguide, the direction of the field of the applied electric field is parallel to the substrate surface along the *z*-axis. The induced refractive index change Δn_z changes the phase in the way similar to Eq. 5.

3.2.2. Mach-Zehnder modulator (MZM)

A typical external modulator is usually in the form of an integrated optical modulator, which includes a waveguide MZI structure that is fabricated on a slab of lithium niobate ($LiNbO_3$). To ensure that light is channeled through the waveguide, the area is typically doped with impurities in order to increase the index of refraction.

Figure 8 shows a typical MZI modulator configuration where the incident light is split by a Y-branch optical splitrer (OS) into the two arms of the interferometer structure. The light traversing the first arm is modulated by a RF signal. As described in the previous section, the applied electric signal induces a change in the index of refraction of the first waveguide and hence the optical signal experiences a phase shift as it propagates through the waveguide. The magnitude of the phase delay is proportional to the applied voltage and is given by [14]:

$$\Delta\theta_{PM} = \pi \frac{V}{V_\pi}, \tag{9}$$

where V is the voltage applied to the modulator and V_π is the half-wave voltage that when applied, causes a phase shift of π. The second portion of the splitted optical signal propagates into the second arm without RF modulation. Then, the two optical signals are recombined at the output by a Y-branch optical combiner (OC). This recombination can be either constructive

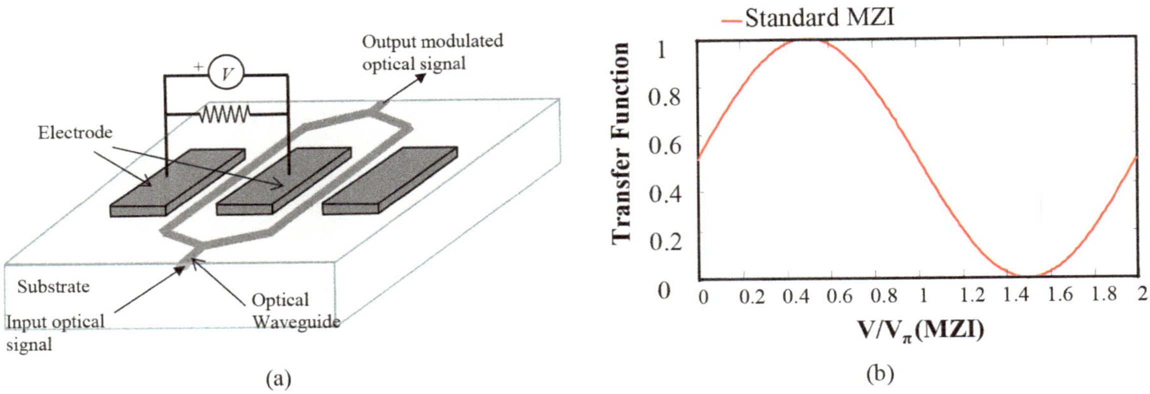

Figure 8. (a) Integrated electro-optic modulation using waveguide Mach-Zehnder Interferometer (MZI) structure, and (b) typical transfer function of a MZI.

or destructive depending on the relative phase difference of the two components of the interfering optical signals. Since the phase of one arm can be controlled by the applied electric signal, the output signal can be modulated. The transfer function of the MZI modulator is given by [73]:

$$T_{\mathrm{MZI}}(V) = \frac{1}{2}\left[1 + \cos\left(\pi\frac{V}{V_\pi} + \theta\right)\right], \tag{10}$$

where θ is the bias phase.

3.2.3. *Single microring resonator (MRR)*

The basic formulation for the physics of microring resonators (MRR) has been extensively discussed [85–86, 90]. We follow this analysis, in which an MRR has a total circumference length equal to d_{MRR} is coupled to a bus waveguide as shown in **Figure 9(a)**. The MRR round trip phase is given as $\theta_{MRR} = \omega \cdot n_0 \cdot d_{\mathrm{MRR}}/c$, where n_0 is the index of refraction of the material, α is the loss factor of the MRR waveguide, and c is the velocity of the light.

In **Figure 9(a)**, the electric fields E_1, E'_1, E_2, and E'_2 are related by the matrix equation which describes their interactions as [91]:

$$\begin{pmatrix} E_2 \\ E'_2 \end{pmatrix} = \begin{pmatrix} \tau & -i\kappa \\ -i\kappa & \tau \end{pmatrix}\begin{pmatrix} E_1 \\ E'_1 \end{pmatrix} \tag{11}$$

where τ is the transmission coupling coefficient of the coupler between the bus waveguide and the MRR while κ is the associated cross coupling coefficient of the coupler. We assume that only a single mode of the MRR is excited, and the coupler is lossless, so that $\tau^2 + \kappa^2 = 1$.

The output electric field E_2, after *one* round trip, can be expressed as [91]:

$$E_2 = \tau E_1 - i\kappa E'_1 = \tau E_1 - (i\kappa)\alpha e^{-i\theta_{\mathrm{MRR}}} E'_2 = \tau E_1 - (i\kappa)^2 \alpha e^{-i\theta_{\mathrm{MRR}}} E_1 = (\tau - \kappa^2 \alpha e^{-i\theta_{\mathrm{MRR}}})E_1 \tag{12}$$

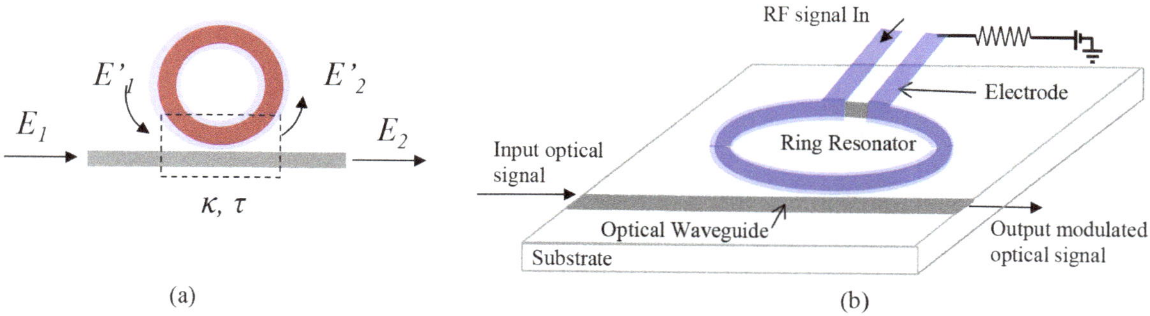

Figure 9. Microring Resonator (MRR) with one bus waveguide, (a) top view showing the field amplitudes at the input and output of the coupling region, (b) perspective view of the MRR with RF electrode.

Here, α is the RR waveguide loss factor (α = 1 indicates that the RR is lossless).

After *two* roundtrips, we have [91]:

$$E_2 = (\tau - \kappa^2 \alpha e^{-i\theta_{\mathrm{MRR}}} E_1 - \kappa^2 \tau \alpha^2 e^{-i2\theta_{\mathrm{MRR}}}) E_1 \tag{13}$$

Repeating this circulation around the MRR n times, output electric field E_2 leads to [91]:

$$E_2 = \left[\tau - \kappa^2 \sum_{n=1}^{\infty} \tau^{n-1} \left(\alpha e^{-i\theta_{\mathrm{MRR}}}\right)^n\right] E_1 \tag{14}$$

And finally, after infinite roundtrips, we have got the output electric field E_2 as [91]:

$$E_2 = \frac{\tau - \alpha e^{-i\theta_{\mathrm{MRR}}}}{1 - \tau\alpha e^{-i\theta_{\mathrm{MRR}}}} E_1 \tag{15}$$

Thus, its corresponding normalized transmission transfer function $T_{MRR}(\theta)$ under steady-state condition is given by [91]:

$$T_{\mathrm{MRR}}(\theta) = \left|\frac{E_2}{E_1}\right|^2 = 1 - \frac{(1-\alpha^2)(1-\tau^2)}{(1-\alpha\tau)^2 + 4\alpha\tau \sin^2(\theta_{\mathrm{MRR}}/2)} \tag{16}$$

Accordingly, the corresponding phase response, Θ_{MRR} and group-delay (GD) [95] response of MRR can be derived as [110, 111]:

$$\Theta_{\mathrm{MRR}} = \pi + \theta_{\mathrm{MRR}} + \arctan\left[\frac{\tau \sin(\theta_{\mathrm{MRR}})}{\alpha - \tau\cos(\theta_{\mathrm{MRR}})}\right] + \arctan\left[\frac{\alpha\tau \sin(\theta_{\mathrm{MRR}})}{1 - \alpha\tau\cos(\theta_{\mathrm{MRR}})}\right] \tag{17}$$

$$GD = -\frac{d}{d\omega}[\Theta_{\mathrm{MRR}}] \tag{18}$$

Eqs. (14) and (16) will be handy in the analysis of the new modulator that we discuss in Section 4.

(A).Passive (steady-state) operation

The relationship between the magnitude of the bus waveguide-to-MRR transmission coupling coefficient (τ) and the RR waveguide loss factor (α) defines the three different coupling conditions of MRR [73, 85, 91, 110–112]. **Figure 10(a)** shows the typical MRR while **Figure 10(b)** depicts its three different operating coupling conditions known as (1) critical-coupled (CC) when $\alpha = \tau$, (2) over-coupled (OC) when $\alpha > \tau$, and (3) under-coupled (**UC**) when $\alpha < \tau$. Their respective intensity and phase responses are shown in **Figures 10(b)** and **(c)**, respectively. One notable observation in **Figure 10(b)** is that the intensity output becomes zero at resonance frequency (normalized frequency $\Omega = 0$) under CC condition. In comparison, the respective dip intensity outputs of OC MRR and UC MRR are always higher than zero. The second significant observation is found in **Figure 10(c)** where the phase response of UC MRR is "opposite in direction" compared with phase responses of both OC MRR and CC MRR. This is significant and is used in many group-delay/chromatic dispersion mitigation approaches. Recall that if waveguide lossfactor $\alpha = 1$, then MRR is ideally lossless and becomes an all-pass filter or phase only filter.

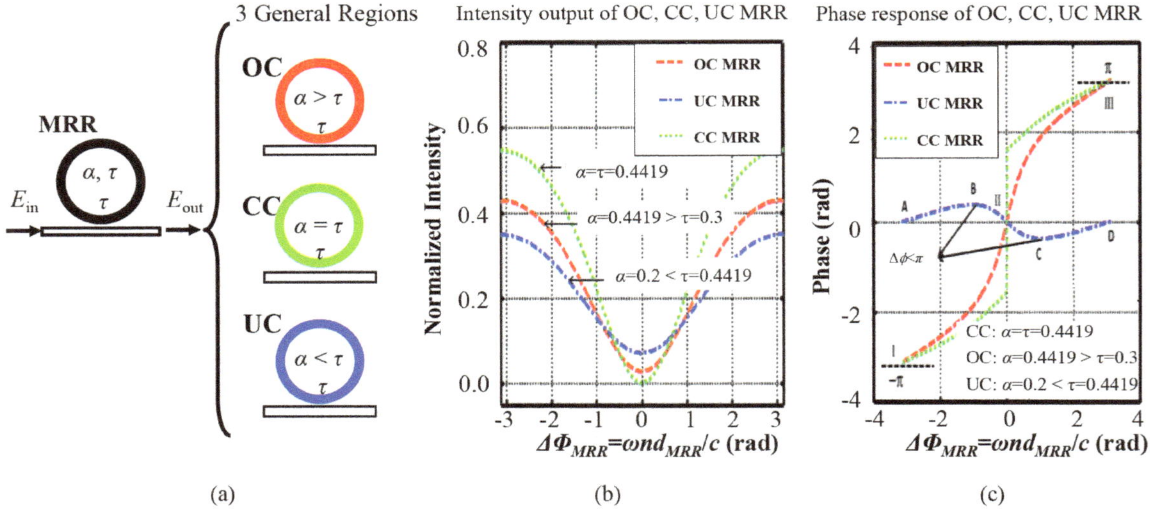

Figure 10. (a)The three different operating coupling conditions of MRR under critical-coupling (CC) at $\alpha = \tau$, over-coupling (OC) at $\alpha > \tau$, and under-coupling (UC) at $\alpha < \tau$ conditions, (b) normalized intensity response for the three operation conditions, and (c) their respective phase responses for the three operation conditions.

(B). Dynamic operation

If a voltage V, is applied to an electro-optic (EO) MRR, a corresponding round-trip phase shift θ_{MRR} induced in the ring resonator is then given as [73]:

$$\theta_{MRR} = \theta_0 + \Delta\beta \cdot d_{MRR}, \tag{19}$$

where $\theta_0 = \omega n\, d_{MRR}/c$ is the static MRR phase response, and

$$\Delta\beta = \frac{\pi n_0^3 r \Gamma V}{\lambda g}. \tag{20}$$

The parameter $\Delta\beta$ is responsible for the dynamic phase change due to applied voltage. Here, r is the electro-optic coefficient of the waveguide, λ is the incident light wavelength, g is the electrode gap and Γ is the overlap integral.

For a sinusoidal voltage, $V=V_0 \sin \omega_m t$ applied to the MRR, then using the same approach that was used to derive Eq. (14), the output amplitude, $E_{\text{out}}(t)$ is given by [77] :

$$E_{out}(t) = \left[\tau - \kappa^2 \sum_{n=1}^{\infty} \tau^{n-1} \alpha^n \exp\left[-i\left(n\theta_{\text{MRR}} + \delta_n \sin\left(\omega_m t - n\phi\right)\right)\right]\right] E_{in}(t) \tag{21}$$

where $\phi = \omega_m/FSR$ with ω_m being the modulation frequency and FSR is the free spectral range of MRR, n is the number of round trip around the ring, and δ_n is the modulation index. The modulation index depends on the electrode structure. For a lumped electrode, the modulation index is [91–92]:

$$\delta_n \sin\left(\omega_m t - n\phi\right) = \int_0^{nL} \Delta\beta \sin\left[\omega_m\left(t + \frac{n_0}{c} z\right) - n\phi\right] dz = \Delta\beta d_{\text{MRR}} \frac{\sin\left(n\phi/2\right)}{\phi/2} \sin\left(\omega_m t - n\phi\right) \tag{22}$$

For a traveling wave electrode, the modulation index is given by [91–92]:

$$\begin{aligned} \delta_n \sin\left(\omega_m t - n\phi\right) &= \sum_{k=0}^{n-1} \int_0^{nL} \Delta\beta \sin\left[\omega_m\left((t + kt_r) - \frac{\Delta n}{c} z\right) - n\phi\right] dz \\ &= \Delta\beta d_{\text{MRR}} \frac{\sin\left(\psi/2\right)}{\psi/2} \frac{\sin\left(n\phi/2\right)}{\sin\left(\phi/2\right)} \sin\left(\omega_m t - \frac{\psi}{2} - \frac{n+1}{2}\phi\right) \end{aligned} \tag{23}$$

Where $t_r = n_0 L/c$ is the optical round-trip time and $\psi = \omega_m \cdot \Delta n \cdot d_{MRR}/c$ is the velocity matching factor. Here, $\Delta n = n_m - n_0$, where n_m is the microwave effective index and n_0 is the optical index.

3.2.4. RAMZI

Figure 11(a) illustrates a RAMZI configuration where MRR is coupled to a MZI in one of its arms. We assumed that the OS and OC of the MZI have balanced (50:50) power split ratios. Then, the transfer function of RAMZI is given as [66–67, 91–92, 98]:

$$T_{RAMZI}^{SS}(\theta) = \frac{1}{2} \left| \left[\frac{1}{\sqrt{2}} |a(\theta_{\text{MRR}})| e^{-iArg[a(\theta_{\text{MRR}})]} + \frac{1}{\sqrt{2}} e^{-i\phi_{\text{MZI}}} \right] \right|^2, \tag{24}$$

where $a(\theta_{MRR})$ is the complex amplitude transfer function of the MRR under steady-state (SS) condition, which is given by Eq. (14) or (15), the phase response of MRR, $\Theta_{\text{MRR}} = \text{Arg}[a(\theta_{MRR})]$ is given by Eq. (17), and ϕ_{MZI} is the DC phase bias between the two arms of the MZI. In Eq. (24), we assumed that the nominal MZI-arm lengths L_1 and L_2 are equal, except for the length due to DC phase bias. If the MRR is lossless ($\alpha = 1$), $\tau = 2 - \sqrt{3}$ and $\phi_{MZI} = \pi/2$, then

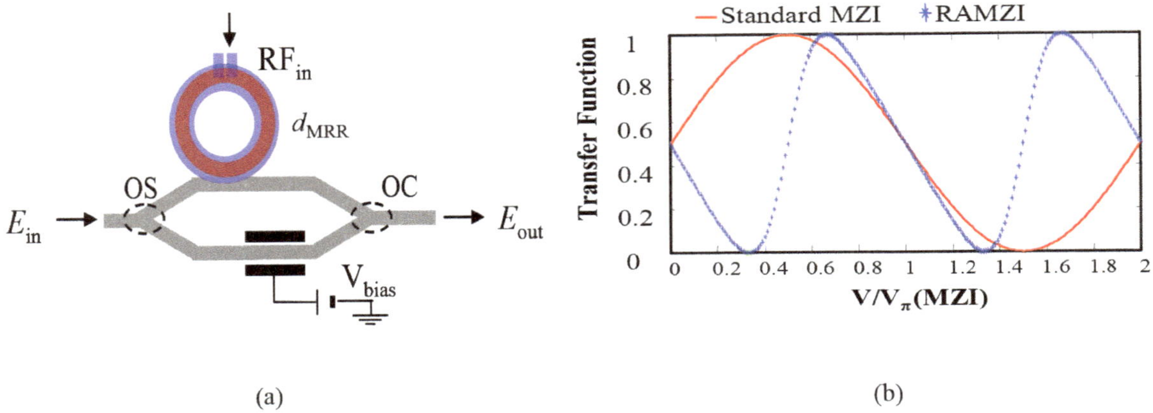

Figure 11. (a)Schematic of RAMZI, and (b) typical transfer function of a RAMZI under steady state condition in comparison with standard MZI.

RAMZI becomes linearized around $\theta = \pi$, which is off-resonance for the MRR [66–67, 91–92, 98]. **Figure 11(b)** shows the typical linearized transfer function of RAMZI in comparison to standard MZI output intensity profile.

On the other hand, under dynamic state (DS) where a modulating sinusoidal RF voltage V is applied to MRR electrode, the corresponding transfer function of RAMZI is given as [91]:

$$T^{DS}_{RAMZI}(\theta) = \frac{1}{2}\left|\frac{1}{\sqrt{2}}|a(\theta_{\mathrm{MRR}})|\exp\left[-i\left(\arg[a(\theta_{\mathrm{MRR}})] + kn_0L_2\right)\right] + \frac{1}{\sqrt{2}}\exp\left[-i(kn_0L_1 + \theta_{\mathrm{bias}})\right]\right|^2, \tag{25}$$

where $\alpha(\theta_{MRR})$ is given by Eq. (21), θ_{bias} is the phase bias of MZI and [91]

$$\theta_{\mathrm{MRR}} = n(kn_0d_{\mathrm{MRR}}) + \Delta\beta d_{\mathrm{MRR}}\frac{\sin(n\phi/2)}{\sin(\phi/2)}\sin\left(\omega_m t - \frac{(n+1)\phi}{2}\right). \tag{26}$$

for a lumped electrode case. The dynamic performance of RAMZI will be shown in Section 4 together with IMPACC performance side-by-side for clear comparison.

3.3. IMPACC

As described briefly in Section 3.1, IMPACC is a generalization of standard MZI modulator and RAMZI. **Figure 12(a)** shows the IMPACC configuration (a) together with its linearized transfer function (b) under its steady-state condition. Here, we assume that both OS and OC have 50:50 optical power split ratios, although later we will relax this restriction and consider arbitrary optical power split ratios namely, P_{OS}:$(1-P_{\mathrm{OS}})$ and P_{OC}:$(1-P_{\mathrm{OC}})$.

Structurally speaking, IMPACC is very much like the RAMZI configuration except for two big differences. First, the passive path-difference, ΔL of the MZI in the RAMZI modulator is now replaced by an active phase modulator (PM) element. Second, the total RF signal driving the electrode in MRR for RAMZI is now split into two portions, by a ratio of F:$(1-F)$, and drives both the electrodes of MRR and PM elements of IMPACC. The two split RF signals are also

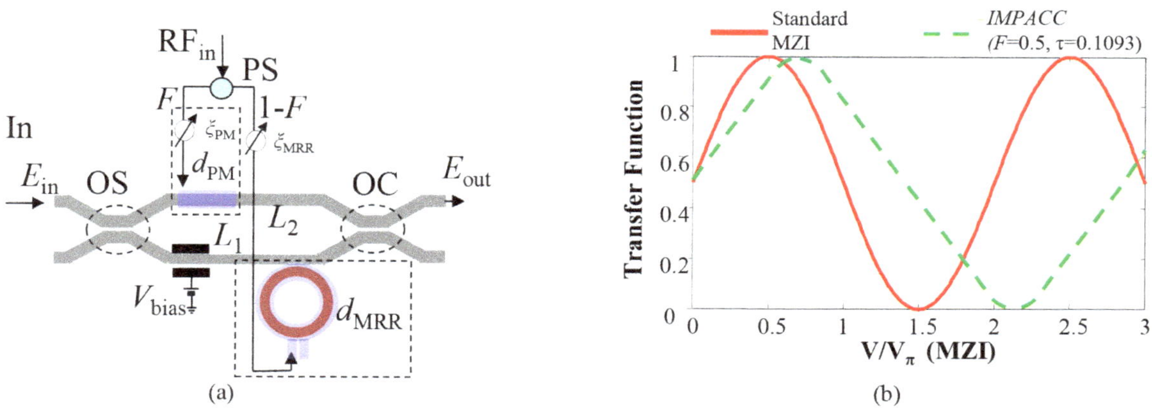

Figure 12. (a) IMPACC configuration, and (b) its transfer function in comparison with standard MZI.

delayed by an amount ξ_{MRR} and ξ_{PM}, respectively. The external control of the RF power split ratio F:(1−F) adds a degree of freedom in the design and serves to drive both the PM and MRR under specific power ratios. The corresponding linearized transfer function of IMPACC under its steady-state condition is shown in **Figure 12(b)**. Here, we assume that both OS and OC have 50:50 optical power split ratios although later in Section 3.3.1, we will relax this restriction by taking optical power split ratios namely, P_{OC}:(1−P_{OC}) to have 50:50 split while OS has arbitrary split of P_{OS}:(1−P_{OS}).

3.3.1. Transfer function

The dynamic transfer function of IMPACC, $T^{DS}{}_{IMPACC}(\theta)$ is given as [104]:

$$T^{DS}_{IMPACC}(\theta) = \frac{1}{2}\Big|\sqrt{p_{MRR}}|a(\theta_{MRR})|\exp\left[-i(\arg[a(\theta_{MRR})] + kn_0L_2)\right] + \sqrt{p_{pm}}\exp\left[-i(\varphi_B + kn_0L_1 + \theta_{pm})\right]\Big|^2 \tag{27}$$

where

$$a(\theta_{MRR}) = \tau - \kappa^2\sum\nolimits_{l=1}^{\infty}\tau^{n-1}\alpha^n\exp\left[-i\theta_{MRR}\right] \tag{28}$$

$$\theta_{MRR} = n(kn_0d_{mrr}) + \sqrt{1-F}\Delta\beta d_{MRR} + \sqrt{1-F}\Delta\beta d_{MRR}\frac{\sin(n\phi/2)}{\sin(\phi/2)}\sin\left(\omega_m t + \xi_{mrr} - \frac{(n+1)\phi}{2}\right) \tag{29}$$

and

$$\theta_{pm} = kn_0d_{pm} + \sqrt{F}\Delta\beta d_{pm} + \sqrt{F}\Delta\beta d_{pm}\sin(\omega_m t + \xi_{pm} - \phi) \tag{30}$$

The parameters p_{MRR} and p_{PM} are the input optical power split ratio coefficients of the input optical splitter (OS) for the MRR and the PM arms, respectively. Note that p_{MRR} = 1 − P_{OS}, p_{PM} = P_{OS}, where P_{OS} was defined in Section 3.3. Also, here we assumed that OC has 50:50

power coupling ratio. The parameters θ_{pm} and $\arg[a(\theta_{MRR})]$ are the phase responses of the PM and MRR respectively, while the parameters d_{PM} and d_{MRR} are the lengths of the PM and MRR, respectively with n being the number of times the beam propagates inside the MRR, ω_{m} is the modulation frequency, $\phi = \omega_{\mathrm{m}}/FSR$, φ_B is MZI optical bias phase, ξ_{MRR} and ξ_{PM}, are the RF signal phase biases injected to the MRR and PM electrodes, respectively, and $\Delta\beta$ is given by Eq. (20). We assume negligible microwave loss and no velocity mismatch.

Unlike RAMZI, IMPACC has an external control of either the RF power split ratio or the RF bias. As we will show in the next section, while both of these modulator designs can result in high SFDR performance, IMPACC has an additional and unique property—an adaptive, inherent compensation capability that can be used to maintain its high SFDR under three nonideal parameter conditions namely: (1) increasing MRR waveguide loss factor, (2) unfavorable MRR coupling operating conditions, and (3) unbalanced input power split ratios of either the optical splitter, OS or optical combiner, OC. This adaptive property provides an additional level of design flexibility and can greatly simplify the overall operation of high-performing modulator.

4. Performance of IMPACC

This section provides details of the linearity performance of IMPACC in terms of SFDR, and compares it with RAMZI and standard MZI modulators. Unlike RAMZI and MZI modulators, IMPACC modulator has two different operational modes, namely: (i) the typical basic mode, and (ii) the unique adaptive mode. As far as we know, IMPACC is the only intensity modulator that has an adaptive mode. In Section 4.1, we focus on its basic operating mode, whereas in Section 4.2 we discuss the adaptive mode.

4.1. Basic operating mode of IMPACC

In the basic mode, IMPACC operates under fixed, predetermined parameter conditions aimed to optimize its performance. This mode of operation is typical in all intensity modulators. Here, we investigate the effects of the different parameter variations (from these optimum conditions) on IMPACC's SFDR performance, as well as its bandwidth capability, to assess its performance limits. These parameters include (i) MRR loss factor α, (ii) τ and α that define the CC, UC, and OC conditions, and (iii) unbalanced power split ratio (e.g., P_{os}:$(1-P_{\mathrm{os}})$) of optical splitters (OS). We also compare IMPACC performance with RAMZI and standard MZI modulators.

4.1.1. Effect of ring resonator's loss on SFDR

Any MRR waveguide has an intrinsic loss, (expressed as loss factor, α in Section 3). Here, we assess its effect on both IMPACC and RAMZI's SFDR performance [104]. **Figure 13(a)** shows the respective SFDRs of both IMPACC and RAMZI under two conditions; (a) an ideal lossless MRR waveguide (e.g., $\alpha = 1.0$), and (b) lossy case (e.g., $\alpha = 0.8$). Note that parameter τ is chosen for both IMPACC and RAMZI to get their respective highest SFDR values. We choose to use $\tau = 0.6355$ for

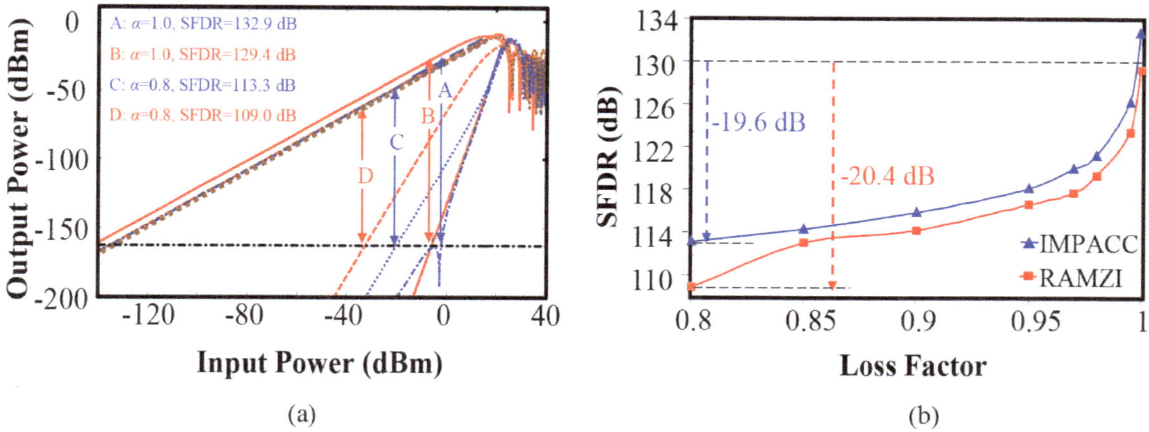

Figure 13. (a) Effect of MRR waveguide loss, α on the SFDRs of IMPACC and RAMZI for ω_m = 1Hz (F, ξ_{PM} are fixed), (A) IMPACC with α = 1, (B) RAMZI with α = 1, (C)IMPACC with α = 0.8, (D) RAMZI with α = 0.8; (b) effect of different values of MRR loss factor on the SFDR values for IMPACC and RAMZI.

IMPACC and the ideal value of τ = 0.2679 for RAMZI. Furthermore, the modulation frequency, ω_m is set to 1 Hz, while the power split ratios of OS and OC are assumed both balanced (50:50). Here, we see that IMPACC has an SFDR value of 132.9 dB under lossless condition (labeled as A) that reduces drastically to 113.3 dB (labeled as C) when loss is introduced. A similar pattern is observed for the RAMZI case, where its SFDR value deteriorates from 129.4 dB for lossless case (labeled as B) to 109.0 dB for lossy case (labeled as D).

Figure 13(b) summarizes the full effect of MRR loss factor α on the SFDR of both IMPACC and RAMZI when the value of α is allowed to vary from 1 to 0.8. We have the following three observations. First, the respective SFDR performances of IMPACC and RAMZI are affected drastically and negatively by the decreasing value of the MRR loss. It is clear from **Figure 13(b)** that the MRR loss value must be minimized within $a < 0.99$ to obtain the higher than 126 dB-Hz SFDR performance. Second, IMPACC always outperforms RAMZI. It exhibits a consistently higher SFDR value (e.g., more than 1 dB) compared than RAMZI for the range $a = 1$ to $a = 0.85$. Third, at the range of a = 0.85 – 0.80, the SFDR performance of RAMZI degrades more than three times compared with IMPACC. The comparative SDFR difference between IMPACC and RAMZI increases from 1.2 dB when a = 0.85 – 4.3 dB with a = 0.8. This implies that IMPACC has a higher parameter tolerance to MRR loss variation especially at higher loss condition.

4.1.2. *Effect of OC, UC, and CC operating conditions on SFDR*

The SFDR performance of any MRR-based devices, including IMPACC and RAMZI, depends highly on the relative ratio between the MRR coupling coefficient strength (τ) and intrinsic loss factor (a) As mentioned in Section 3, there are three MRR regions of operation namely; (i) over-coupling (OC) condition, when $a > \tau$, (ii) critical coupling (CC) when $a = \tau$, and (iii) under-coupling (UC) when $a < \tau$.

Here, we selected the three regions with the following parameters; (i) OC condition (τ = 0.6364, α = 0.7364), (ii) CC condition (τ = 0.6364 and α = 0.6364), and (iii) UC condition (τ = 0.6364 and

α = 0.5364). Note again that we choose τ = 0.6364 for IMPACC and the ideal value of τ = 0.2679 for RAMZI. Furthermore in order to simplify the analysis and gain better understanding of the different parameter effects, we assumed an ideal, balanced 50/50 power split ratio for both the input optical splitters (OS) and output optical combiner (OC). The effect on nonideal 50:50 split ratio of OS will be discussed in the next section. Moreover, we assumed that IMPACC and RAMZI are driven at one of its resonant modulation frequency of 23 GHz.

Briefly speaking, the SFDR performance of both IMPACC and RAMZI under OC condition is shown in **Figure 14(a)**. IMPACC has superior linearity with SFDR =130 dB which is far better than RAMZI (e.g., 110 dB). This superior SFDR value is maintained (e.g., 129.7 dB) under CC condition as shown in **Figure 14(b)**. Lastly, the SFDR performance of both IMPACC and RAMZI under UC condition is also shown in **Figure 14(c)**. Under UC condition, both IMPACC and RAMZI are no better than the typical nonlinearized MZI modulator having SFDR value in the range of 110–113 dB. However, we can state that IMPACC's SFDR value (e.g., 113 dB) is still 3 dB higher when compared with RAMZI (e.g., 110 dB). Thus, the takeaway from **Figure 14** is that the huge SFDR advantage of IMPACC to RAMZI exists only when MRR is operated either in OC or CC condition.

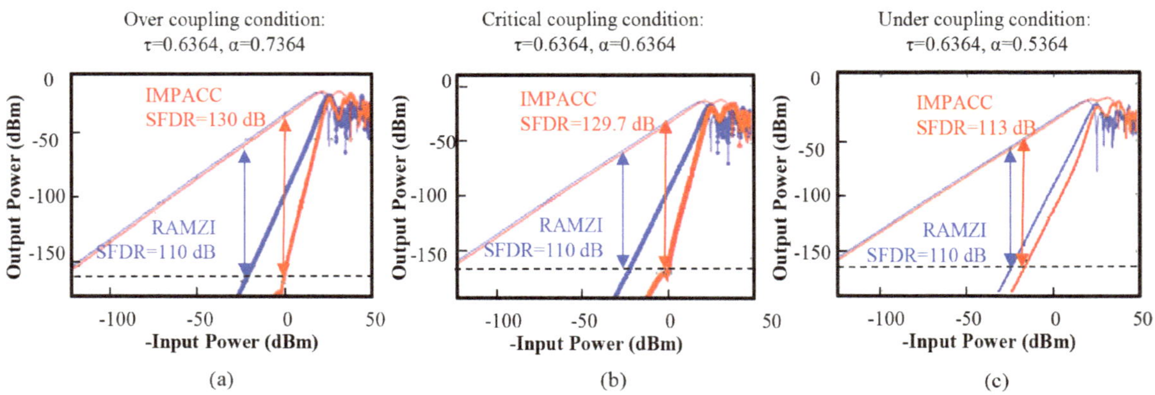

Figure 14. SFDR performance at (a) over-coupling, (b) critical-coupling and (c) under-coupling condition for 23 GHz modulation frequency (analysis assumes a resolution BW of 1 Hz).

4.1.3. Effect of unbalanced optical coupler on SFDR

This section studies the effect on nonideal, unbalanced split ratios of the optical splitter (OS) or optical combiner (OC) on the SFDR performance of IMPACC for three different MRR operating conditions (OC, UC, CC) [110]. To simplify the analysis, we limit our study to unbalanced OS by maintaining OC to have 50:50 power split ratio. Note that we use the same parameters as found in Section 4.1.2 with τ = 0.6364 for IMPACC and the ideal value of τ = 0.2679 for RAMZI.

4.1.3.1. Over-coupled (OC) IMPACC

Figure 15(a) shows the effect of the unbalanced power split ratio (e.g., 55:45) of OS on SFDR performance of IMPACC. The SFDR value drops by 6 dB from 130 dB-Hz under balanced condition to 124 dB-Hz under unbalanced case. **Figure 15(b)** depicts the SFDR performance for various split ratios. It is plotted as a function of the offset from the ideal optical splitter balanced condition (0 deviation corresponds to an ideal power split ratio of 50:50) to a very unbalanced condition (e.g., 0.20 corresponds to power split ratio of 70:30 and −0.20 corresponds to power split ratio of 30:70). One conclusion from **Figure 15(b)** is that in order to maintain a SFDR value of above 124 dB-Hz, we need to ensure that the OS power split ratio deviation should be no greater than ± 0.08 or no more than 58:42 or 42:58 power split ratio. Note that the results of **Figure 15(b)** were obtained without changing any external IMPACC parameters (e.g., RF power split ratio-F or the RF phase bias-ξ_{PM}) except the coupling ratio OS.

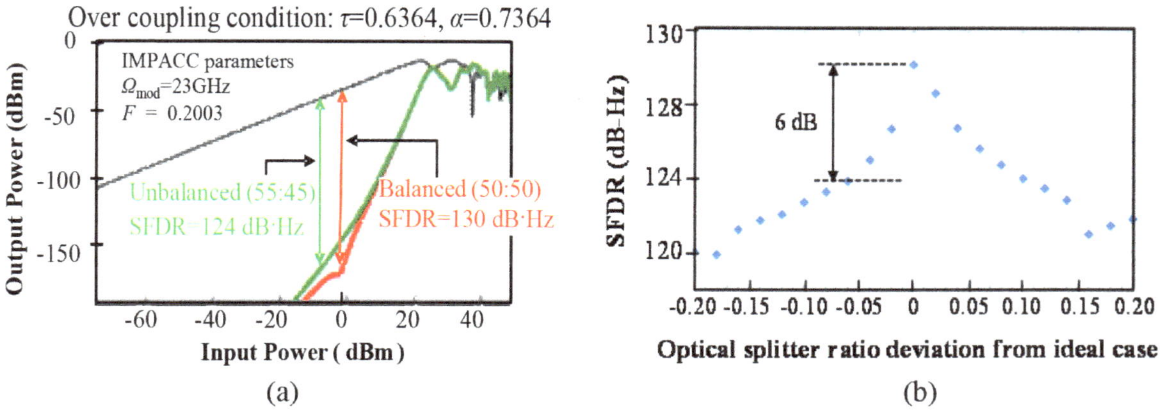

Figure 15. (a) SFDR degradation due to particular unbalanced optical power splitter (55:45) when MRR is operated under over-coupling condition, (b) SFDR performance for different unbalanced optical power split ratios.

4.1.3.2. Critical-coupled (CC) and under-coupled (UC) IMPACC

Figure 16 shows IMPACC's SFDR performance under (a) CC and (b) UC conditions for balanced (i.e., 50:50) and unbalanced (e.g., 55:45) cases. We see that the SFDR value of the OC condition degrades by 11 dB from balanced case of 129 dB-Hz to 118 dB-Hz for unbalanced case. Compared with the OC case (**Figure 15a**) where the SFDR drops only by 6 dB, the CC condition implies higher sensitivity to OS split ratio deviation. On the other hand, **Figure 16(b)** illustrates that OS split ratio deviation has no effect in the balanced or unbalanced cases for UC condition. However, IMPACC has only low SFDR value of 113dB-Hz.

4.1.4. SFDR versus modulation bandwidth response of IMPACC

We know that the modulation bandwidth of all ring resonator based modulators, such as RAMZI and IMPACC is typically limited by the MRR's free-spectral range (FSR) [91, 102, 104–105]. Here, we describe the SFDR-vs-modulation bandwidth capability of both IMPACC and RAMZI, and assess their respective performance limits. We evaluate the SFDR-vs-modulation bandwidth under lossless case (A), and under the three MRR operating conditions (B).

4.1.4.1. MRR lossless case (OC Condition)

Here, we assume that the loss factor a in the MRR waveguide, for both the IMPACC and the RAMZI, is lossless or equal to 1. Although, this is not a realistic case, it is done to focus on the

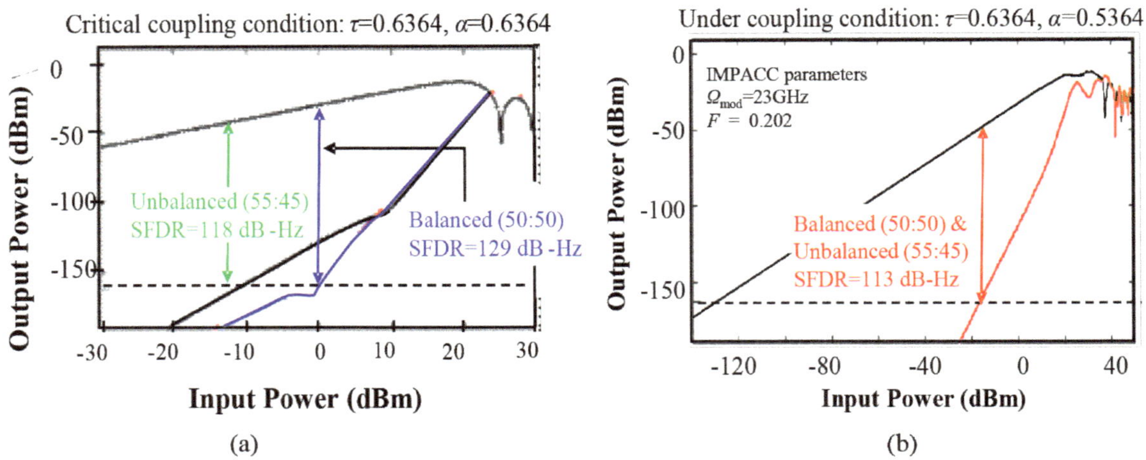

Figure 16. SFDR of IMPACC at over-coupling condition (a), and under-coupling condition (b) for both unbalanced and balanced scenarios (modulation frequency 23 GHz, $\xi_{PM} = 1.3893\pi$, $F = 0.2008$).

bandwidth capability and understand its performance limits. Later in the next subsection, we consider the effect of MRR loss. Here, we use $\tau = 0.208$ for IMPACC, $\tau = 0.268$ (or $\tau = 00.2679$) for RAMZI [90–91, 95, 104], MRR RF phase bias $\xi_{MRR} = 0$, and a PM RF bias control sensitivity ξ_{PM} of 0.2° [105]. The two τ values have been selected such that they provide maximum SFDR value for each case.

First, **Figure 17(a)** shows the SFDR-vs-Modulation frequency of IMPACC (blue line), RAMZI (red line), and MZI modulator (green line) for modulation frequencies up to 75 GHz. IMPACC has an increasingly higher SFDR, (e.g., 3.7 dB to 13.7 dB) compared with RAMZI and standard MZI modulators for these increasing resonant frequencies. It is important to note that the resonance enhancement in IMPACC is accomplished without resorting to a smaller ring size. Smaller ring size not only makes the fabrication more challenging, but would also introduce larger resonator waveguide losses.

Next, we identify two frequency regions of operation: (i) the resonance region at around ~23 GHz (shown in **Figure 17(b)** and ~70GHz (shown in **Figure 17(c)**), and (ii) the non-resonance region at around 56 GHz (shown in **Figure 17(d)**) [105]. The resonance region is defined as the region where the SFDR is greater than that obtained from an ideal MZI, which has a relative flat response as a function of frequency. These resonant regions occur at multiples of the FSR of the modulator (e.g., 23.3 GHz, 69.9 GHz).

As shown in **Figure 17(a)**, IMPACC provides an increasingly higher SFDR, (e.g., 3.7 dB to 13.7 dB) compared with RAMZI and MZI at these resonant frequency regions. **Figures 17(b)** and **(c)** depict the respective IMPACC's SFDR performance when the central RF modulation frequencies are set to 23 GHz and 69.9 GHz, respectively. We note that the modulation linewidth around the central frequency 23 GHz is wider compared with case of central frequency of 69.9 GHz. On the other hand, in the non-resonance region, (defined as the region between the resonance regions), the SFDR is typically less than that obtained from the ideal MZI as shown

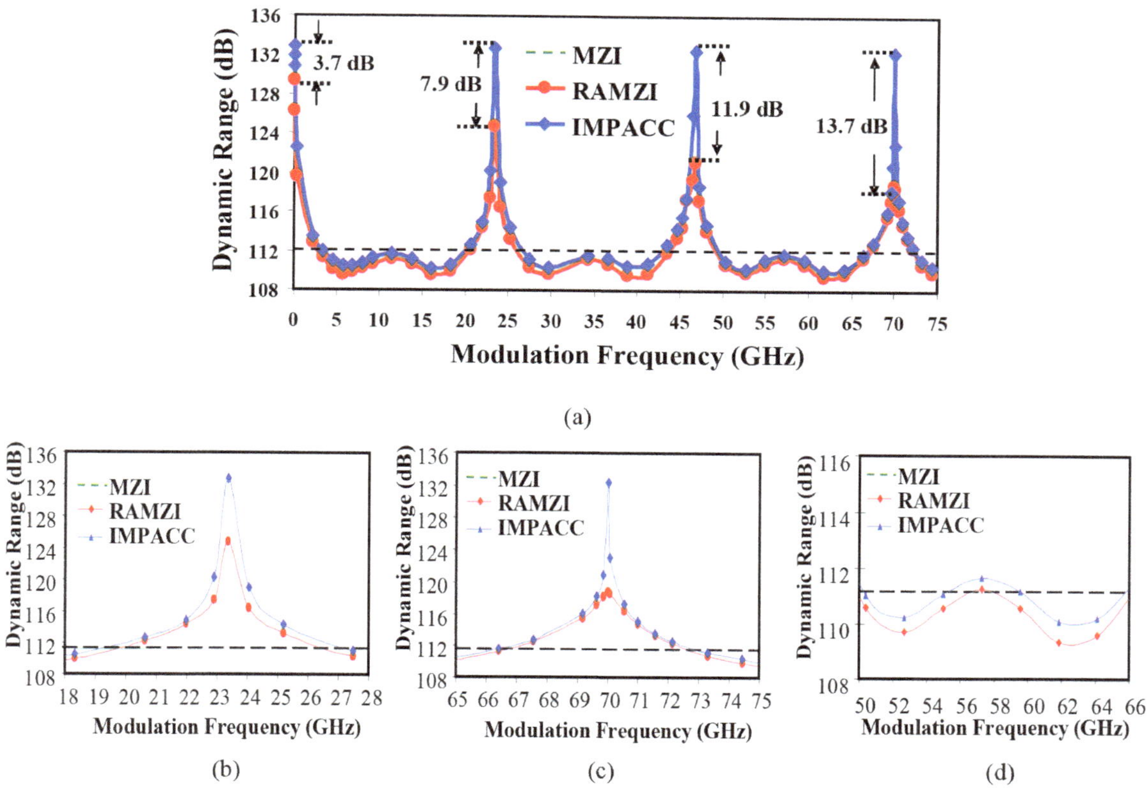

Figure 17. (a) Frequency response for the IMPACC, RAMZI, and MZI, (b) resonant region around a central modulation frequency of 23.3 GHz, (c) resonant region around a central modulation frequency of 70 GHz, (d) non-resonance region around 58 GHz. (analysis assumes a 1 Hz resolution bandwidth).

in **Figure 17(d)**. However, the IMPACC still shows an improved SFDR performance by 0.5 to 2.0 dB compared to RAMZI.

4.1.4.2. MRR lossy cases (OC, CC and UC Conditions)

Figure 18 depicts the SFDR performance of both IMPACC and RAMZI under over-coupled (OC), critical-coupled (CC), and under-coupled (UC) conditions. **Figure 18(a)** shows the OC IMPACC (τ = 0.6364, α = 0.7364) and OC RAMZI (τ = 0.2679, α = 1) as a function of the RF modulation frequency. As we have described earlier, in the regions close to the resonance peak (e.g., ~23 GHz), IMPACC clearly outperforms RAMZI, which operates under ideal condition by 10–20 dB.

The case of CC IMPACC is shown in **Figure 18(b)**. IMPACC also outperforms RAMZI within or in the resonant peak regions. Outside the resonant regions, both IMPACC and RAMZI have nearly the same performance. Lastly, **Figure 18(c)** shows IMPACC performance versus RAMZI in the case of UC condition at the same resonance region as above (e.g., modulation frequency ~23 GHz). In regions close to the resonance peak, the IMPACC still outperforms the ideal over-coupled RAMZI, but now the maximum difference is much smaller (e.g., ~4 dB). We note that

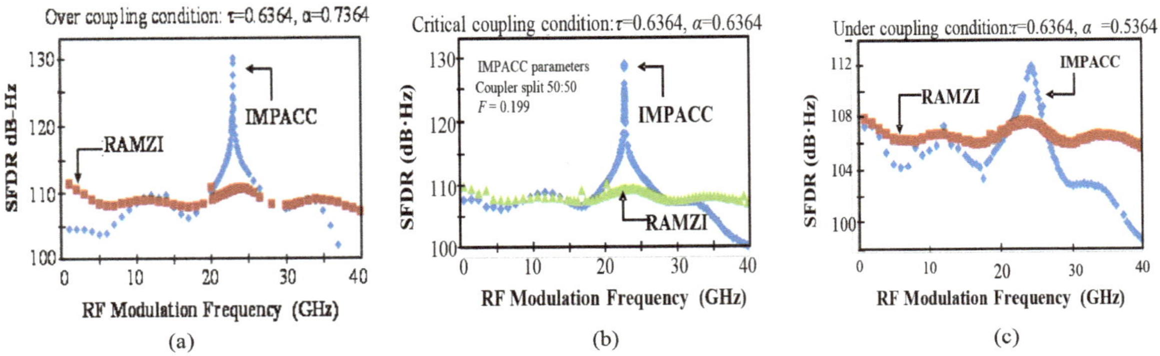

Figure 18. (a) SFDR comparison of IMPACC and RAMZI at (a) over-coupled condition, (b) critical-coupling condition, and (c) under-coupling condition for modulation frequency 23 GHz, $\xi_{PM} = 1.3893\pi$, $F = 0.2008$.

in all these situations, we did not optimize the performance of IMPACC by changing its external parameters (e.g., F or ξ_{PM}).

4.2. Adaptive operating mode of IMPACC

IMPACC's adaptive operating mode is all about using dynamically the two built-in, adjustable, and externally controllable two parameters to maintain IMPACC's 130 dB SFDR performance under unfavorable parameter conditions such as: (i) intrinsic RR loss, (ii) parameter variation of τ value, and (iii) unbalanced OS power split ratio. The unavoidable parameter variations can be due to (a) environmental temperature change, (b) device aging, and (c) fabrication tolerance and errors. The externally controllable parameters used for mitigation purposes are the RF power split ratio F, and the RF phase shift, ξ_{PM}. Here, we set the other RF phase shift, ξ_{MRR} equal to 0, for simplicity. Using again these two external parameters, we show that IMPACC can also extend its SFDR-vs-modulation bandwidth capability beyond that of RAMZI [91, 106–108, 110]. Overall, the adaptive mode of IMPACC allows the link design to be flexible with excellent results.

4.2.1. Extending parameter tolerance to ring resonator loss

The ability to adjust the control parameters F and ξ_{PM} can be exploited to compensate for the negative effect of loss factorα of MRR. In Section 4.1.1, we showed that the SFDR performance of both IMPACC and RAMZI can deteriorate by as much as ~20 dB when the RR loss factor (e.g., $\alpha < 1$) increases. This effect is reproduced in **Figure 19(a)** for IMPACC and RAMZI with fixed F values. **Figure 19(a)** also shows the SFDR performance for IMPACC with dynamic or adaptive F parameter value. The two cases with fixed F parameter value are similar to the one shown in **Figure 13(b)** and discussed in Section 4.1.1.

The result of this adaptive mode is to maintain the SFDR value to higher than 130 dB by dynamically changing the required F value as a function of the MRR loss factorα as shown in **Figure 19(b)**. It shows the range of required changes in the value of F to achieve the SFDR performance shown in **Figure 19(a)**. This unique ability of IMPACC overcomes the negative effect of the intrinsic insertion loss due to MRR. This feature is not feasible in RAMZI or any

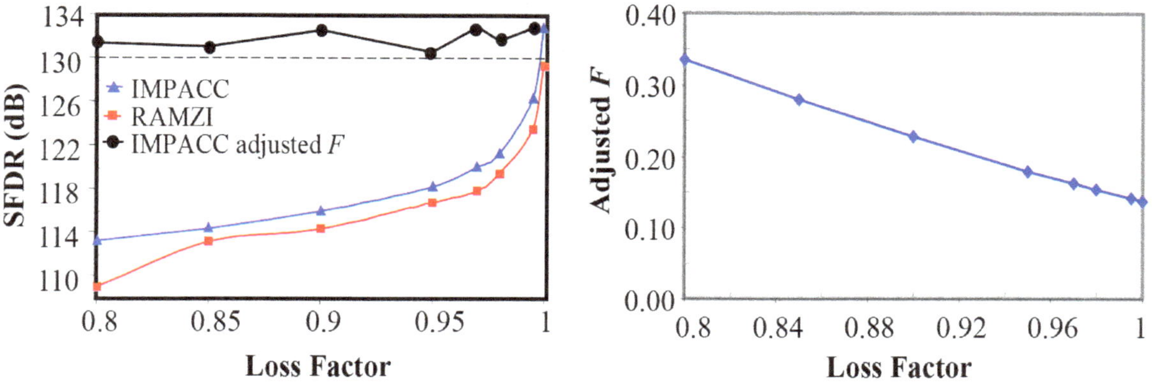

(a) (b)

Figure 19. (a) IMPACC compensates for MRR intrinsic losses to obtain SFDR values above 130 dB, (b) RF power split ratio adjustments required to achieve high SFDR (e.g., >130 dB) performance under MRR with exhibited insertion losses.

previously reported linear intensity modulators. It makes the link design and engineering more flexible.

4.2.2. Extending parameter tolerance to unbalanced split ratio of OS

4.2.2.1. Over-coupled (OC) IMPACC and critically-coupled (CC) IMPACC

In this section, we highlight IMPACC's ability to compensate the negative effect due to unbalanced optical splitter performance [110, 113]. **Figure 20(a)** shows the SFDR performance of OC IMPACC (τ = 0.6364, α = 0.7364) for unbalanced OS (55:45; green line), balanced (50:50; uncompensated; red line), and unbalanced (55:45; blue line) but compensated by adjusting the F parameter.

This is emphasized in **Figure 20(a)** where the original drop of 6 dB in the SFDR performance, when the optical splitter is nonideal (55:45), can be compensated to return to SFDR value equal to 129 dB by adjusting the external parameters, F and ξ_{PM} to 0.2003 and 1.307π, respectively.

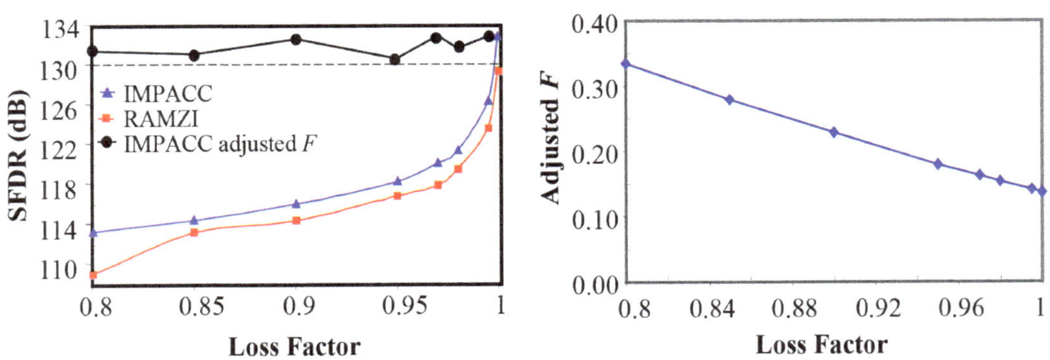

(a) (b)

Figure 20. SFDR comparison of (a) over-coupled and (b) critical-coupling condition for RAMZI and IMPACC for both unbalanced and balanced scenarios (modulation frequency 23 GHz, ξ_{PM} = 1.3893π, F = 0.2008).

Hence, clearly, IMPACC's inherent compensation capability is quite effective and able to overcome this limitation. **Figure 20(b)** depicts the SFDR performance of IMPACC at a power balanced for the case of CC condition (i.e., 50:50), unbalanced condition (e.g., 55:45) without compensation and with compensation for the case of CC condition. In the unbalanced condition, the SFDR drops 11 dB from 129 to 118 dB. Nevertheless, IMPACC's inherent compensation capability enables one to increase the SFDR back up to 129 dB for the unbalanced (55:45) case under CC by adjusting the F and ξ_{PM} parameters.

4.2.2.2. Under-coupled (UC) IMPACC

In UC condition, the inherent compensation capability of IMPACC is severely limited and cannot increase the SFDR of IMPACC for the unbalanced case. Adjusting the external parameters,F or ξ_{PM} do not improve the SFDR value. Similar to previous Section 4.1.3, we observed that the IMPACC SFDR performance (e.g., 111 dB) is still higher than that of RAMZI (e.g., 108 dB) but the difference is noticeably smaller compared with the over- and critical-coupled conditions with SFDR of around 129–130 dB.

4.2.3. Extending SFDR versus modulation bandwidth response

As discussed earlier in Section 4.1.3, all ring resonator based modulators (e.g., RAMZI and IMPACC) are limited by the FSR of the ring resonator [88, 91–92, 94,102,104–108]. This is graphically represented by the narrow Lorentzian shape of its SDFR profile as depicted in **Figures 17** and **18**. This limitation is troublesome, especially for wideband RF link applications. Fortunately, the IMPACC has significant advantage compared with RAMZI and other linear modulators that relaxes this built-in RF modulation bandwidth limitation due to MRR.

IMPACC can broaden the SFDR profile at its peak value by dynamically adjusting both F and ξ_{PM} for any given modulation frequency band. **Figure 21** shows this remarkable feature for a frequency of operation from 16 GHz to 20 GHz, where the IMPACC can maintain a SFDR of >130 dB. For the results presented in **Figure 21**, we have used λ = 1550 nm and τ = 0.208, and both the RF phase, ξ_{PM} and power split, F parameters were changed in other to maintain the

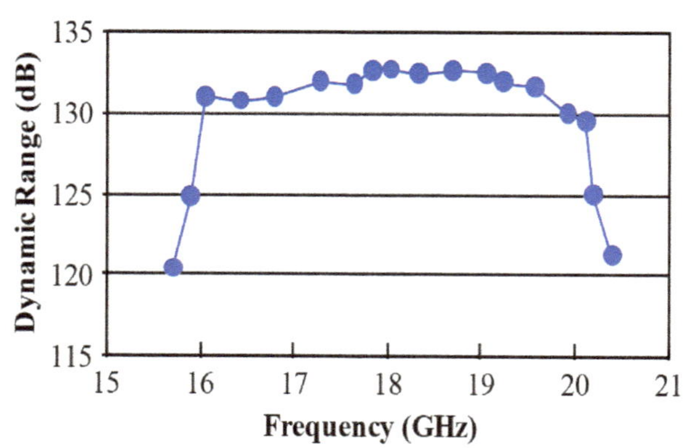

Figure 21. Changing the power split ratio F:(1−F) and RF bias, phase ξ_{PM}, of RF signal can achieve a flat SFDR response over a range of ~20% of around a central frequency (λ = 1550 nm).

high SFDR. We are not aware of any MRR-based modulator designs that can accomplish such a dynamically adjustable wide frequency response using externally controlled parameter.

5. Conclusion

In summary, electro-optic modulators are critical part of optical communications. They will continue to play an important role in fiber optic links. However, an essential requirement is that they have linear response with a high dynamic range of input RF powers of multiple tones. We have shown that IMPACC can operate with highly linear response with peak SFDR value of 132 dB.

Another important requirement in the link design is the inherent versatility of the device or link to compensate for unexpected changes or unavoidable parameter variations due to environmental temperature change, device aging, and fabrication errors. In this regard, IMPACC has a unique place among previously reported linear intensity modulators since it is the only linear modulator design, as far as we know, that has an inherent compensation capability. We demonstrated this inherent feature by compensating various nonideal and often detrimental effects in the modulator from RR waveguide loss and unbalanced optical inputs to various coupling conditions that could lower manufacturing tolerance and degrade linearity performance if not mitigated. The ability to maintain high SFDR (e.g., >130 dB) under these conditions makes IMPACC a viable candidate for many high-bandwidth RF FO-link applications, and well positioned as ideal linear intensity modulator for software-defined MPLs

Acknowledgements

NM would like to thank partial support from the European Committee under program FP7-MC-CIG 333829.

Author details

Benjamin Dingel[1,2]*, Nicholas Madamopoulos[3,4] and Andru Prescod[5]

*Address all correspondence to: bbdingel@stny.rr.com

1 Nasfine Photonics Inc., Painted Post, NY, USA

2 Ateneo Innovation Center, Ateneo De Manila University, Quezon City, Philippines

3 Department of Aeronautical Science, Hellenic Air Force Academy, Dekeleia, Greece

4 Department of Electrical Engineering, The City College of City University of New York, USA

5 ManTech International Corporation, Arlington, VA, USA

References

[1] Elbers J-P, Autenrieth A. From static to software-defined optical networks. In: 16th International Conference on Optical Network Design and Modeling (ONDM), 2012. IEEE; 2012

[2] Gringeri S, Bitar N, Xia TJ. Extending software defined network principles to include optical transport. IEEE Communications Magazine. 2013;**51**(3):32–40

[3] Amaya N, et al. Software defined networking (SDN) over space division multiplexing (SDM) optical networks: Features, benefits and experimental demonstration. Optics Express. 2014;**22**(3):3638–3647

[4] Chen X, Zhang Y. Intelligence on optical transport SDN. International Journal of Computer and Communication Engineering. 2015;**4**(1):5

[5] Kreutz D, et al. Software-defined networking: A comprehensive survey. Proceedings of the IEEE. 2015;**103**(1):14–76

[6] Thyagaturu AS, et al. Software defined optical networks (SDONs): A comprehensive survey. IEEE Communications Surveys & Tutorials. 2016;**18**(4):2738–2786

[7] Lannoo B. Radio-over-fibre for ultra-small 5G cells. In: 17th International Conference on Transparent Optical Networks (ICTON). 2015. pp. 1–4. DOI: 10.1109/ICTON.2015.7193591

[8] Waterhouse R, Novack D. Realizing 5G: Microwave photonics for 5G mobile wireless systems. IEEE Microwave Magazine. 2015;**16**(8):84–92

[9] Way WI. Broadband Hybrid Fiber/Coax Access System Technologies. San Diego, CA: Academic Press; 1998

[10] Al-Raweshidy H, Komaki S, editors. Radio over Fiber Technologies for Mobile Communications Networks. Norwood, MA: Artech House; 2002

[11] Llorente R, Beltrán M. Radio-over-fibre Techniques and Performance. Rijeka, Croatia: INTECH Open Access Publisher; 2010

[12] Liu C, et al. Key microwave-photonics technologies for next-generation cloud-based radio access networks. Journal of Lightwave Technology. 2014;**32**(20):3452–3460

[13] Ackerman E, Cox III C, Riza N, editors. Analog Fiber Optic Links. SPIE Milestone Series. Vol. MS-149; 1998

[14] Chang W, editor. RF Photonics Technologies in Optical Fiber Links. Cambridge, UK: Cambridge University Press; 2002

[15] Vilcot A, Cabon B, Chazelas J, editors. Microwave Photonics from Components to Applications. New York, NY: Kluwer Academic Publishers; 2003

[16] Cox III CH. Analog Optical Links: Theory and Practice. Cambridge, UK: Cambridge University Press; 2004

[17] Ji PN. Software defined optical network. In: 11th International Conference on Optical Communications and Networks (ICOCN), 2012. IEEE; 2012

[18] Channegowda M, et al. Experimental demonstration of an OpenFlow based software-defined optical network employing packet, fixed and flexible DWDM grid technologies on an international multi-domain test bed. Optics Express. 2013;**21**(5):5487–5498

[19] Simeonidou D, Nejabati R, Channegowda MP. Software defined optical networks technology and infrastructure: Enabling software-defined optical network operations. In: Optical Fiber Communication Conference and Exposition and the National Fiber Optic Engineers Conference (OFC/NFOEC); 2013. IEEE; 2013

[20] ONF Solution Brief. OpenFlow-enable Transport SDN. May 2014

[21] Capmany J, Novak D. Microwave photonics combines two worlds. Nature Photonics. 2007;**1**(6):319–330

[22] Marpaung D, et al. Integrated microwave photonics. Laser & Photonics Reviews. 2013;**7**(4):506–538

[23] Capmany J, Munoz P. Integrated microwave photonics for radio access networks. Journal of Lightwave Technology. 2014;**32**(16):2849–2861

[24] Guan B, et al. CMOS compatible reconfigurable silicon photonic lattice filters using cascaded unit cells for RF-photonic processing. IEEE Journal of Selected Topics in Quantum Electronics. 2014;**20**(4):359–368

[25] Iezekiel S, et al. RF engineering meets optoelectronics: Progress in integrated microwave photonics. IEEE Microwave Magazine. 2015;**16**(8):28–45

[26] Bowers JE. Integrated microwave photonics. In: 2015 International Topical Meeting on Microwave Photonics (MWP). IEEE; 2015

[27] Xu X, Bosisio RG, Wu K. Analysis and implementation of six-port software-defined radio receiver platform. IEEE Transactions on Microwave Theory and Techniques. 2006;**54**(7):2937–2943

[28] Ibrahim S, et al. Demonstration of a fast-reconfigurable silicon CMOS optical lattice filter. Optics Express. 2011;**19**(14):13245–13256

[29] Pérez D, Gasulla I, Capmany J. Software-defined reconfigurable microwave photonics processor. Optics Express. 2015;**23**(11):14640–14654

[30] Zhuang L, et al. Programmable photonic signal processor chip for radiofrequency applications. Optica. 2015;**2**(10):854–859

[31] Pérez D, et al. Reconfigurable lattice mesh designs for programmable photonic processors. Optics Express. 2016;**24**(11):12093–12106

[32] Wei W, et al. Software-defined microwave photonic filter with high reconfigurable resolution. Scientific Reports. 2016;**6**

[33] Capmany J, Gasulla I, Pérez D. Microwave photonics: The programmable processor. Nature Photonics. 2016;**10**(1):6–8

[34] Khan MH, et al. Ultrabroad-bandwidth arbitrary radiofrequency waveform generation with a silicon photonic chip-based spectral shaper. Nature Photonics. 2010;**4**(2): 117–122

[35] Burla M, et al. Integrated waveguide Bragg gratings for microwave photonics signal processing. Optics Express. 2013;**21**(21):25120–25147

[36] Pant R, et al. On-chip stimulated Brillouin scattering for microwave signal processing and generation. Laser & Photonics Reviews. 2014;**8**(5):653–666

[37] Shin H, et al. Control of coherent information via on-chip photonic–phononic emitter–receivers. Nature Communications. 2015;**6**. DOI: 10.1038/ncomms7472

[38] Wang J, et al. Reconfigurable radio-frequency arbitrary waveforms synthesized in a silicon photonic chip. Nature Communications 2015;**6**. DOI:10.1038/ncomms6957

[39] Liu W, et al. A fully reconfigurable photonic integrated signal processor. Nature Photonics. 2016;**10**(3):190–195

[40] Wang J, Long Y. A Chip-Scale Microwave Photonic Signal Processing. Rijeka, Croatia: INTECH Open Access Publisher; 2017

[41] Cox CH, et al. Limits on the performance of RF-over-fiber links and their impact on device design. IEEE Transactions on Microwave Theory and Techniques. 2006;**54**(2): 906–920

[42] Iezekiel S, editor. Microwave Photonics: Devices and Applications. Chichester, West Sussex, UK: John Wiley & Sons, Ltd; 2009

[43] Urick Jr VJ, Mckinney JD, Williams KJ. Fundamentals of microwave photonics. In: Chang K, editor. Wiley Series in Microwave and Optical Engineering. Hoboken, New Jersey: John Wiley & Sons, Inc.; 2015

[44] Urick VJ, Diehl JF, Draa MN, McKinney JD, Williams KJ. Wideband analog photonic links: Some performance limits and considerations for multi-octave implementations. In: Nelson RL, Prather DW, Schuetz C, editors. Proceedings of the SPIE, RF and Millimeter-Wave Photonics II; 2012. Vol. 8259:825904

[45] Bridges W, Shaffner J. Distortion in linearized electrooptic modulators. IEEE Transactions on Microwave Theory and Techniques. 1995;**43**(9):2184–2197. DOI: 10.1109/22.414563

[46] Ridder RM, Korotky S. Feedforward compensation of integrated optic modulator distortion. In Proceedings of the Technical Digest Optical Fiber Communications Conference, San Francisco, CA; 1990

[47] Chiu Y, et al. Broad-band electronic linearizer for externally modulated analog fiber-optic links. IEEE Photonics Technology Letters. 1999;**11**(1):48–50

[48] Sadhwani R, Basak J, Jalali B. Adaptive electronic linearization of fiber optic links. In: Optical Fiber Communication Conference. Optical Society of America; 2003

[49] Sadhwani R, Jalali B. Adaptive CMOS predistortion linearizer for fiber-optic links. Journal of Lightwave Technology. 2003;**21**(12):3180–3193

[50] Killey RI, et al. Electronic dispersion compensation by signal predistortion using digital processing and a dual-drive Mach-Zehnder modulator. IEEE Photonics Technology Letters. 2005;**17**(3):714–716

[51] Säckinger E. Broadband Circuits for Optical Fiber Communication. John Wiley & Sons; 2005

[52] Moon H, Sedaghat R. FPGA-based adaptive digital predistortion for radio-over-fiber links. Microprocessors and Microsystems. 2006;**30**(3):145–154

[53] Lee T-K, et al. Theoretical analysis and realization of optoelectrical predistortion optical transmitter for the simultaneous suppression of IM3 and IM5 signal. Optics Communications. 2012;**285**(10):2697–2701

[54] Lam D, et al. Digital broadband linearization of optical links. Optics Letters. 2013;**38**(4):446–448

[55] Schmidta C, Kottkea C, Jungnickela V, Freunda R. High-speed digital-to-analog converter concepts. In: Proceedings of the SPIE; 2017, January, Vol. 10130, pp. 101300N-1

[56] Lee G-W, Han S-K. Linear dual electroabsorption modulator for analog optical transmission. Microwave and Optical Technology Letters. 1999;**22**(6):369–373

[57] Betts GE, et al Gain limit in analog links using electroabsorption modulators. IEEE Photonics Technology Letters. 2006;**18**(19):2065–2067

[58] Xie XB, et al. Analysis of linearity of highly saturated electroabsorption modulator link due to photocurrent feedback effect. Optics Express. 2007;**15**(14):8713–8718

[59] Kogelnik H, Schmidt RV. Switched directional couplers with alternating $\Delta\beta$. IEEE Journal of Quantum Electronics. 1976;**12**:396–401

[60] Thaniyavarn S. Modified 1×2 directional coupler waveguide modulator. Electronics Letters. 1986;**22**:941–942

[61] Tavlykaev RF, Ramaswamy RV. Highly linear Y-fed directional coupler modulator with low intermodulation distortion. Journal of Lightwave Technology. 1999;**17**:282–291

[62] Zhang X, et al. Highly linear broadband optical modulator based on electro-optic polymer. IEEE Photonics Journal. 2012;**4**(6):2214–2228

[63] Li J, et al. Optical sideband processing approach for highly linear phase-modulation/direct-detection microwave photonics link. IEEE Photonics Journal. 2014;**6**(5):1–10

[64] Zheng X-P, et al. All-optical signal processing for linearity enhancement of Mach–Zehnder modulators. Chinese Science Bulletin. 2014;**59**(22):2655–2660

[65] Chen J, Zhu D, Pan S. Linearized phase-modulated analog photonic link based on optical carrier band processing. In: 25th Wireless and Optical Communication Conference (WOCC); 2016. IEEE; 2016

[66] Tazawa H, Steier W. Linearity of ring resonator-based electro-optic polymer modulator. Electronics Letters. 2005;**41**(23):1297–1298

[67] Tazawa H, Steier WH. Analysis of ring resonator-based traveling-wave modulators. IEEE Photonics Technology Letters. 2006;**18**(1):211–213

[68] Song M, Zhang L, Beausoleil RG, Willner AE. Nonlinear distortion in a silicon microring-based electro-optic modulator for analog optical links. IEEE Journal of Selected Topics in Quantum Electronics. 2010;**16**(1):185–191

[69] Fegadolli W, Oliveira JEB, Almeida VR. Highly linear electro-optic modulator based on ring resonator. Microwave Optics Technology Letters. 2011;**53**(10):2375–2378

[70] Ayazi A, Baehr-Jones T, Liu Y, Lim AE-J, Hochberg M. Linearity of silicon ring modulators for analog optical links. Optics Express. 2012;**20**(12):13115–13122

[71] Chen L, Chen J, Nagy J, Reano RM. Highly linear ring modulator from hybrid silicon and lithium niobate. Optics Express. 2015;**23**(10):13255–13264

[72] Hosseinzadeh A, Middlebrook CT. Highly linear dual ring resonator modulator for wide bandwidth microwave photonic links. Optics Express. 2016;**24**(24):27268–2727924

[73] Yariv A, Yeh P. Optical Waves in Crystals-Propagation and Control of Laser radiation. New York, NY: John Wiley and Sons; 1984

[74] Johnson LM, Roussell HV. Reduction of intermodulation distortion in interferometric optical modulators. Optics Letters. 1988;**13**(10):928–930

[75] Johnson LM, Roussell HV. Linearization of an interferometric modulator at microwave frequencies by polarization mixing. IEEE Photo Technology Letters. 1990;**2**(11):810–811

[76] Djupsjobacka A. A linearization concept for integrated-optic modulators. Photonics Technology Letters. 1992;**4**(8):869–871

[77] Ho KP, Kahn JM. Optical frequency comb generator using phase modulation in amplified circulating loop. IEEE Photonics Technology Letters. 1993;**5**(6):721–725

[78] Betts GE, O'Donnell FJ. Optical analog link using a linearized modulator. In: IEEE LEOS Annual Meeting, 31 Oct.–3 Nov. 1994; Vol. 2, pp. 278–279

[79] Betts GE. Linearized modulator for suboctave-bandpass optical analog links. IEEE Transactions Microwave Theory Technology. 1994;**42**(12):2642–2649

[80] Sabido DJM, Tabara M, Fong TK, Lu CL, Kazovsky LG. Improving the dynamic range of a coherent AM analog optical link using a cascaded linearized modulator. IEEE Photonics Technology Letters. 1995; **7**(7):813–815

[81] Burns WK. Linearized optical modulator with fifth order correction. IEEE Journal of Lightwave Technology. 1995;**13**(8):1724–1727

[82] Betts GE, O'Donnell FJ. Microwave analog optical links using suboctave linearized modulators. Photonics Technology Letters. 1996;**9**(8):1273–1275

[83] Loayssa A, Alonso M, Benito D, Garde MJ. Linearization of electro-optic modulators at millimeter-wave frequencies. In: Proceedings of the LEOS '99 - 12th Annual Meeting. 8–11 November 1999; Vol. 1, pp. 275–276

[84] Ackerman EI. Broad-band linearization of a Mach-Zehnder electro-optic modulator. IEEE Transactions on Microwave Theory and Techniques. 1999;**47**(12):2271–2279

[85] Yariv A. Universal relations for coupling of optical power between microresonators and dielectric waveguides. Electronics Letters. 2000;**36**:321–322

[86] Okamoto K. Fundamentals of Optical Waveguides. Burlington, MA: Academic Press; 2000

[87] Rengand N, Shpantzer I, Achiam Y, Kaplan A, Greenbalatt A, Harston G, Cho PS. Military Communications Conference. IEEE MILCOM 2003. 13–16 Oct. 2003;2:1208

[88] Xie X, Khurgin J, Kang J, Chow FS. Linearized Mach-Zehnder intensity modulator. IEEE Photonics Technology Letters 2003;**15**(4):531–533

[89] Dingel B. Ultra-linear, broadband optical modulator for high performance analog fiber link system. In: IEEE International Topical Meeting on Microwave Photonics 2004 (MWP'04); 4–6 Oct. 2004. pp. 241–244

[90] Schwelb O. Transmission, group delay, and dispersion in single-ring optical resonators and add/drop filters – A tutorial overview. Journal of Lightwave Technology. 2004;**22**: 1380–1394

[91] Tazawa H. Ring resonator based electro-optic polymer modulators for microwave photonics applications [Ph.D. thesis]. Los Angeles, CA: University Southern California; 2005

[92] Tazawa H, Steier W. Bandwidth of linearized ring resonator assisted Mach-Zehnder modulator. IEEE Photonics Technology Letters. 2005;**17**(9):1851–1853

[93] Dingel B. Linear optical modulator. US Patent 6943931, Sept. 2005

[94] Dingel BB, Madabhushi R, Madamopoulos N. Super-linear optical modulator technologies for optical broadband access network: Development and potential. In: SPIE Proceeding 6012, Optical Transmission Systems and Equipment for WDM Networking IV, October 2005; paper No. 32, Boston

[95] Chen WY, Van V, Herman WN, Ho PT. Periodic microring lattice as a bandstop filter. IEEE Photonics Technology Letters. 2006;**18**(19):2041–2043

[96] Madamopoulos N, Dingel B. Performance analysis of the ultra-linear optical intensity modulator. In: Proceedings of the SPIE 6390, Optical Transmission Systems and Equipment for Networking V, October 2006; paper No. 15, Boston

[97] Van V, Herman W, Ho PT. Linearized microring-loaded mach-zehnder modulator with RF gain. IEEE Journal of Lightwave Technology. 2006;**24**(4):1850–1854

[98] Tazawa H, Kuo YH, Dunayevskiy I, Luo J, Jen A, Fetterman H, Steier W. Ring resonator-based electrooptic polymer traveling-wave modulator. IEEE Journal of Lightwave Technology. 2006;**24**(9):3514–3519

[99] Oliveira JEB, Sakamoto BFR, dos Santos Fegadolli W. Similarities between birefringent gires tournois interferometer and double ring assisted Mach Zehnder. In: European Microwave Conference, Munich; 9–12 Oct 2007. pp. 1330–1333

[100] Zhu G, Liu W, Fetterman HR. A broadband linearized coherent analog fiber-optic link employing dual parallel Mach–Zehnder modulators. IEEE Photonics Technology Letters. 2009:**21**(21):1627–1629

[101] Prescod A, Dingel B, Madamopoulos N. Super-linear modulator with extended bandwidth capability for broadband access applications. In: Proceedings of the SPIE OPTO 2009, Vol. 7234, San Jose, CA; Jan. 24–29, 2009

[102] Prescod A, Dingel BB, Madamopoulos N. Traveling wave optical modulator with high dynamic range (<<130 dB) at large modulation frequencies. In: Proceedings 2009 IEEE Sarnoff Symposium, Session: Optical Communications and Networking, Princeton, NJ; March 30–April 1, 2009. pp. 1–5

[103] Prescod A, Dingel B, Madamopoulos N. An ultra-linear modulator with inherent SFDR compensation capability. In Proceedings of the OSA Integrated Photonics Research, Silicon and Nano Photonics (IPR), July 27, 2010, Monterey, California, USA

[104] Prescod A, Dingel BB, Madamopoulos N, Madabhushi R. Effect of ring resonator waveguide loss on SFDR performance of highly linear optical modulators under suboctave operation. Photonics Technology Letters. 2010;**22**(17):1297–1299

[105] Dingel B, Prescod A, Madamopoulos N. Inherent RF linearized bandwidth broadening capability of an ultra-linear optical modulator. In: Proceedings of the SPIE 7958, Broadband Access Communication Technologies IV; January 2011; paper 7958–3, San Francisco, CA

[106] Dingel B, Madamopoulos N, Prescod A, Madabhushi R. Analytical model, analysis and parameter optimization of a super linear electro-optic modulator (SFDR > 130 dB). Optics Communications. 2011;**284**(24):5578–5587

[107] Madamopoulos N, Dingel B, Prescod A. Interferometric modulator with phase-modulating and cavity-modulating components (IMPACC) for high linearity microwave applications. In Proceedings of the ICPS 2013, The International Conference on Photonics Solutions, Pattaya City, Thailand, May 26–28, 2013 (Co-sponsored by OSA, SPIE, IEEE)

[108] Madamopoulos N, Dingel BB, Prescod A. Review and challenges of high frequency cavity-assisted electro-optic modulators. In: Wireless Optical Communication Conference, WOCC 2014, Newark, NJ, USA; 2014

[109] Jiang W, et al. A linearization analog photonic link with high third-order intermodulation distortion suppression based on dual-parallel Mach–Zehnder modulator. IEEE Photonics Journal. 2015;**7**(3):1–8

[110] Dingel B, Madamopoulos N, Prescod A, Madabhushi R. Power balancing effect on the performance of IMPACC modulator under critical coupling (CC), over coupling (OC), and under coupling (UC) conditions at high frequency. Proceedings of the SPIE Broadband Access Communication Technologies. 2012;**VI**:828208

[111] Ye B, Dingel BB, Cui W. Minimalist-design, high-functionality, micro-ring resonator-based optical filter with narrow linewidth and low group delay using Looped Back Over- and Under-coupled Resonator (LOBOUR). Proceedings of the SPIE. Broadband Access Communication Technologies VII. 2013;8645:86450T

[112] Dingel B, Ye B, Cui W, Madamopoulos N. High-performance, minimalist-design, microring resonator-based optical filter with reduced low group delay and simplified center-wavelength control. International Conference on Photonics Solutions 2013. Thailand; 2013

[113] Dingel B, Prescod A, Madamopoulos N, Madabhushi R. Performance of ring resonator-based linear optical modulator (IMPACC) under critical coupling (CC), over coupling (OC), and under coupling (UC) conditions. In: IEEE Photonics Annual Society Meeting; 2011

5

Single-Photon Avalanche Diodes in CMOS Technologies for Optical Communications

Edward M.D. Fisher

Additional information is available at the end of the chapter

Abstract

As optical communications may soon supplement Wi-Fi technologies, a concept known as visible light communications (VLC), low-cost receivers must provide extreme sensitivity to alleviate attenuation factors and overall power usage within communications link budgets. We present circuits with an advantage over conventional optical receivers, in that gain can be applied within the photodiode thus reducing the need for amplification circuits. To achieve this, single-photon avalanche diodes (SPADs) can be implemented in complementary metal-oxide-semiconductor (CMOS) technologies and have already been investigated in several topologies for VLC. The digital nature of SPADs removes the design effort used for low-noise, high-gain but high-bandwidth analogue circuits. We therefore present one of these circuit topologies, along with some common design and performance metrics. SPAD receivers are however not yet mature prompting research to take low-level parameters up to the communications level.

Keywords: single-photon, avalanche, diodes, visible, light, communications, receivers

1. Introduction

Optical technologies are routinely used in inter-continental communications, where both data rates and transmission lengths are high, or in fast networking applications such as data centres [1, 2]. At the opposite end of the distance scale, interconnects in microelectronics are moving towards photonics. This is due to the bandwidth restrictions of metals on silicon ICs. By using free-space links for personal computing, optical communication may offer an alternative to wireless standards, i.e. visible-light communications (VLC) applications [3, 4]. The radio spectrum is becoming increasingly crowded, with the UK's frequency allocation tables

(UKFAT) showing severe restrictions on wireless bandwidth, especially as 4G mobile internet is expanded towards 5G. Ultimately, the radio spectrum is limited at the high-GHz and low-THz bands by atmospheric attenuation, principally the molecular resonances of water and oxygen. There can be no doubt that as a mature technology, optical communication has become ubiquitous and is growing as bandwidth demands rise [5]. However, with that ubiquity and growth, the total electrical power, and indeed the materials and complexity of systems, is also increasing at an extraordinary annual rate.

The design of optical communication systems is highly application dependant [1, 2, 6]. The high-speed, long-distance intercontinental links can use complex, power-hungry and expensive transmitters (Tx) and receivers (Rx). This is principally as the energy per bit (J/b) and cost per bit ($/b) can be kept low through long-term installations with high throughput and multiple end-users. As a direct comparison, hardware for applications such as VLC to mobile phones and personal computers, must be simple, low-power and cheap [3, 4]. This is especially true as smart-phones or personal computers imply a single end-user and a bandwidth low enough for simple or minimal forward error correction (FEC) schemes. It also implies battery operation with an expectation of long battery life and short product lifetimes (e.g. the 2-year average for mobiles). In terms of costs, VLC transmitters and receivers must: (i) be comparable to other system level components, i.e. memory, processor or display, and (ii) be suitable for high volume production, i.e. > 100,000 units.

1.1. Basic receiver operations and research directions

Taking a high-level approach, all optical communication receivers must perform eight basic operations [1, 2, 6]:

1. Convert incoming optical signals into an electrical signal, usually electrical current.
2. Amplify and convert the current signal into a form that can be processed easily.
3. Equalize and/or modify the signal to increase bandwidth or remove unwanted artefacts.
4. Demodulate the incoming signal, depending on its transmission modulation.
5. Sample the processed signal to recover the transmitted digital bit stream.
6. Output the signal in a form suitable for the end-user's application.
7. Optionally recover the clock embedded within the data stream, and
8. Optionally perform error correction tasks.

The role of the receiver front-end covers only tasks #1 to #6, however as part of the sampling operation many receivers also perform a clock and data recovery (CDR) operation [1], i.e. task #7. As error correction (task #8) requires a memory buffer, such tasks are separated from a receiver front-end and incorporated within encoding and protocol specific units within the overall system. The above front-end and back-end split can be viewed conveniently with respect to Ethernet technologies. The front-end Tx/Rx is often denoted as the 'PHY' as it relates to physical transmission and reception, it is also viewed as the physical layer in the Open Systems Interconnect (OSI) model. At the back-end, a Medium Access Controller (MAC) is used to implement the OSI data link layer, and contains both (i) transmit and receive buffers and (ii) framing, addressing and protocol level functions. Within the optical communication physical layer community, there are a number of key performance metrics that are driving innovation and the exploration of the current limiting factors [1, 2, 6]. These overall performance metrics are given below with current areas of innovation.

- Demand for increased bandwidth and transmission speeds
  - *Increase in bits per symbol through increased modulation complexity*
  - *Increase in channels per physical link though multiplexing in the time, frequency, phase or wavelength domains*
  - *Increased use of channel equalization techniques at both Tx and Rx*
- Reduction in energy usage and increased energy efficiency
  - *Decrease in the energy per bit*
  - *Trends towards quantum limited transmission (minimum optical power)*
  - *Increased transmission spectral efficiency for given real-world channel*
  - *Increased receiver sensitivity through (i) low-noise amplification, or (ii) optical and/or electrical coherent reception either homodyne or heterodyne detection*
- Decreased manufacture and design costs, and
  - *Increased Tx or Rx levels of CMOS integration, i.e. single-chip solutions*
  - *Decreased use of rare earth or high-cost materials*
  - *Minimization of bulky (for a particular CMOS node) analogue circuitry*
  - *Decreased design costs through Tx/Rx standardization*
- Decrease in bit errors and noise immunity
  - *Increased optical Tx extinction ratio*
  - *Decreasing Tx and Rx electrical noise*
  - *Decrease in inter-symbol interference*

1.2. Key receiver performance metrics

Before discussing the target application, a number of key receiver performance metrics need to be defined. These are (i) basic parameters at the communications level and (ii) measurements that can be made to assess the quality of a receiver, or its fitness for a particular application.

1.2.1. The data rate

The data-rate, R_D, measured in bits per second (b/s) is a measure of the serialized data throughput. As technology progresses, the throughput that is required at personal, commercial, national and international levels is increasing [5]. The data rate is intrinsically linked to the transmitter, channel and receiver bandwidth, BW, as each must pass all required frequency components in order to allow accurate, low-error decoding of the received signal back into binary data. In a communications receiver, the sampling point is chosen to be within the centre of the data symbol, thus allowing some protection from distortions on a signal's rising and falling edges. This leads to a lower frequency optimum bandwidth, BW_{-3dB}, which is related to the data rate [6], see Eq. (1). This is a mid-way point between (i) a low bandwidth that reduces input noise but produces low-pass inter-symbol interference, and (ii) a high bandwidth that captures all of a signal without distortion but also captures a large noise bandwidth and unnecessarily increases design specifications. For a 10 ns (100 Mb/s) data symbol, this would suggest a receiver bandwidth of 66.7 MHz. This is significantly below the knee frequency (Eq. (2)) of 250 MHz for a 2 ns rise time, T_R (10–90%) signal, i.e. 20% of a 10 ns 100 Mb/s data rate symbol.

$$BW_{-3dB} = \frac{2R_D}{3} \tag{1}$$

$$F_{KNEE} = \frac{0.5}{T_R} \tag{2}$$

1.2.2. The bit error rate

The bit error rate (BER) is defined as the number of single-bit errors within a continuous bit stream [1, 2, 6]. Data links at 100 Mb/s to 10 Gb/s may utilize error correction and are thus specified with 'native', no-FEC, error rates at 1×10^{-9}, i.e. a single bit error within a transmission of one billion bits. For links with slower data rates or where the application incurs significant channel distortions (such as VLC), the use of FEC may be a requirement. For these cases, the designer aims for the best native BER performance but as this cannot be guaranteed due to unknown factors impinging on the communications channel, a worst case maximum BER is chosen. For many FEC algorithms there is a limit as to effective recovery, at an approximate BER level of 1×10^{-3}. The BER is both a specification of the link and a measurable performance parameter. As an example, a BER of 1×10^{-6} to 1×10^{-9} may be specified as part of a protocol standard, but a measured value of 1×10^{-6} at a particular incident optical power may also become improved at a higher optical power.

There are two critical issues for the BER, (i) signal noise and (ii) inter-symbol interference. Assuming the use of non-return-to-zero (NRZ) on-off-key (OOK) modulation, the error rate is related to the probabilities that the noise of a 'one' or 'zero' cross a threshold, N_{th}. As both zeros and one are subject to noise, both have amplitude distributions about their means, (N_0 and N_1), these will have standard deviations of σ_0 and σ_1 respectively. The threshold and bit error rate can be approximated, for Gaussian noise profiles by Eqs. (3) and (4).

$$N_{th} = \frac{\sigma_0 N_1 + \sigma_1 N_0}{\sigma_0 \sigma_1} \tag{3}$$

$$BER = \frac{1}{2} erfc\left(\frac{N_1 - N_{th}}{\sigma_1 \sqrt{2}}\right) \tag{4}$$

Inter-symbol interference (ISI) is the bleed-through of symbols directly before and after, interfering with the reception of a bit. If we view the system as a low-pass filter of bandwidth, BW_{-3dB}, rectangular pulses become smeared in time with exponential rise and fall times [1]. As the signal is the supposition of multiple transitions the amplitude at the sampling point becomes corrupted by these tails. If the bandwidth decreases, these tails become longer in comparison to a bit period. Ideally, all ISI should have settled prior to sampling, however ISI acts to decrease the region within a symbol where robust sampling can occur.

1.2.3. Receiver sensitivity

For communications link budgets, it is useful to establish the receiver sensitivity [1]. This is defined as the incident optical power necessary for a receiver to reach the specified BER at the specified data rate, and is measured in dBm [1]. To ensure communication through a variety of signal attenuation factors, a high receiver sensitivity is required. We would therefore choose a receiver with a sensitivity of −20 dBm (0.01 mW) above a receiver with −5 dBm (0.32 mW). This would be particularly true if we knew the optical power was likely to be low at 0.05 mW due to perhaps low transmitter power and optical filtering. The sensitivity is dependent not only on the physical optical sensitivity, but also the received signal as a proportion of the receiver noise and inter-symbol interference. Thus, for a high-sensitivity receiver, (i) optical efficiency and electrical gain need to be high, while (ii) noise, inter-symbol interference and sampling phase-noise need to be low [1, 6].

1.2.4. The quantum limit

The receiver sensitivity can easily be reformulated in terms of the number of photons required per bit [1]. As this can be generalized to a theoretical receiver, we come across the theoretical ideal known as the quantum limit, QL. An ideal receiver would have zero noise (zero bits given by zero photon arrivals) and would receive single-photons with 100% efficiency. As photon arrivals follow Poisson statistics, the QL can be given by Eq. (5), where T_B is the symbol duration and $\gamma(T_B)$ denotes the average number of photons received per symbol. For a BER of 1×10^{-9}, approximately 20 photons are needed per symbol, which at 100 Mb/s ($T_B = 10$ ns), is 2×10^9 photons per second. For 650 nm (red) light this is 0.611 nW (−62.14 dBm).

$$BER_{QL} = \frac{1}{2} e^{-\gamma(T_B)} \tag{5}$$

1.2.5. Energy per bit and power consumption

While the power dissipation of a receiver is a product of the supply voltage and receiver current, receivers with different data rates are often compared using the energy per bit, measured in Joules per bit (J/b) [1, 6]. As data rates are stretching beyond Gb/s, the energy per bit and therefore the total energy usage become significant issues [5]. Amplification and equalization circuitry often incur large energy costs; hence effort is being concentrated on efficient implementation. While reduced voltage supplies for integrated circuits (ICs) helps reduce power consumption, analogue design becomes more complex. Coupled with the bulkiness of robust analogue circuitry in advanced CMOS nodes (<45 nm), there is a trend towards denser digital circuitry in place of analogue where this is achievable, (i.e. trends of delay-locked loops (DLLs) replacing phase-locked loops (PLLs)). A significant open question is how to achieve high receiver sensitivity while obtaining a low energy per received bit? As transmitter efficiency and overall power/bit should also be included in the transmission energy budget of data, it is in our interest to reduce the power required of the most inefficient sub-system, along with improving its efficiency.

1.2.6. Integration and silicon area usage

The degree of receiver circuit integration, directly impacts system level costs and performance. Older receivers required separate ICs for the basic receiver tasks discussed in Section 1.1, however increased integration allows higher bandwidths, lower power dissipation and significantly reduced system level design times [7]. As an example, integrating a CDR circuit with the receiver removes high-bandwidth, accurate traces from a printed circuit board (PCB). It also reduces the power that must be dissipated by the input and output pads of multiple ICs, and drastically improves jitter [1]. Likewise, by integrating a photodiode in the same substrate as a CMOS receiver, the capacitance and inductance of bond pads and wires between an external photodiode and IC can be removed and thus achieving higher bandwidth. As multi-channel and multi-port ICs become more prevalent, some circuit blocks can be shared, thereby increasing area and energy efficiency. While requiring increased configurability, and difficulty in testing and verification, there is significant drive towards reduction in the number of input/output (I/O) pads within a system, hence integration of sub-systems onto a single IC.

1.3. Target application: visible-light communications

Visible light communication is a free-space method emerging as an alternative to local wireless systems [3, 4, 8, 9]. It is intended for data rates between 100 Mb/s and 1 Gb/s and has recently been ratified by the institute of electrical and electronic engineers (IEEE) (IEEE P802.15.7) [10]. A significant thrust is the dual use of energy-efficient light emitting diode (LED) technologies for illumination and communications [3, 10] (**Figure 1**). The advantages of VLC are principally

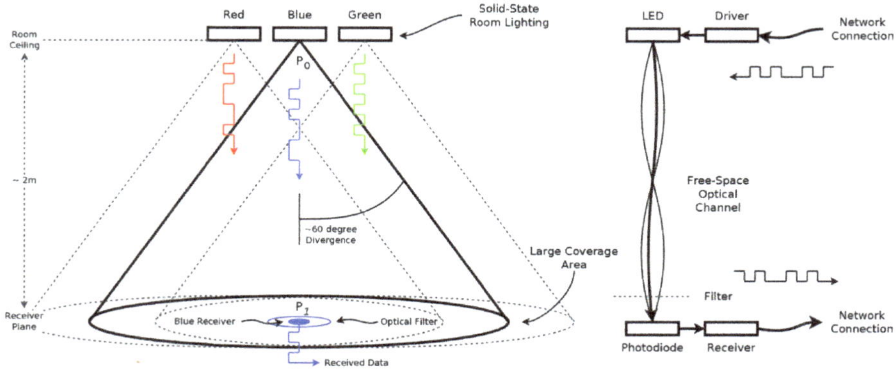

Figure 1. Visible optical communications and room lighting using red, green and blue modulated, LEDs. The wide divergence angle, presents an issue as the received optical power is low.

that the visible band is: (i) of wide bandwidth, (ii) is un-licensed and (iii) immune to electromagnetic interference from existing wireless technologies [1, 2, 6]. The primary target for end-users are mobile devices such as smart-phones and laptops where VLC can be used to supplement wireless systems for high bandwidth applications [3, 4] such as real-time, high-quality media streaming [5]. It is projected that as (i) the frequency allocation tables become increasingly dense and (ii) transmitters and receivers become increasingly difficult to produce at the high GHz radio levels, that some form of optical link—above the THz gap—will be required [3].

Despite advances in VLC, the application has a number of inherent issues that require application specific receiver designs. Firstly, the divergence angle of LEDs—particularly when wide-area lighting is needed—produces a large, diffuse illuminated area [3, 4]. When using a receiver IC of limited size, as per mobile systems, the receiver will capture a small proportion of the total transmitted optical power. Secondly, as optical sources are modulated at speeds close to their native bandwidth, both the average optical power and the modulation extinction ratio (the difference between high and low amplitudes) become smaller. Finally, in order to utilize the wide optical bandwidth, some form of optical filter is needed to perform wavelength multiplexing, i.e. multiple data channels on different colours. This has two implications, firstly that for ideally white illumination (wide-band), the receiver will receive a signal strength which is a narrow-band subset of wavelengths. A three-wavelength system (e.g. red, green, and blue, RGB), will naturally imply optical powers approximately one third of the total optical power. The second implication is that the filter itself will reduce optical power as they are neither perfect within their pass-band (signal attenuation) and have finite stop-band rejection (inter-colour interference and therefore finite separation). The effect of these issues, is that optical power incident on a receiver will be attenuated, while still being modulated at high data rates and with the user expecting both low error rates and low power consumption.

1.4. Research aims: extreme receiver sensitivity

We can consider VLC, as involving an inherent communications link budget issue, i.e. multiple attenuation factors necessitate reciprocal receiver gains. Further, the future requirements for low energy-per-bit communications would suggest an overall reduction in the link budget. The difficulty in such a scenario is twofold:

- First, the gain required in the communication budget may be extreme, especially if Gb/s links are required with multiple wavelength multiplexed channels, wide-area coverage and compact optically simple receivers.
- Second, as optical powers decrease and electrical gains increase, the signal-to-noise ratio (SNR) becomes a significant hurdle for robust detection. Johnson noise may limit analogue gain, especially if cooling is unachievable. Photon shot-noise may also become significant, without the possibility of using increased optical power at the transmitter (room lighting limited).

The work within this chapter therefore focuses on receiver designs with extreme sensitivity and high bandwidths to overcome the inherent attenuation factors of VLC, see [11, 12] and references therein. The overall method of using the avalanche multiplication of photo-generated carriers may use extra electrical power at the receiver—principally through increased biasing voltages. But, as this removes the need for analogue amplification circuitry, represents a more economical use of both electrical power and silicon area.

2. High-sensitivity optical to electric conversion

When electromagnetic radiation is incident upon a material, there are three processes that can lead to photon interaction, and thus the detection of that photon [13]. These are the (i) photo-electric effect [14], (ii) Compton scattering and (iii) pair production. Taking the high-energy processes first, we can begin to discount processes in order to arrive at the process whereby optical, and in particular visible, sensors operate.

- In Compton scattering, the incident photon is scattered by an atomic electron. It imparts some energy to the electron, meaning that the photon energy is reduced, and thus the wavelength becomes longer. Compton scattering therefore does not destroy the photon. As noted in [13], this effect is small for energies "below tens of KeV", i.e. 1 KeV is 1.24 nm and 10 KeV is 0.124 nm, and therefore can be discounted from visible applications within the band 400–700 nm.
- In pair-production, the photon energy is high enough to result in the production of an electron-positron pair, i.e. the electron's anti-matter counterpart. The photon energy must therefore be higher than $E_p = 2m_e c^2$, where m_e is the rest-mass of an electron. As this is 1.02 MeV, i.e. 0.0012 nm, this process occurs only for X-ray and Gamma-ray interactions, and thus can be discounted from any visible light application [13, 14].

Elimination therefore leaves the photoelectric effect—which is subdivided (within textbooks, patents, company websites, whitepapers and journals) into the external and internal photoelectric effects. It is therefore the only physical process suitable for optical to electrical conversion within visible applications [13, 14].

2.1. The photoelectric effect: converting light

In both photoelectric processes, a photon is absorbed by an atom, thus destroying the photon. If the photon energy is sufficient to overcome the work function of the material, a bound-free transition takes place whereby an electron is promoted from an outer electron orbital and is ejected from the surface [13, 15]. Remaining energy is accounted for by the kinetic energy of the electron as a free particle [15]. This is the external photoelectric effect as photo-electrons physically leave the material. The photon energy is given by Eq. (6), where h is the Plank constant, c is the speed of light, λ is the wavelength and ν is the optical frequency.

$$E_p = \frac{hc}{\lambda} = h\upsilon \quad (6)$$

In contrast, for semiconductors with band-gap, E_G, between the valance, E_V, (i.e. outer-orbital bound electrons) and conduction electrons, E_C, (i.e. delocalized cloud of electrons) [16], a photon of energy greater than the band-gap ($E_p > E_G$) promotes an electron from the valence to the conduction band. As the absence of an electron in a valence state is described as a hole, the internal photoelectric effect produces an electron-hole pair [16]. This is a bound-bound or intrinsic transition [14, 15]. The electron is still ejected from the atom; however, not from the surface. If two electrodes are placed on the material with a slight potential gradient, or if that potential exists due to a p-n doped junction, the electron-hole pair are separated and drift due to their relative charges. With many photo-generated carriers within the material due to many incident photons, the bulk conductivity of the material increases [16], allowing a photocurrent to flow through an external circuit. Photons of high energy are highly likely to cause band to band transitions, however as the wavelength increases towards a photon energy close to the band-gap, the likelihood of transition decreases, given by the absorption coefficient, α. This leads to a long-wavelength cut off, λ_C, given by Eq. (7). For silicon, this is 1.1 μm, where the absorption coefficient is 1×10^1 cm^{-1}, whereas at 400 nm it is 1×10^5 cm^{-1}.

$$\lambda_C = \frac{1.24}{E_G} \tag{7}$$

The internal photo-electric effect is therefore the mechanism whereby semi-conductor materials can be utilized for the detection of light. As materials have various band-gap energies, different materials can be used to detect optical energy with different wavelengths [16]. Typically, less than one electron-hole pair is produced per absorbed photon, fundamentally limiting the internal quantum efficiency and the spectral responsivity measured in Amperes per Watt (A/W). The optically-induced current, I_p, flowing through a photodiode is given by Eq. (8) [16], assuming a detector thickness much larger than the light penetration depth, $(1/\alpha)$. q is the electronic charge, P_{OPT}, is the incident optical power, μ_n is the electron mobility. η is the quantum efficiency, ε is the electric field within the photoconductor, L is the distance between the contacts, and τ is the carrier lifetime.

$$I_P = q\left(\eta \frac{P_{OPT}}{hv}\right)\left(\frac{\mu_n \tau \varepsilon}{L}\right) \tag{8}$$

Within CMOS technologies, photodiodes are fabricated using p- and n- type dopants to form a p-n junction [14, 16]. As the average depth of absorption changes with photon wavelength, the depth of the junction is chosen, if possible, to maximize the received photocurrent. The width of the p-n junction is also critical in this; however, it also has implications for photodiode bandwidth. The response speed is restricted by three phenomena. Firstly, the capacitance of the p-n junction (dictated by the junction width). Secondly, the time delay of carriers generated outside of the junction, diffusing into the junction. And thirdly, the drift or transit time, t_r, of the carriers within the junction [1, 6, 16]. The transit time is given by Eq. (9), where L can be replaced by the junction width, W, as the electrode contacts are often fabricated on the top and bottom of the junction within planar technologies.

$$t_r = \frac{L}{\mu_n \varepsilon} \tag{9}$$

Eqs. (8) and (9) indicate both limits and opportunities with respect to receiver design. If the junction width or photoconductor length are reduced, the response speed increases, although this becomes limited if a thin junction width leads to a high capacitance and therefore a slow photodiode resistance/capacitance (RC) time constant. Reducing the width also reduces the photocurrent, requiring a high receiver gain to compensate. If the electric field is increased, both the photocurrent and diode bandwidth increase. It is here we encounter the fundamental limit and trade-off within high-speed, high-sensitivity receiver design. That transit time, and therefore bandwidth behaves in a reciprocal manner to photocurrent magnitude. As the electric field looks to be beneficial, this is a clue to guide research in the field.

2.2. The Avalanche gain mechanism: photo-carrier multiplication

A central challenge within receiver design is how to obtain the required amplification [16], or sensitivity. This must be done without undue use of (i) amplifier circuits (thus silicon area and power usage) or (ii) wide and therefore slow p-n junctions [1, 6]. Avalanche multiplication used in avalanche photodiodes (APDs), has been previously investigated for communications [1, 16–19]. This moves amplification into the diode providing initial gain without circuitry, and by increasing ε, allows both a larger photocurrent and shorter transit time.

The multiplication process uses an increased reverse bias voltage, V_R [16], creating an energy difference of $q(V_{bi} + V_R)$, where V_{bi} is the built-in potential between the p-side and n-side regions. The resultant increase in electric field, ε, accelerates a free carrier, labelled 1 in **Figure 2**, to a kinetic energy, E_K, sufficient to overcome the ionisation energy, E_G, of the material [16] (Eq. (10)). Upon a collision between a photo-carrier and the crystal lattice, the accelerated carrier ionises another carrier. An electron-hole pair, labelled 2 and 2′, is generated with those carriers then accelerated by the electric field, causing further ionisation [16, 17, 20]. This continues exponentially, creating an avalanche of carriers within the depletion region.

$$E_K = (1.5)\, E_G \tag{10}$$

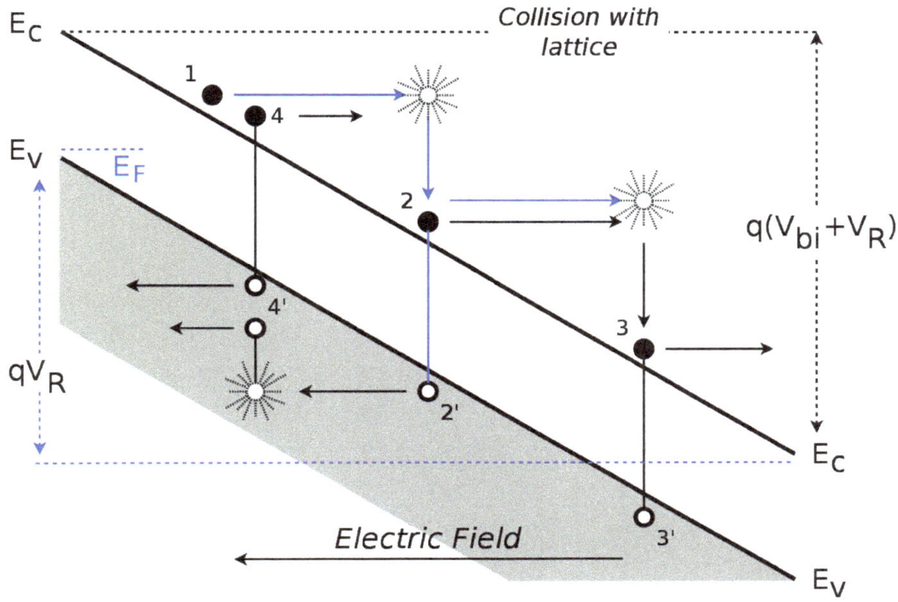

Figure 2. Band diagram showing reverse bias avalanche multiplication. When the electric field is elevated above the ionisation energy level, an accelerated carrier imparts a significant kinetic energy to a bound electron upon collision with the lattice. Adapted from Ref. [16], pp. 79 and 97.

APDs are biased such that multiplication achieves a constant generation rate, and thus constant gain [21] but not run-away avalanche leading to junction breakdown ([22] and references therein). The diode produces a current dependant on the incident photon flux with the gain and depletion region width being dependant on the bias [16, 21]. Some structures can be biased further into reverse bias, giving larger gains and greater sensitivities. In exceptional cases, run-away multiplication is used for specialised diodes called single-photon avalanche diodes (SPADs) [20, 22–25].

The use of avalanche gain, can be advantageous for sensitivity, as a gain of $M = 10–1000$, can be accomplished within the diode without the need of analogue amplification [1, 6]. Avalanche gain, is thus useful, however it comes with some disadvantages. Firstly, the higher bias voltage must be generated, necessitating additional power supply circuits or on-chip charge pumps. Secondly, multiplication is inherently random leading to increased noise. Finally, there is a gain-bandwidth trade off formed by the persistence of the avalanche process, whereby if the optical power decreases quickly, the avalanche caused by the previous high-optical power state takes some time to subside. To progress sensitivity and speed, through use of avalanche multiplication, circuits mitigating some of the above disadvantages are needed. For example, could we switch to smaller, lower power digital circuits and utilize very high multiplication factors by reducing the impact of multiplication noise?

2.3. Single-photon avalanche diodes: history and operation:

The Geiger region lies at the extreme end of the reverse photodiode range, beyond the linear avalanche gain region but prior to breakdown of a guard ring surrounding it ([11, 16, 21, 23, 25], and references therein). Initial research into avalanche behaviour, centered on microplasmas [21, 25]. These are small breakdown regions, corresponding to silicon defects [21, 25]. The historical study of microplasmas, resulted in artificial microplasmas [25] with guard rings to force a known breakdown region. These artificial structures later became known as Geiger-mode avalanche photodiodes (GM-APDs) or single-photon avalanche diodes (SPADs) [11, 23]. As they allow the avalanche mechanism to run-away, breaking down the p-n junction, their gain factors can be greater than 1×10^6, i.e. a single-photon, yielding a single electron-hole pair is able to produce a sizable avalanche photocurrent.

The Geiger region provides long periods of quiescent operation at a voltage of $V_{OP} = V_{BD} + V_{EX}$, where V_{EX} is the excess bias applied above the breakdown voltage, V_{BD} [20–23]. Upon the arrival of a photon, the device transitions from the quiescent point (**Figure 3**), to the I-V curve, with the avalanche quickly building up (10–100 ps). The current flows through any external series resistances, causing a significant voltage drop, designed to be of magnitude V_{EX} over that resistance. This reduction in voltage, (i) is sensed, indicating a single-photon event and (ii) reduces the voltage across the SPAD p-n junction [22, 23]. The reduced voltage, decreases the kinetic energy of carriers and halts the avalanche, a process known as quenching. The lower current flow allows the SPAD to recharge to $V_{BD} + V_{EX}$. There are two principal operating modes, passive quenching with passive reset (PQPR) and active quenching with active reset (AQAR) [23]. The passive circuit uses a large ~100 kΩ resistance (typically implemented using transistors), R_Q, and optionally a smaller 50 Ω sense resistor. The circuit recharges with a time constant given by $R_Q C_{SPAD}$, where C_{SPAD} is formed by (i) the p-n junction

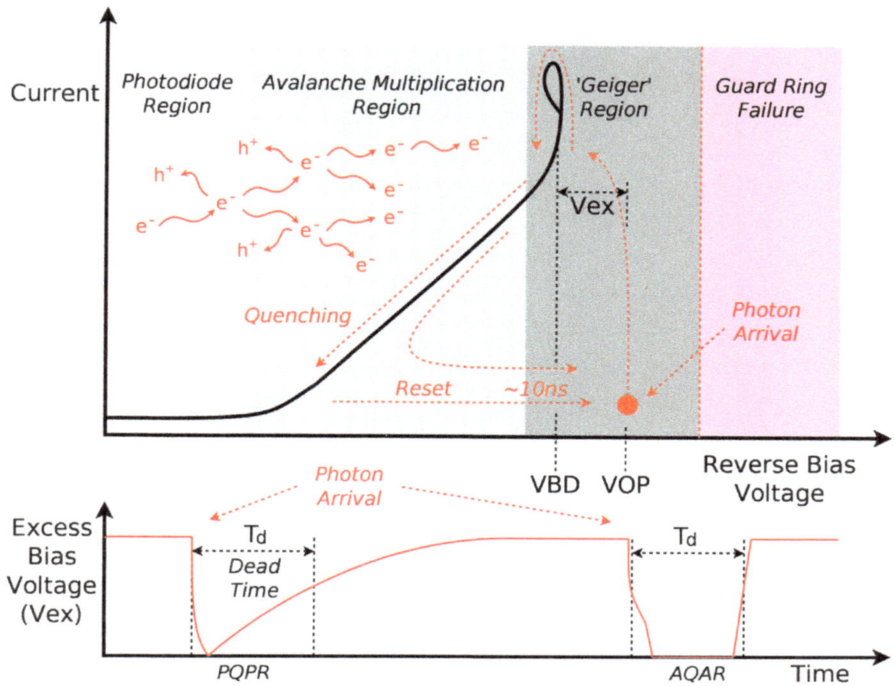

Figure 3. Top: the bias regions and I-V curve of a SPAD, showing the avalanche, quench and reset cycle. Bottom: the voltage-time pulses for both passive and active quenching circuits. Notice the steep, well defined edges of the AQAR pulse shape. The dead-time T_d is shown for both.

capacitance, and any parasitic capacitances on the diode (**Figure 4**). In the active case, the change in voltage triggers a quenching transistor which pulls V_{EX} to zero. After a short delay a separate low on-resistance transistor resets the device to V_{EX}.

In **Figure 4**, both positive-going and negative-going PQPR SPAD circuits are shown, along with the parasitic anode to substrate (C_{AS}) and cathode to substrate (C_{CS}) capacitances. The transistor gate voltage, DT_{ctrl}, can control the on-resistance, allowing adaptation of the SPAD dead-time, T_d. The n-type metal-oxide-semiconductor (NMOS) transistor circuit is preferable

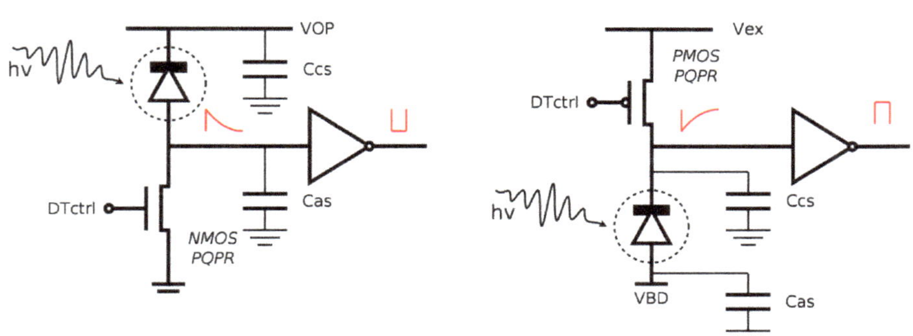

Figure 4. Two PQPR circuit topologies. Left: positive-going circuit with an NMOS quenching transistor and a small anode parasitic capacitance on the diode moving node. Right: negative-going circuit with a PMOS transistor, but with a larger cathode to substrate parasitic capacitance.

as (i) C_{AS} is typically smaller than C_{CS} and (ii) NMOS transistors can be integrated within a SPAD array without the deep N-well required for p-type metal-oxide-semiconductor (PMOS) devices.

It is with circuits such as this, that high-sensitivity receivers can be realized [11, 12]. One circuit approach would be to combine avalanche gain exceeding that used in APDs, with topologies known to give high sensitivity. The SPAD circuit can be used to replace the continuous-time photodiode, trans-impedance amplifier (TIA) and feedback network, Rf, circuitry (**Figure 5**, left). A small digital inverter, necessary to prevent loading on the SPAD, replaces any post-amplification, while the resulting full logic swing can be used with a digital counter (**Figure 5**, right). This discrete time "direct-to-digital" approach allows all gain to be implemented within the diode, removing complex analogue circuitry, and drastically reducing area requirements. Electrical power is expended within the SPAD due to a total voltage of V_{OP}, and the avalanche current flow per detection, however analogue amplifier circuits typically require continuous bias currents for each amplification stage branch [1, 6].

2.4. Single-photon avalanche diodes: key performance parameters

Before discussing SPADs within optical receivers, there are several key performance parameters that must be discussed. As the field is still under active investigation, there are as yet no definitive methods to bring low-level parameters up to the communications level, particularly the bit error rate (BER). This contrasts with mature photodiode receivers where the receiver BER can be derived [1, 2, 6]. Taking the properties of a single SPAD first, we encounter non-ideal behaviour away from true single-photon counting [23]. For receivers, the concept of the receiver power penalty within the link budget [6], can be used to estimate the effects, to first order, of various offsets and non-idealities. The power penalty, expressed in decibels (dB), models the adjustment in optical power required to compensate for an effect.

1. SPADs produce avalanche events despite the absence of light, this is the so-called dark count rate (DCR), measured in counts per second (cps) or Hertz (Hz) [20, 21, 23, 24]. This is related to (i) thermal generation of carriers, and (ii) tunnelling of carriers predominantly due to traps within the band structure. The DCR has steadily improved through (i) academic investigation of guard ring structures such as retrograde guard rings [26], (ii) modelling and understanding of the transient behaviour of the junction itself [27] and (iii) CMOS process cleanliness, and can now be as low as 1 Hz [27]. A per-SPAD rate such as this is now effectively ideal in

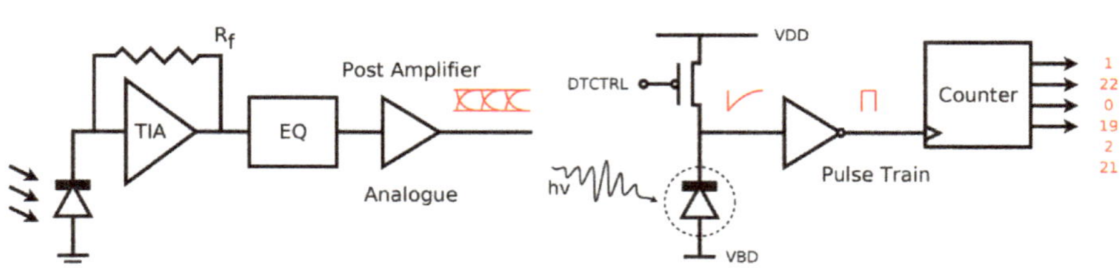

Figure 5. Left: a continuous-time conventional photodiode receiver, and right: a discrete-time replacement with single-photon sensitivity, where the SPAD and inverter allows optical to electrical conversion into a voltage (i.e. trans-impedance) with full logic swing magnitudes.

comparison to high (>100 Mb/s) data-rate communications. The DCR limits the optical photon flux at which reliable single-photon detection can be achieved. Therefore, the incident optical power would need to be increased by a power penalty, PP_{DCR}. This is given by Eq. (11), where DCR_{PC} is the DCR per channel per symbol period as a proportion of the required number of photons per symbol, similar to the voltage threshold offset power penalty ([1], p. 71). As the DCR and symbol period decrease, this penalty becomes negligible. For a DCR_{PC} of 10% of the required photons (~20), the PP_{DCR} is 0.79 dB.

$$PP_{DCR} = 10\log_{10}\left(1 + \frac{2\,DCR_{PC}}{100}\right) \tag{11}$$

2. A SPAD—like a photodiode or avalanche diode—does not detect light with 100% efficiency. The photon detection efficiency (PDE), typically some 20–40%, is linked to three processes, (i) the internal quantum efficiency [16], (ii) the wavelength specific absorption depth and the width and depth of the p-n junction, and (iii) the avalanche turn-on probability [20, 23, 25]. This latter factor is a product of the kinetic energy achievable before a carrier recombines, and is thus dependent on the electric field. In combination with the DCR, this limits the lower optical power that can be reliably detected. A power penalty, PP_{PDE}, to account for the finite detection efficiency can be calculated using Eq. (12). For a SPAD with 20% PDE, this is 7 dB.

$$PP_{PDE} = -10\log_{10}\left(\frac{PDE}{100}\right) \tag{12}$$

The PDE has increased through commercial R&D into SPAD-dedicated micro-lenses. However, it has proved difficult to improve the internal quantum efficiency as increased electric fields correspondingly increase the DCR, while wider depletion regions require modifications to dopants that may not be an option for a multi-customer CMOS foundry. Correspondingly, the drive for high-PDE has led to the optimisation of the layer stack above the silicon surface, including reduced metallisation, thinner nitride layers and anti-reflection coatings, or the use of back-side illuminated structures. The combination of these design options has promoted foundries such as ST Microelectronics and TowerJazz to develop dedicated optical CMOS processes for SPAD implementation.

3. The dead time, T_d, is a discharge and recharge period, during which the bias falls below the normal $V_{EX} + V_{BD}$ [20, 21, 23]. No photon arrivals can be detected, although for PQPR as the voltage exponentially recharges there is an increasing probability of an avalanche event, and thus the dead time can be extended, an effect called paralysis [28, 29]. A short dead time is advantageous as it maximizes the detections per second, C_{MAX} (Eq. (13)). Unfortunately, an arbitrarily short dead-time, cannot be achieved as second order effects become more pronounced [23]. The dead time has been reduced within the literature, principally through the use of AQAR, and the use of smaller diodes. For AQAR, the maximum count rate and the dead time required for a target minimum number of photons per second, $N_{PHOTONS}$, for a number of SPADs, N_{SPADS} and target number of channels, N_{CHANS}, are given by Eqs. (13) and (14) respectively. For passive quenching, both equations are multiplied by $1/e$ [28, 29].

$$C_{MAX} = \frac{N_{SPADs}}{N_{CHANs}\, T_d} \tag{13}$$

$$T_d = \frac{N_{SPADs}}{N_{CHANs} \lfloor N_{PHOTONs} \rfloor} \tag{14}$$

4. Without a long dead-time, there is a small but finite probability of a secondary avalanche, denoted as an after-pulse, ap [20, 23]. This is related to the release of a charge carrier from a trap within the device. The number of trapped carriers is dependent on the charge flow during an avalanche, hence reducing SPAD capacitance is beneficial [22, 24]. The number of traps is a function of process cleanliness and device structure, while the trap lifetime, and hence the approximate required dead-time is a function of temperature. A minority-carrier effect, may limit the lower-bound of the dead-time with respect to after-pulsing, and hence fundamentally limits reductions in dead-times [22].

5. The Poisson distribution of photon arrivals is complemented by a similar timing variation in photon detection [20, 26]. This event timing jitter is a product of two contributions. Firstly, the statistics of avalanche build up [30], lateral spreading of the avalanche throughout the junction volume and the carrier transit time [31]. Secondly a lengthy, but low-probability jitter tail is caused by carriers deeper within the substrate diffusing towards the p-n junction. The use of silicon-on-insulator (SOI) CMOS processes may offer some mitigation, however a significant trend for jitter reduction has been SPADs of small diameter and the tailoring of the SPAD doping.

Arrays of SPADs in CMOS planar processes also have a number of non-ideal behaviours and limitations. As the SPAD dead-time or IC application may necessitate the use of multiple diodes, array level limitations have a large impact on final receiver performance [11].

6. The fill factor (FF) is the ratio of the optically receptive area to the total device area and is a key parameter in obtaining the goal of a high sensitivity receiver. For a square array of circular SPADs, the fill factor (FF_{SQ}) is given by Eq. (15). Where A_{ACT} is the total optically active area, A_{ARR} is the total square array area and N_{SPAD} is the number of SPADs within the array. D_R is the radius of the per-SPAD active area, D_G is the guard ring width, and D_S is the guard-ring to guard-ring separation. The fill factor significantly alters the receiver sensitivity as photons incident on inactive regions are not counted but non-the-less are included in link budgets. This leads to a fill-factor power penalty, PP_{FF}, given by Eq. (16) [11]. The fill factor is often limited by the diode geometries available for a given noise or capacitance specification [24], the complexity integrated onto the receiver and the CMOS process design rules [22, 24].

$$FF_{SQ} = \frac{A_{ACT}}{A_{ARR}} = \frac{\pi D_R^2 N_{SPAD}}{\left[2 D_R \sqrt{N_{SPAD}} + (2 D_G + D_S)\left(\sqrt{N_{SPAD}} - 1\right)\right]^2} \tag{15}$$

$$PP_{FF} = -10 \log_{10}\left(\frac{FF}{100}\right) \tag{16}$$

Prior to 2010, the fill-factor was often limited to 1–2% [11]. This necessitates a factor of 50–100 increase in optical power (PP_{FF} = 17 dB), to return the detected photons per data symbol to the ideal ($FF = 100\%$) theoretical value. Despite restrictions [24], the fill factor has been increasing through a number of techniques. N-well sharing, out-of-array electronics, reduced quenching circuit areas, NMOS-only logic, increased diode radii with retained low-noise, use of clustered mini arrays, non-circular geometries and micro-lenses, have all lead to fill factors reaching recent maximums of 67% [32] and 70% [33]. With a fill factor such as 70%, the optical power would need to be increased by a factor of 1.43 ($PP_{FF} = 1.55$ dB). To date, the fill-factor has been a particular focus of both academic and commercial R&D with several companies becoming heavily invested.

7. Detector cross talk, which is often at the 1–2% level in SPAD arrays using out of array circuitry, is induced by secondary photons in the substrate [34]. These are produced through radiative recombination during an avalanche. In-array circuitry has also been recognised as a contributor [22]. Therefore, while area within the array should be used to maximise the functionality to area cost ratio, high-speed circuitry, clock buffers and digital supplies should be placed outside the active areas. Cross talk can be reduced with increased SPAD spacing; however, this negatively impacts the fill factor.

8. As with all sensors, the temporal response departs from the ideal as characterized by the step and impulse response [11]. Upon the reception of a "zero" to "one" bit-level transition (a positive step assuming OOK), the ideal would be for the received transient to be limited by the channel. The step response however may cause a finite rise time and ripple that must settle before the output can be sampled. Likewise, upon the reception of a "zero" after a "one", the impulse response limits the speed at which the receiver settles to the new level. Upon either transition, the system must wait before sampling, and thus the symbol rate is fundamentally limited (discussed further later).

3. Circuit approaches and topologies

To date there are three distinct circuit topologies used within SPAD-based receiver designs. Two of the topologies use the innate full-logic-swing of the SPAD to produce an all-digital approach with either a sampled integrating or continuous time output. The third topology again aims for a continuous time output, but does so using the SPAD to switch current steering circuits. Each topology has advantages and disadvantages; however, the largest impactor is the level of integration within a CMOS technology as a function of the receiver functions discussed in Section 1.1. Along with these topologies, pure analogue, parallel connections have been formed into so called multi-pixel photon counters (MPPCs), these differ from SPAD arrays as the diode is not used to create a full logic level voltage. A positive feasibility study of MPPC use for communications can be found in [35].

3.1. Digital count summation: digital synchronous discrete time

One of the first all-digital multi-SPAD arrays custom designed for optical communications [12], aimed to replicate a synchronous integration mode-receiver [1]. A receiver of this type is presented in Section 3.2 as an example SPAD-based receiver. Integration prior to signal sampling allows: (i) the output signal amplitude to be increased, i.e. allowing a doubled integration time, will approximately double the output amplitude, and (ii) noise which has a period shorter than the integration window is reduced. As SPADs produce a discrete time pulse, integrating signals is equivalent to pulse counting [11].

The topology is shown in **Figure 6A**, where the pulses from a SPAD are counted digitally using a ripple or synchronous counter. An integration window (**Figure 6B**) shorter than the symbol period can be used (i) to allow rejection of photon counts caused by the rise and fall periods of the input, (ii) to reject DCR events outside of the window, and (iii) to stop, sample and reset the digital counter. The addition of multiple high-speed synchronous data streams from multiple 'pixels', (formed from a SPAD and a counter), can be performed easily and with low latency. A multi-diode approach therefore allows multiple detection events during periods equivalent to the SPAD dead-time. A SPAD array output, in the form of a wide bit-depth synchronous digital stream can then either be (a) output to allow array diagnostics and signalling formats such as pulse amplitude modulation (M-PAM) or (b) thresholded on-chip allowing recovery of single-bit serial data streams. As the readout clock can be adjusted in phase relative to the transmitter clock, the position of the integration window can be chosen to minimize the BER, and its width can be optimized to maximize received signal amplitude.

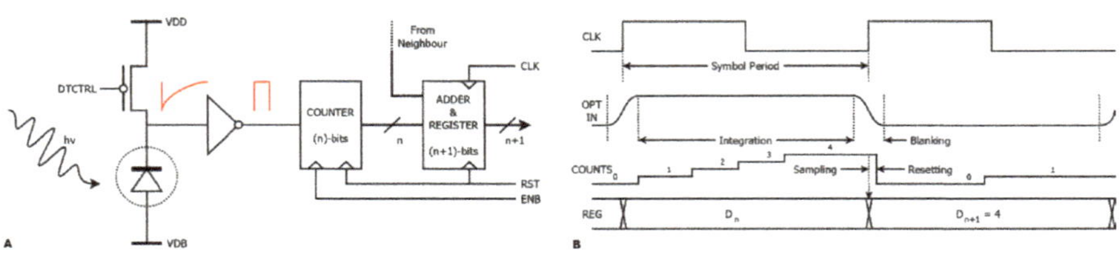

Figure 6. (A) The circuit topology with a binary counter per SPAD with synchronous binary addition of multiple SPAD counts within a defined integration window. (B) Timing diagram showing an input optical symbol, with an integration period blanking off the low-high and high-low transitions.

3.2. XOR-tree summation: digital asynchronous discrete time

To decrease the required logic, negated AND and negated OR (NAND/NOR) trees were investigated for combining multiple SPADs in small, higher fill factor mini- silicon photomultipliers (SiPMs) ([11] and references therein). However, to operate at high counting rates, the SPAD pulse (10–30 ns) must be reduced to <1 ns using per-SPAD pulse shortening. This is still smaller than per-SPAD digital counters, however it prompted the investigation of exclusive-OR (XOR) trees for mass digital pulse summation of larger SPAD arrays [36, 37] (**Figure 7A**). The XOR truth table lends itself readily to a summation task as sequential leading edges of SPAD pulses (without pulse shortening), simply toggle the output (**Figure 7B**). The limitation of this is the rate at which the final XOR within a tree can toggle its output.

As XOR gate standard cells occupy a smaller area than the digital logic required per SPAD in both digital summation and pulse-shortened NAND/NOR summation, the fill factor of the circuit can be increased significantly, in the case of [36], a fill factor of 43% was achieved. Indeed, such a receiver has demonstrated 4-PAM VLC, with highly discernible Poisson-limited levels,

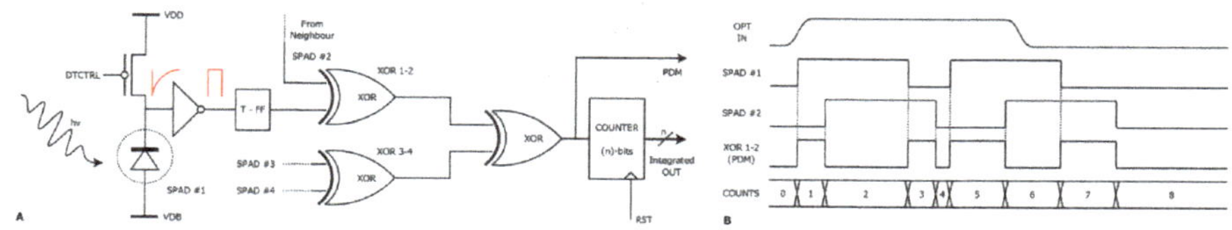

Figure 7. (A) The XOR summation of pulses, without pulse shortening, of four SPADs. The toggling of the output is a form of pulse density modulation; however, a binary counter can be used for pulse integration during a symbol. (B) Timing diagram of the XOR circuit for events on SPADs 1 and 2.

albeit with a low data rate [38]. As noted by the authors, this topology can be combined with (a) the digital summation approach, (b) on-chip modulation decoding logic, (c) signal processing logic, and (d) the adaption of the XOR tree and number of SPADs per channel. This would allow receiver optimization for power usage and optical sensitivity [38].

3.3. Current steering summation: analogue continuous time

In **Figure 8**, the current summation topology is shown [39], this uses the SPAD to switch currents onto output nodes shared by multiple SPADs, IQ– and IQ+. When the SPAD toggles, the current path changes due to the opening or closing of transistors MS1 and MS2. Transistor MCS could be implemented outside of the array, and transistors MS1 and MS2 can be far smaller than large digital circuits such as synchronous carry look-ahead counters. The authors therefore hoped for a high fill factor, however the use of PMOS devices limited fill factor through the SPAD N-well spacing rules. A fill factor of 3% was achieved however this was not optimized. A deeper issue is that the topology includes an inherent signal attenuation, whereby a full-logic swing on the SPAD (V_{EX} = 1.6 V) and inverters results in a maximum per-SPAD steered current of 100 uA. With a value of 300 Ω for RO+ and RO– this yields 30 mV per SPAD. Larger trans-impedance values could be used (e.g. 1 kΩ, yielding 0.1 V per SPAD), however trans-impedance bandwidth decreases as this resistance increases, i.e. $BW_{-3dB} = (2\pi R_O C_D)^{-1}$ [1]. It seems a shame to insert (i) a signal attenuation, and (ii) the transform {SPAD avalanche current to SPAD voltage to readout current to readout voltage}, into the topology to read out the SPAD array and communication signals.

While summation using Kirchhoff's current law for many SPADs is more efficient in terms of area and power than a multi-bit digital summation tree, and thus a lower energy per bit,

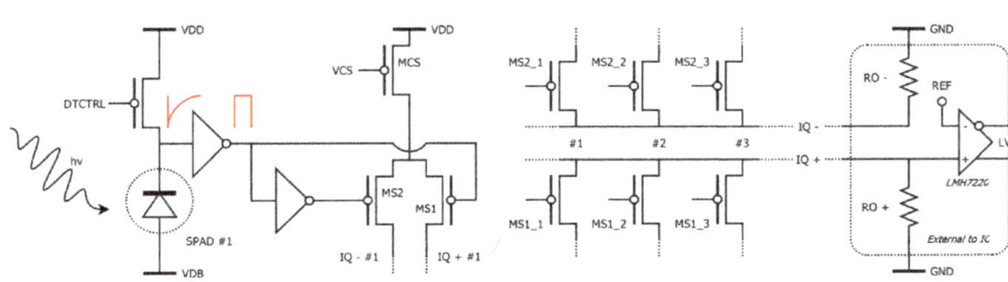

Figure 8. A current summation topology where a SPAD steers current through transistors MS1 or MS2. Multiple outputs can be added using Kirchhoff's current law on nodes IQ– and IQ+. Off-chip trans-impedance resistances RO– and RO+ are used to convert the current to a continuous-time voltage.

this readout structure introduces noise through transistor mismatch and the Johnson noise of the trans-impedance resistances. As this noise acts upon the voltage levels of "zeros" and "ones" at the analogue LMH7220 comparator, the steered current per SPAD, and thus the per-SPAD voltage has a fundamental limit, especially if the receiver is intended for M-PAM modulation schemes. This is an additive readout noise in combination with both Poisson photon arrival and noise induced by dead-time, DCR and after-pulsing effects. Further, to compare the energy per bit and the receiver area, modelling would need to include the power dissipated and area of (i) both RO resistors, (ii) the comparator, (iii) the field programable gate array (FPGA) serial-deserial (SERDES) input pads and (iv) the latching flip-flop. This would be the case as this yields the same synchronous full-logic level digital bit as the previous topologies, albeit single- rather than multi-bit. While the output is asynchronous with samples taken when the data eye is maximally open, with no blanking periods SPADs breakdown due to photons within the transitions of the optical source, thus preventing them from contributing once the optical signal has reached its correct level. This would be critical for reducing the standard deviations, σ_1, σ_2 and σ_3, upon M-PAM levels with means N_1, N_2 and N_3. Modulated OOK data streams have been received with this topology with 125 Mb/s at a BER of $1 \times 10^{-3.5}$ achievable with a dead time of 8.8 ns and before decision feedback equalization [39, 40]. We must note however, that to be fair all receivers should be compared without equalisation, although equalisation represents a logical subsequent signal processing step for low BER communications.

3.4. An exemplar receiver IC

To demonstrate the concepts and to investigate issues inherent in the design of such receivers, the digital summation topology has been investigated using a 32 × 32 array of 8 μm SPADs in 130 nm imaging CMOS (**Figure 9**) [11, 12]. This does use a hybrid approach with NAND/NOR signal addition prior to the main digital parallel summation tree, but does so mildly using 'super-Pixels' formed from four SPADs (**Figure 10**).

The design, discussed in detail within [11, 12], operates as an integration-mode receiver, with two blanking periods of 500 ps at the start and end of a symbol. These screen out SPAD events during the transitions of an optical source, and thus decreasing the variability of counts within an integration period. The digital summation tree allows clocking at 100 MHz, culminating in either a single channel 14-bit output or a 9-bit 16 channel output operating at 25 MHz per channel. As the receiver was intended as a test-harness for various concepts, the array treated the fill-factor and PDE as fixed offsets that could be improved in subsequent generations. Rather than designing to maximize known performance metrics, the array was designed to test transient issues along with unknown factors important for both communications and high-sensitivity, high sample-rate metrology.

By passing full logic swing synchronous parallel data, rather than (a) fast toggling pulse density modulated data, (b) asynchronous ripple-counter data or (c) currents requiring a trans-impedance, the receiver requires no circuitry prior to its test data being captured by an FPGA FIFO, or being thresholded. The receiver, summarized in **Table 1** (see [11] for results/modelling), represents a test harness with dedicated modes for array characterization.

Figure 9. Micrograph of the 1024 SPAD digital summation integration-mode receiver with the SPAD inset. This has a fill factor of 2.4%, and implements up to 16 'Data Channels' each of 64 SPADs.

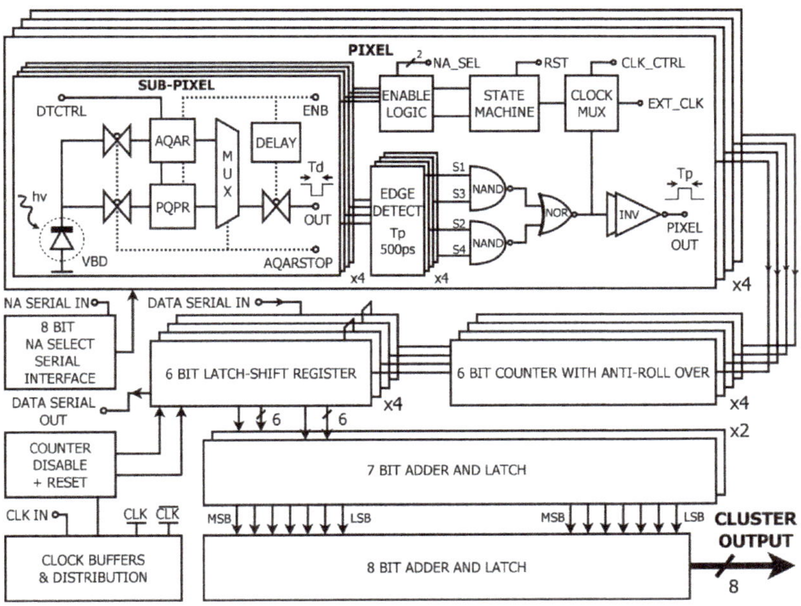

Figure 10. Block diagram of the architecture for 16 SPADs, showing per-SPAD active and passive quenching, dead time control via the analogue voltage DTCTRL, and pulse shortening. A binary counter (6 bit) is used per group of four SPADs and NAND/NOR signal addition is used.

Parameter	Value	Units	Notes
CMOS process	130 nm Imaging	—	90 nm metallization
Peak PDE	20	%	VEX = 1.2 V at 450 nm
SPAD VBD	−13.05	V	—
Dead Time	5–250	ns	Active or passive
Pulse shortening	500	ps	For NAND/NOR Sum
SPAD VEX	1.35	V	Typically, 1.25 V
Fill factor	2.42	%	Exc. I/O Pads
# SPADs	1024	—	Circular
SPAD diameter	8	μm	Active area
Quenching	AQAR or PQPR	—	Adjustable
Digital VDD	1.2 and 3.3	V	Core/IO
Pixel counter bits	6	Bit	Per 4 SPADs
Output width	14 or 9	Bit	Single- or Multi-Channel
# Channels	1 or 16	—	Configurable
Die size	2.4 × 2.1	mm	~5 mm^2
Clock frequency	108 100	MHz	Pad-Limited (Max) (Typical)
Median DCR	2.5	kHz	16°C, V_{EX} = 1.35
Max count rate	65	Gphoton/s	~10 ns Dead time
After-pulsing	0.9	%	1 μs window
Energy per bit	370	pJ/Hz	Estimated
Sensitivity estimate	−31.7	dBm	100 Mb/s BER = 1 × 10^{-9} 100% extinction ratio
Sensitivity estimate calibrated	−28.5	dBm	Inc. LED extinction ratio reduction

Table 1. Performance details for the 32 × 32 digital summation integration-mode receiver.

4. Results

This section will present experimental results from the example digital summation integration-mode receiver. The sensitivity curve or photon transfer curve will be shown and a model for this will be presented. The step response will also be shown as it makes a direct impact on communications performance. This will also be modelled illuminating the choices a designer has with respect to the SPAD dead-time, number of SPADs, digital-counter read-out rate and maximum modulated input power. Finally, experimental communications results will be shown, demonstrating both feasibility and several issues we have discussed.

4.1. Photon transfer curve and optical sensitivity

In **Figure 11**, the photon transfer curve (PTC) is shown. This plots the detected photon events per second against the input optical power. The N_a modes describe a time multiplexing approach, which while beneficial for the temporal response, will not be discussed here (see [11]). The N_a = 4 mode is of interest as it turns on all SPADs within the summation. At the centre of the curve, the receiver has a linear response, however it saturates at both the low and high levels due to the combined DCR and the active or passive quenching maximum count rate, C_{MAX}, respectively. Passive quenching shows both the presence of the $1/e$ factor in its maximum counting rate, and the issue of paralysis whereby the count rate starts to decrease at high optical powers. For communications, the receiver must operate within the linear region, as nonlinearities act to modify the extinction ratio of the detected signal, and thus negatively impacts the BER.

For design of SPAD-based receivers, active quenching appears to be advantageous, however as it requires extra circuitry per-SPAD, the choice of active or passive quenching must be taken in view of top level communications performance metrics. By reducing the dead-time, a higher count rate is achievable, while an increased number of SPADs has the same effect but comes with increased total DCR. It is advantageous to reduce the DCR through both innovation and by turning high-DCR SPADs off (as shown on **Figure 11**) as this extends the lower optical power limit and pushes the receiver closer to being quantum limited.

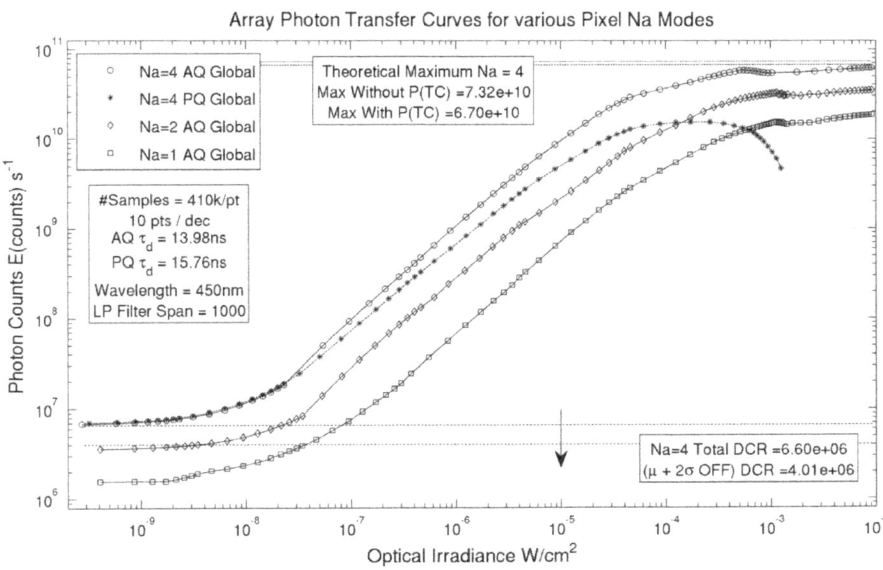

Figure 11. The photon transfer curve of photon counts per second against optical input power. The curve is linear within the centre; however, saturation occurs due to the combined DCR of all SPADs within the summation and the active or passive quenching maximum count rate, C_{MAX}.

The dynamic range can be calculated as: (i) the ratio of maximum (i.e. C_{MAX}) to minimum (i.e. $\sum(DCR)$ observable signals (approx. 80 dB in Figure), or (ii) the ratio of the maximum to the noise of the dark count floor rather than its absolute value, i. e. $\sqrt{\sum(DCR)}$ (approx. 149 dB). If we accept that between two modulated amplitude levels, a suitable gap must be left in order to achieve the required BER, the size of the linear region gives an estimate as to the theoretical number of M-PAM levels that could be implemented. With the required offsets, due to the finite fill factor and photon detection efficiency known through measurement of the PTC, other power penalties can be assessed and ranked as a guide to future design.

4.1.1. A competitive count model of SPAD avalanche initiations

The PTC can be modelled using competitive interaction, matching the measured experimental data (**Figure 12**). The arrival of a photon, (assuming AQAR), predicates loss of a photon, dark count or after-pulse if one occurs during the dead time. This creates competition at high count rates, in which (i) photons prevent dark counts or after-pulses, (ii) dark counts prevent photon counts or after-pulses, and (iii) after-pulses prevent photon counts or dark counts.

A simple model, describing the number of output counts as a function of input photons is given in Eq. (17). Where FF is the fill factor, DCR is the dark count rate, ΔT represents the time that counts are collected and $P(ap)$ is the after-pulsing probability. $PDE(\lambda)$ is the photon detection efficiency at a particular wavelength and $\psi(\lambda)$ is the number of photons, of wavelength λ, incident per second. This does not account for the dead-time, although it could be combined with hybrid paralysable models of dead-time limited counting [28, 29]. This would modify the output counts, N_{counts}, but would not include the concept of competition due to events other than photon arrivals. The model assumes that photons, dark counts and after-pulses can be separated, which while achievable with careful experimentation, cannot be performed during normal operation. The question therefore arises of what is the DCR level when the receiver is being illuminated, knowing that some DCR events will not be detected as the SPAD has already broken down? As noted within [23], a photon arrival and an after-pulse, or a dark count and an after-pulse are not statistically independent.

$$N_{counts} = [1+P(ap)](PDE(\lambda)\psi(\lambda)FF+DCR)\Delta T \tag{17}$$

As high counting rates prevent the separability, independence and static contribution assumptions being used, the concept of an enable time, T_{enb}, as a proportion of ΔT, can be introduced, i.e. an avalanche induced dead time will remove a period, T_d, from ΔT. The time in which a SPAD is in its receptive state, is then given by Eq. (18), and is the first step towards a count competitive model. The values ψ_{act} and DCR_{act} are the actual numbers of photon and dark detections, however these are intermediate output values in Eq. (18) as they must be corrected for count competition [11].

$$T_{enb} = \Delta T - [(1+P(ap))(\psi_{act}+DCR_{act})(T_d)] \tag{18}$$

A regression can be made backwards to the input values for photons, ψ_0, dark counts, DCR_0, and after-pulses, AP_0, that can be calculated from the incident optical power, fill factor or

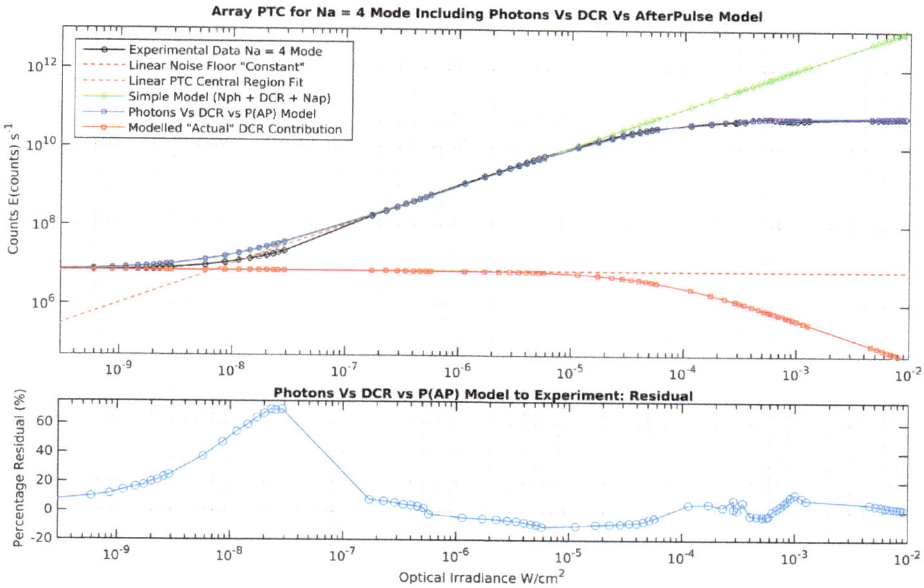

Figure 12. Count competitive and simple addition models for the photon transfer curve. The fitting parameters are: λ = 450 nm, *PDE* = 23%, *FF* = 2.4%, *P(ap)* = 1%, T_d = 16 ns, DCR = array average DCR. Two linear fits are included to both the central region of the experimental PTC and the constant noise floor. The competition implies a reduced DCR as the PTC saturates, this is also included.

carrier generation rate etc. The final counts N_{counts}, including competitive interaction and the finite dead time, is then given by Eq. (19).

$$N_{counts} = \left(\frac{\Delta T}{1+(1+P(ap))(\psi_0+DCR_0)(T_d)}\right)(\psi_0+DCR_0) + \left(\frac{\Delta T}{1+(1+P(ap))(\psi_0+DCR_0)(T_d)}\right)^2(\psi_0+DCR_0)P(ap) \quad (19)$$

In **Figure 12**, this model shows a low overall residual error, while incorporating the saturation at the level of C_{MAX}. The deviation from experimental values in the region of 1×10^{-8} W/cm^2 is due to small inconsistencies in the measurement approach as the optical power approached the limits of the optical power meter. The power of the model however, is that the instantaneous DCR or after-pulsing counts can be estimated given count competition rather than assuming their calculated, or stand-alone measurement values. For example, DCR_0 may be measured at a mean of 100 cps, however through the introduction of some photon detections and the corresponding loss in T_{enb}, the actual detected dark count rate, DCR_{act} may instead be 90 cps, i.e. 10 thermal generation events were ignored as they happen to coincide with SPAD dead-times caused by photon events. The value of DCR_{act} is therefore plotted showing that as a SPAD approaches count saturation, the number of detected dark events falls. This effect must therefore be included in any assessment of count noise at a particular optical power, as would be needed for the BER of OOK or M-PAM of finite extinction ratio.

4.2. Step and impulse response

The step and impulse responses play a key role in the effectiveness of communication. In [11, 12], the OOK communications experiments are severely impacted by both (i) a step response initial peak, ripple and lower than expected settled count rate and (ii) an impulse response slow decay. Both act to close the data eye, thereby increasing the BER. For the step response, the inter-symbol interference effect is directly linked to the dead-time of SPADs as a function of the array read-out-rate, and has also been observed experimentally in [39, 40].

Figure 13 shows a large change in the peak and ripple as the input power increases [11]. For the 5 μW step, the peak is 29% above the steady state. This will force a separate peak in histograms of the data amplitude levels depending on the data sequence. The following data point shows a value 10% below the steady state, indicating that a received 'one' at this point would have an artificially lowered amplitude. In comparison, lower optical powers such as the 0.5 μW step, give peaks only 10% above the steady state, followed by a minimal trough. When combined with the observation that the lower light level responses have a steady state closer to the initial, (and correct incident), number of photons, there will be a corresponding reduction in the splitting of the N_1 data distribution and an improved BER (explored further in [11]). As the steady state is below the initial number of counts, the relative decrease in steady state as optical power increases acts to decrease the received extinction ratio of the modulated OOK or M-PAM signal, thereby increasing errors.

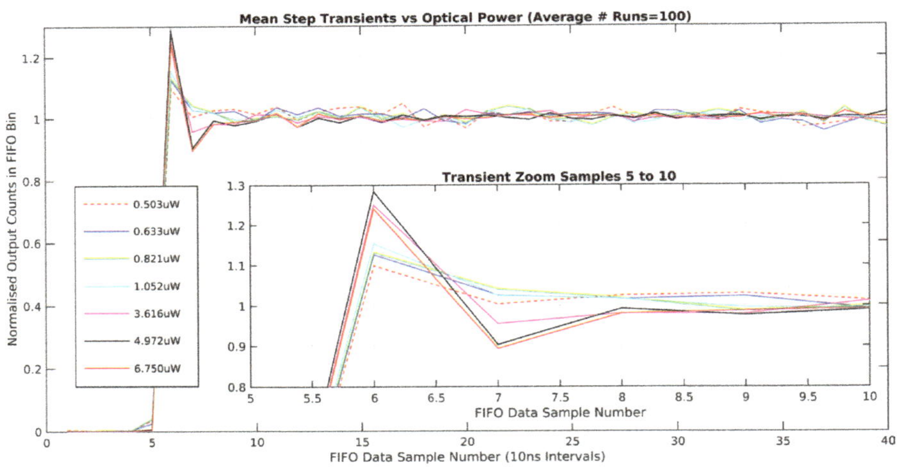

Figure 13. Normalised average step response over a sweep of optical power. The peak and ripple in the number of counts per 10 ns bin, is highlighted in the inset zoom (50 ns) after the start of the laser step.

4.2.1. A detector pooling model of SPAD array step response:

The SPAD-array step response can be modelled allowing optimisation of future designs. Initially, the receiver has a full complement of SPADs (e.g. N_{array} = 1024). At some subsequent time, a number of diodes have broken down, giving a number at the output. These have been removed from the pool, and hence the effective total efficiency of the array, (i.e. the PDE and array fill-factor for an array of N_{array} SPADs), η_{det}, for subsequent photon arrivals reduces as there is a new number of available SPADs, N_{avail}. The removal of SPADs from the pool, and the reduction in detection probability, leads to decreased output counts despite continuous input levels. Sometime later, the SPADs that were removed will be recharged back into the available pool, bringing the effective detection efficiency back towards unity.

Eq. (20) is an initial recursive discrete time model for the number of SPADs available at a time step, n, where $C_{in}(n)$ is the expected number of input counts (photons, dark counts and after-pulses). The model makes several assumptions, however fitting to experimental results shows that: (i) the assumptions hold for the 32 × 32 digital summation integration mode receiver and (ii) these can be tackled in subsequent revisions of the model. The model assumes that the SPADs break down at the start of a discrete time step, and that the array readout frequency giving rise to these discrete time steps is initially equal to the SPAD dead time. This latter assumption is expanded within the modelling below to include dead times integer multiples of the readout period, i.e. $2n$, $3n$ etc. It should be noted that the quantity, $PDE_0FF_0C_{in}(n)$ models the number of SPAD detections received by the SPAD array within the discrete period and thus can be combined with the count competitive model (Eq. (19)).

$$N_{avail}(n) = N_{avail}(n-1) - \frac{PDE_0\,FF_0\,N_{avail}(n-1)\,C_{in}(n-1)}{N_{array}} + \frac{PDE_0\,FF_0\,N_{avail}(n-2)\,C_{in}(n-2)}{N_{array}} \tag{20}$$

The steady state and the reduction of the extinction ratio due to this effect, can be estimated by Eq. (21), (the 'alpha model'), or a simple model (Eq. (22)) using the initial input counts (see [11]). The 'alpha model' uses the long-term average input counts, $\langle C_{in}\rangle$, expected for a light level, i.e. N_{counts} from Eq. (19), along with a fitting factor, α, which has been found through minimisation of the residual to be 0.725 i.e. approximately $1/\sqrt{2}$.

$$C_{out}(n) = \langle C_{in}\rangle\left[1 - \frac{\langle C_{in}\rangle}{N_{array}}\left(1 + \alpha\frac{\langle C_{in}\rangle}{N_{array}}\right)\right] \tag{21}$$

$$C_{out}(n) = \langle C_{in}\rangle\left[1 - \frac{\langle C_{in}\rangle}{N_{array}}\right] \tag{22}$$

The discrete time model (Eq. (20)) is (i) uncalibrated, (ii) departs from noisy experimental results and (iii) is not yet combined with the steady state model of Eq. (21). Despite this it can be used as an indication as to future modelling and device requirements. The critical point of both time and steady state models is that they allow preliminary investigation into the splitting of the OOK N_1 distribution that is seen clearly in experimental communication results (next section). In **Figure 14**, an array of 1024 SPADs receives a step change in optical power. The input is fixed at an example 100 photons per symbol as ultimately the BER must remain static (Eq. (5)) [1, 6]. With the dead time fixed at 10 ns, the discrete time 'readout' period is changed as a proxy for different data rates. As the data symbol or 'readout' period becomes shorter than the dead time, there is a severe increase in the ripple, the number of symbols prior to SPADs recharging into the detector pool and a decrease in the steady state. While readout periods of 20 ns for the 10 ns dead time produce minimal decreases in the effective detection efficiency and step-response ripple, we must remember that there may be a limit to how short the dead time can be. As designs push the data rate in the first instance, a dead-time equal to the readout or symbol period is preferable.

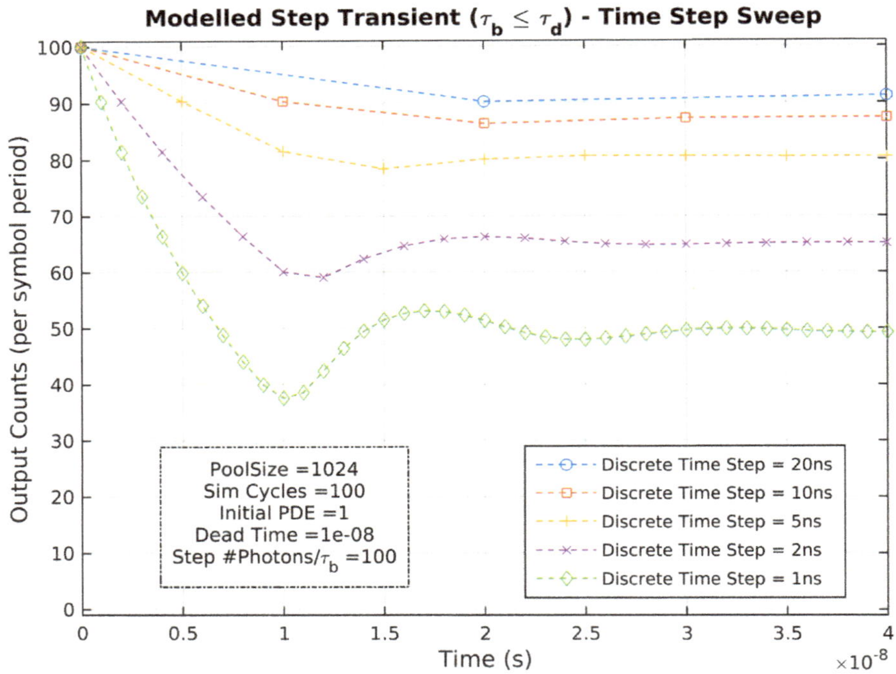

Figure 14. Modelled (uncalibrated) step response with an input step of 100 photons per symbol, a pool of 1024 SPADs and a SPAD dead time of 10 ns. The discrete time step or array readout speed is swept as a proxy for data rates between 50 Mb/s and 1 Gb/s.

In **Figure 15**, a readout period of 1 ns is chosen to highlight observations within the model. The step is held at 100 photons per symbol and the dead time is again set to 10 ns, however here the size of the array is changed from 2048 to 256 SPADs. It is possible to decrease the severity of the step response by increasing the number of diodes. However, clearly it is the ratio of the expected counts per symbol to the pool size that is critical. Reduce the number of diodes too much and the step response ripple becomes the limiting factor for both communications (BER) and optical metrology (for 100 photons per sample, the steady state of the 256 SPAD array is approx. 20 counts per sample). For completeness, the model, (once calibrated), can be used to estimate the number of diodes necessary, at a specified dead time (10 ns) and symbol period (1 ns), for a step response ripple and steady state that is within the Poisson noise of the expected input count level. For 100 counts per symbol, the standard deviation, ($\sqrt{100} = 10$), leads to ~9750 SPADs required for a steady state of 90.

These parameter sweeps, although chosen to visually highlight SPAD array issues, point to three heuristic design rules:

- First, that the dead-time should be equal to the data rate, (10 ns for 100 Mb/s etc.),
- Second, that the number of SPADs should be high in comparison to the expected photon counts per data symbol, and
- Third, if T_d is limited, the number of diodes can be increased to reduce the detrimental effects on the step response if we wish to operate at a data rate higher than $1/T_d$.

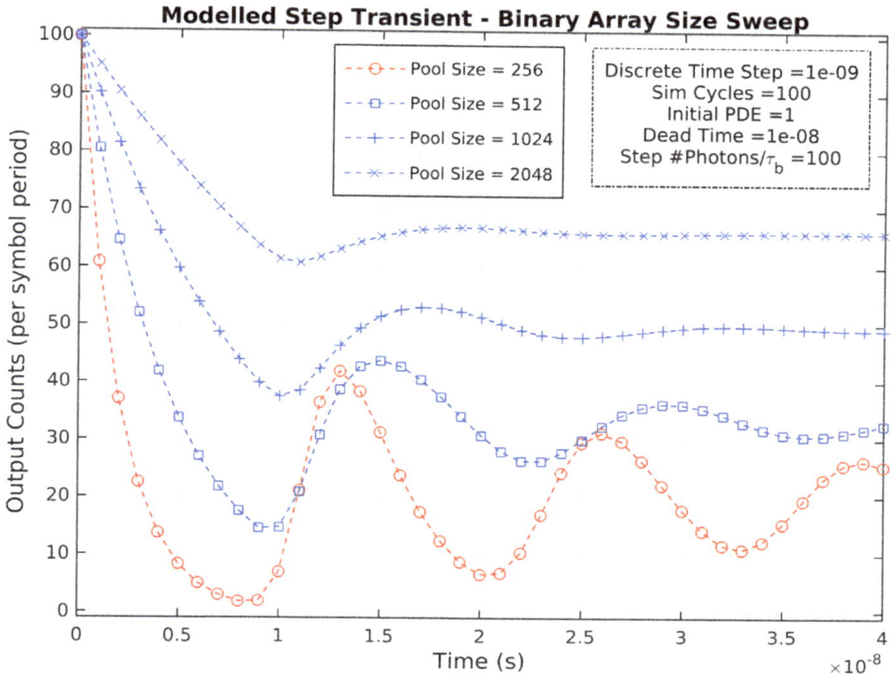

Figure 15. Modelled (uncalibrated) step response with an input step of 100 photons per symbol, a SPAD dead time of 10 ns and an array readout period of 1 ns (for illustration purposes). The number of SPADs in the pool is swept highlighting that if a dead time equal to the symbol period is not possible, the number of SPADs must be large in comparison to the expected photon arrivals per symbol.

Of course, these rules are a product of only the step response model. Increasing the number of diodes within the array may have a detrimental effect on the array fill factor. Hence research must concentrate on models that once combined, can be used for overall receiver optimisation with respect to communication level metrics (sensitivity, BER, data rate, etc.).

4.3. Communication results and feasibility

Early communications results have been achieved using the 32 × 32 SPAD-based integrating receiver, demonstrating that such receivers are feasible. **Figure 16** shows the eye diagram of a 50 Mb/s on-off key non-return-to-zero data stream. The transmitter in this case is a Gallium-Nitride 34 μm diameter 450 nm micro-LED, with a bandwidth of 220 MHz. The data eye is open at this optical power (20 μW/cm^2) and data rate, however several issues become apparent.

- Firstly, upon a 'zero' to 'one' transition, there is a large peak 75–100 counts higher than the main N_1 level (approx. 250). This can be directly attributed to the step response discussed above, and can be seen clearly near sample 110 in the time traces of **Figure 17**. Indeed, it leads to a split N_1 distribution, with the severity of the splitting of the ideally two distributions (N_0 and N_1) being evident in the count histogram.
- Secondly, the impulse response, (measured in [11]), creates a slow 'one' to 'zero' transition. This is clearly visible near time sample 90 (**Figure 17**). For some high frequency transitions, the impulse response prevents the transmitted 'zero' from obtaining the low count average

(approx. 45) as the impulse response is still decaying when the next 'one' bit is transmitted (see near sample 150). This therefore leads to the splitting of the N_0 distribution.

While excellent agreement is shown between the transmitted and received pseudo random bit streams (**Figure 17**), the closure of the eye can be attributed to (a) the steady state of the step response, (b) the bleeding of previous 'one' bits into subsequent 'zero' bits due to the impulse response, and (c) the finite extinction ratio of the micro-LED transmitter. The 32 × 32 SPAD receiver, as a testing harness, has provided crucial details for the future development of SPAD-based receivers in that clearly, the transient response which

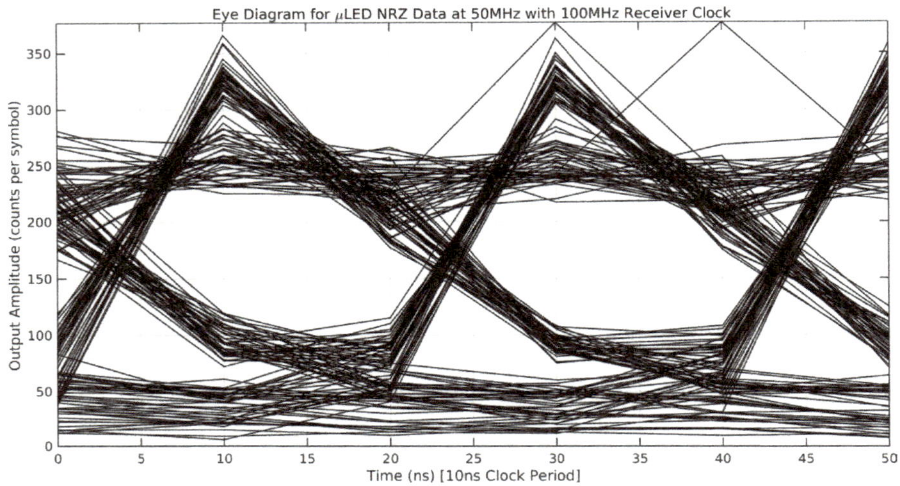

Figure 16. A measured eye diagram for 50 Mb/s OOK-NRZ data. The dead time was approx. 13.5 ns.

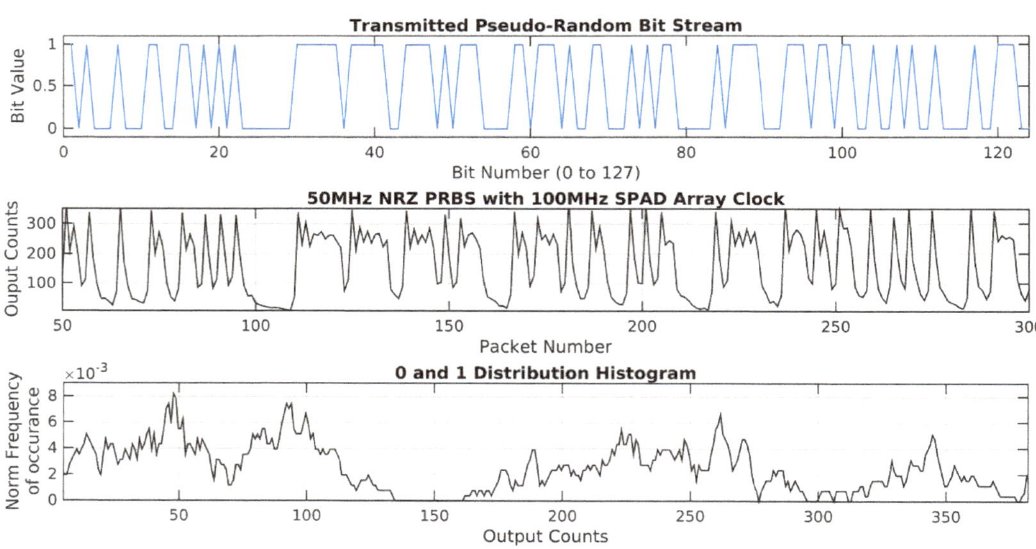

Figure 17. The transmitted OOK NRZ data stream, a time trace of the SPAD-receiver output and the histogram of received amplitudes. A sampling threshold at 150 counts shows separability of the N_0 and N_1 distributions, however the impulse and step responses have acted to split the distributions and close the data eye. With a reduced impulse and step response, the transmitter should render Poisson-limited distributions near 50 and 350 counts.

was only evident once empirically measured, must be addressed for low bit error rate communications.

5. Conclusions

SPAD-based optical receivers increase the prospective methodologies for applications requiring extremely high optical sensitivity. This chapter has covered the specifics of both optical detection using the internal photoelectric effect, and how avalanche multiplication can be extended to run-away avalanche producing full logic-level voltage pulses with single-photon sensitivity. The application, visible light communications, highlights that such receivers require further study before all noise sources and all circuit topologies are understood fully with respect to the higher-level communication performance metrics.

Three SPAD receiver topologies were introduced, the digital summation, the XOR summation and a current-summation. The performance metrics of both (i) optical communications (data rate, bit error rate, energy per bit etc.) and (ii) single SPAD and SPAD array circuits (fill factor, PDE, DCR), were also discussed. At present, there is little receiver theory to link low-level SPAD metrics with higher-level performance, although some work as already been done. For example, both the SPAD array fill factor and photon detection efficiency are known to directly impact optical sensitivity, and appropriate modelling has been developed. The receiver BER has also been investigated with respect to the SPAD dead time and the output photon-countingstatistics of a digital summation SPAD-receiver. To date however, there is little to link the SPAD after-pulsing probability with the BER, although one would assume its inherent delay would act as a form of ISI.

One requirement for future research and commercialization, would be the use of application specific design methodologies fully linked to higher level communications performance. How should the designer balance the SPAD fill factor and dead-time for example, when attempting to optimize a receiver for a required data rate and sensitivity? At present, too few high to low performance relations exist to allow such an interplay to be optimized. Likewise, if the application speed requires the dead-time to be traded off with after-pulsing, the effect that after-pulsing has needs to be quantified to allow comparison with dead-time ISI effects.

Overall, SPAD-based receivers are an active field of research, aiming to provide optimized designs with both extreme sensitivity and high-speed. Data-rates and energy per bit will likely remain modest and far removed the highest speed links of today. However, such receivers may offer options in (i) room-lighting VLC applications, supplementing existing Wi-Fi for download intensive tasks such as high-quality, real-time video streaming, and (ii) niche applications where optical attenuation is significant enough to warrant increased receiver gain and a higher energy per bit.

6. Student exercise questions

1. Calculate the quantum limit number of photons required per symbol, assuming an ideal, noiseless receiver, at a data rate of 50 Mb/s, a BER of 1×10^{-6}, and a wavelength of 600 nm.

2. Calculate the minimum number of SPADs within a receiver array, for the conditions in question #1, if all SPADs have a dead time of 10 ns.

3. If the SPADs are circular, and laid out in a square array, calculate the approximate array fill factor (10 μm SPAD diameter), assuming (i) no CMOS circuitry is incorporated within the array and (ii) that D_s and D_g are 2 μm and 5 μm respectively.

Author details

Edward M.D. Fisher

Address all correspondence to: e.fisher@ed.ac.uk

Institute of Digital Communications (IDCOM), The School of Engineering, The University of Edinburgh, Scotland, UK

References

[1] Razavi B. Design of Integrated Circuits for Optical Communications. 1st ed. McGraw-Hill, New York, USA; 2003.

[2] Binh LN. Digital Optical Communications. 1st ed. CRC Press, Boca Raton, Florida, USA; 2008.

[3] Hanzo L, Haas H, Imre S, O'Brien D, Rupp M, Gyongyosi Y. Wireless myths, realities, and futures: From 3G/4G to optical and quantum wireless. Proceedings of the IEEE. 2012;**100**:1853-1888 (Special Centennial Issue)

[4] Elgala H, Mesleh R, Haas H. Indoor broadcasting via white LEDs and OFDM. IEEE Transactions on Consumer Electronics. 2009;**55**(3):1127-1134.

[5] Cisco Systems Inc. Cisco Visual Networking Index: Global Mobile Data Traffic Forecast Update, 2016-2021 White Paper [Internet]. 09/02/17 [Updated: 09/02/17]. Available from: http://www.cisco.com/c/en/us/solutions/collateral/service-provider/visual-networking-index-vni/mobile-white-paper-c11-520862.pdf [Accessed: 24/02/17]

[6] Sackinger E. Broadband Circuits for Optical Fiber Communication. Wiley, New Jersey, USA; 2005.

[7] Carusone A, Yasotharan H, Kao T. Progress and trends in multi-Gbps optical receivers with CMOS integrated photodetectors. In: Custom Integrated Circuits Conference (CICC); IEEE, San Jose California, USA; 2010:1-8.

[8] Zeng L, O'Brien D, Minh H, Faulkner G, Lee K, Jung D, Oh Y, Won ET. High data rate multiple input multiple output (MIMO) optical wireless communications using white LED lighting. IEEE Journal on Selected Areas in Communications. 2009;**27**(9):1654-1662.

[9] O'Brien D, Le Minh H, Zeng L, Faulkner G, Lee K, Jung D, Oh Y, Won ET. Indoor visible light communications: challenges and prospects: free-space laser communications VIII. Proceedings of the SPIE. 2008;**7091**:709106-709109.

[10] IEEE Computer Society—WG802.15—Wireless Personal Area Network (WPAN) Working Group. P802.15.7 —IEEE Approved Draft Standard for Short Range Wireless Optical Communication Using Visible Light [Internet]. April 2011 [Updated: Rev. P802.15.7]. Available from: https://standards.ieee.org/develop/project/802.15.7.html

[11] Fisher E. A parallel, reconfigurable single-photon avalanche diode array for optical communications [Thesis]. The School of Engineering, The University of Edinburgh, Edinburgh, Scotland, UK; 2015.

[12] Fisher E, Underwood I, Henderson R. A reconfigurable single-photon-counting integrating receiver for optical communications. IEEE Journal of Solid-State Circuits. 2013;**48**(7): 1638-1650.

[13] Henley E, Garcia A. Sub-Atomic Physics. 3rd ed. World Scientific, Toh Tuck Link, Singapore; 2007.

[14] Durini D, editor. High Performance Silicon Imaging: Fundamentals and Applications of CMOS and CCD Sensors. 1st ed. Elsevier Ltd, Woodhead Publishing (for Elsevier), Cambridge, UK; 2014.

[15] Roychoudhuri C, Kracklauer AF, Creath K, editors. The Nature of Light: What is a Photon?. 1st ed. CRC Press, Boca Raton, Florida, USA; July 2008. 452 p.

[16] Sze SM. Semiconductor Devices: Physics and Technology. 2nd ed. Wiley and Sons, New Jersey, USA; 2001.

[17] Campbell J. Recent advances in telecommunications avalanche photodiodes. Journal of Lightwave Technology. 2007;**25**(1):109-121.

[18] Yang B, Schaub J, Csutak S, Campbell J. 10 Gbps all silicon APD optical receiver. Lasers and Electro-Optics Society. The 15th Annual Meeting of the IEEE. 2002;**2**:681-682.

[19] Takahashi S, Tajima A, Tomita A. High-efficiency single photon detector combined with an ultra-small APD module and a self-training discriminator for high-speed quantum cryptosystems. In: Proc. 13th Micro-Optics Conference, Japan. 2007

[20] Ghioni M, Gulinatti A, Rech I, Zappa F, Cova S. Progress in silicon single-photon avalanche diodes. IEEE Journal of Selected Topics in Quantum Electronics. 2007;**13**(2):852-862.

[21] Haitz RH. Mechanisms contributing to the noise pulse rate of avalanche diodes. Journal of Applied Physics. 1965;**36**(10):3123-3131.

[22] Webster E. Single-photon avalanche diode theory, simulation, and high performance CMOS integration [Thesis]. The School of Engineering: University of Edinburgh; 2013.

[23] Cova S, Ghioni M, Lacaita A, Samori C, Zappa F. Avalanche photodiodes and quenching circuits for single-photon detection. Applied Optics. 1996;35(12):1956-1976.

[24] Richardson J, Webster E, Grant L, Henderson R. Scaleable single-photon avalanche diode structures in nanometer CMOS technology. IEEE Transactions on Electron Devices. 2011;**58**(7):2028 -2035.

[25] Goetzberger A, McDonald B, Haitz R, Scarlett RM. Avalanche effects in silicon p-n junctions. II. Structurally perfect junctions. Journal of Applied Physics. 1963;**34**(6):1591-1600.

[26] Richardson J, Grant L, Henderson R. Low dark count single-photon avalanche diode structure compatible with standard nanometer scale CMOS technology. IEEE Photonics Technology Letters. 2009;**21**(14):1020 -1022.

[27] Webster E, Richardson J, Grant L, Henderson R. Single-photon avalanche diodes in 90 nm CMOS imaging technology with sub-1 Hz median dark count rate. Proceedings of the International Image Sensor Workshop. 2011:262-265.

[28] Lee S, Gardner R. A new G-M counter dead time model. Applied Radiation and Isotopes. 2000;**53**(4-5):731-737.

[29] Neri L, Tudisco S, Musumeci F, Scordino A, Fallica G, Mazzillo M, Zimbone M. Note: dead time causes and correction method for single photon avalanche diode devices. Review of Scientific Instruments. 2010;**81**(8): 086102-1 to 086102-3

[30] Blazej J, Prochazka I. Avalanche photodiode output pulse rise-time study. In: SPIE 7355, Photon Counting Applications, Quantum Optics, and Quantum Information Transfer and Processing II; May; SPIE Headquarters: Bellingham, Washington, USA; 2009. pp. 73550M-11.

[31] Spinelli A, Lacaita A. Physics and numerical simulation of single-photon avalanche diodes. IEEE Transactions on Electron Devices. 1997;**44**(11):1931-1943.

[32] Vilella E, Alonso O, Montiel A, Vila A, Dieguez A. A low-noise time-gated single-photon detector in a HV-CMOS technology for triggered imaging. Sensors and Actuators A: Physical. 2013;**201**(1):342-351.

[33] C. Niclass, M. Soga, H. Matsubara and S. Kato, "A 100m-range 10-frame/s 340×96-pixel time-of-flight depth sensor in 0.18μm CMOS," 2011 Proceedings of the ESSCIRC (ESSCIRC), Helsinki, 2011, pp. 107-110. doi: 10.1109/ESSCIRC.2011.6044926

[34] Rech I, Ingargiola A, Spinelli R, Labanca I, Marangoni S, Ghioni M, Cova S. Optical crosstalk in single photon avalanche diode arrays: a new complete model. Optics Express. 2008;**16**(12):8381-8394.

[35] Zhang G, Yu C, Zhu C, Liu L. Feasibility study of multi-pixel photon counter serving as the detector in digital optical communications. Elsevier Optik. 2013;**124**(22):5781-5786. DOI: http://dx.doi.org/10.1016/j.ijleo.2013.04.060

[36] Dutton N, Gnecchi S, Parmesan L, Holmes A, Rae B, Grant L, Henderson R. A Time-correlated single-photon-counting sensor with 14GS/s histogramming time-to-digital converter. In: Technical Digest International Solid-State Circuits Conference (ISSCC); 22 Feb; San Francisco. All IEEE references share the same IEEE headquarters location of: New York, USA; 2015. pp. 1-3.

[37] Gnecchi S, Dutton N, Parmesan L, Rae B, Pellegrini S, McLeod S, Grant L, Henderson R. Digital silicon photomultipliers with OR/XOR pulse combining techniques. IEEE Transactions on Electron Devices. 2016;**63**(3):1105-1110.

[38] Almer O, Dutton N, Abbas T, Gnecchi S, Henderson R. 4-PAM visible light communications with a XOR-tree digital silicon photomultiplier. In: Summer Topicals Meeting Series (SUM); 13 July; IEEE; 2015. pp. 41-42.

[39] Chitnis D, Collins S. A SPAD-based photon detecting system for optical communications. Journal of Lightwave Technology. 2014;**32**(10):2028-2034.

[40] Chitnis D, Zhang L, Chun H, Rajbhandari S, Faulkner G, O'Brien D, Collins S. A 200Mb/s VLC demonstration with a SPAD based receiver. In: Summer Topicals Meeting Series (SUM), 2015; July; All IEEE Publisher Locations are New York, USA; 2015. p. 226-227.

6

Novel MT/MPO Single-Mode Multifiber Connector Technologies for Optical Fiber Communications

Mitsuru Kihara

Additional information is available at the end of the chapter

Abstract

This chapter reports the latest mechanically transferable (MT) and multifiber push-on (MPO) multifiber connector technologies. Low insertion-loss and high return-loss angled physical contact (APC)-MPO single-mode multifiber connectors are developed. For these multifiber connectors to achieve a high performance, it is necessary to reduce deviations of all components and to control fiber offset due to shear force caused by compressing two obliquely ended MT ferrules. MT single-mode 84-fiber connectors are developed to realize higher density multifiber connectors. We have developed a novel high-density optical cable joint that has a 24-connector array consisting of new 84-fiber MT connectors. The new cable joining technique significantly shortens the time needed to join the optical fiber cables. Novel optical fiber switches are developed on the basis of MT multifiber connector technology. The switches use a manual guide-pin slide method in special guide holes in a novel multifiber array. Because high precision multifiber connectors need to be measured and inspected, a new inspection technique for MT ferrules is developed and equipment for it is fabricated. The fabricated inspection equipment can measure MT ferrules with high speed, cost effectiveness, and high accuracy. These single-mode multifiber connectors can be used as key technologies for advanced optical fiber communication systems.

Keywords: optical connector, insertion loss, return loss, physical contact, fiber-to-the-home, data center

1. Introduction

The number of subscribers to broadband services in Japan now exceeds 38 million, and about 29 million subscribers were using fiber-to-the-home (FTTH) services as of September 2016 [1]. The FTTH networks use many various single-mode optical fiber connection technologies.

One such connection is a field mountable optical fiber connector (FMC) or field installable connector (FIC), and another type is a manufactured optical fiber connector. The mechanically transferable (MT) or multifiber push-on (MPO) connectors, which belong to the manufactured connectors, are particularly expected to be used in upcoming FTTH networks because these multifiber connectors are superior to single-fiber connectors with high-density optical fiber cable joints [2–6].

Optical fiber communication systems have been used in data centers in the USA. Multimode single-fiber connectors such as a little connector (LC) were mainly used previously, and multimode MPO connectors are also used in current data centers. In addition, high precision MPO connectors with single-mode fibers are expected to be used for communicating more information over long distances in huge data centers.

This chapter reports the latest MT/MPO multifiber connector technologies. Current MT/MPO connectors used in Japan and the USA are explained in Section 2, and our developed innovative single-mode multifiber connectors are reported in Section 3. The low insertion-loss and high return-loss angled physical contact (APC)-MPO single-mode multifiber connectors are described in Section 3.1. Next, MT single-mode 84-fiber connectors for realizing higher density multifiber connectors are explained in Section 3.2. Novel optical fiber switches based on MT multifiber connector technology are reported in Section 3.3. These connectors need to be measured and inspected for high precision multifiber connectors. In Section 4, a new inspection technique for MT ferrules and equipment using the technique are introduced. These single-mode multifiber connectors can be used as key technologies for advanced optical fiber communication systems.

2. Overview of MT/MPO multifiber connectors

Figure 1 shows the configuration of a typical FTTH network in Japan. It is mainly composed of an optical line terminal (OLT) in the central office, underground and aerial optical fiber cables, and an optical network unit (ONU) inside a customer's home and building. The network requires various fiber connections at office, outdoor, and home sites. With the fiber connections at aerial and home sites in particular, field installable connectors or field mountable connectors and mechanical splices are used to fit the best wiring depending on aerial conditions and room arrangement. Field assembly small (FAS) connectors and field assembly (FA) termination connectors are field installable connectors [7–9]. In contrast, manufactured connectors, such as miniature-unit (MU) coupling optical fiber and single-fiber coupling (SC) optical fiber connectors, are used in central offices and homes. MT connectors are also used in central offices for multifiber ribbon joints. MPO connectors are used in customers' buildings. These MT and MPO connectors are particularly expected to be used in the upcoming FTTH networks with high-density optical fiber cable joints.

Optical fiber communication systems have been used in data centers in the USA. Previously, multimode single-fiber connectors such as an LC connector were mainly used, and multimode MPO connectors are also used in current data centers. This is because multifiber

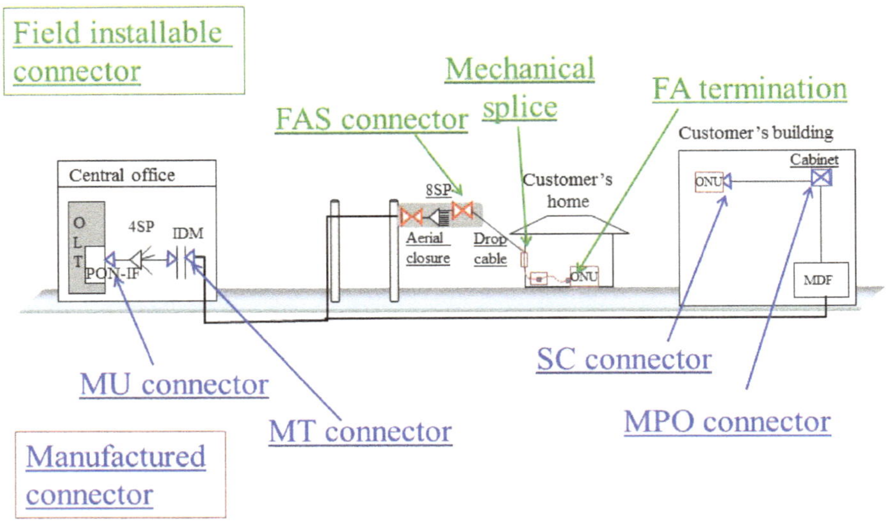

Figure 1. Typical FTTH network and various optical fiber connections.

connectors are superior to other single-fiber connectors with high-density optical fiber cable joints. **Figure 2** shows examples of optical fiber wiring (a) with LC single-fiber connectors and (b) with MPO multifiber connectors. Optical fiber wiring with the single-fiber connectors is slightly complicated, whereas that with the multifiber connectors is very simple and well ordered. The MPO multifiber connectors are consequently expected to be used in data center networks with high-density optical fiber cable joints. In addition, high precision MPO connectors with single-mode fibers are expected to be used for communicating more information over long distances in huge data centers. **Figure 3** shows the structure of the MT multifiber connector [2, 3], which consists of two plastic ferrules with two guide holes, two guide pins, and a clamp spring. The multifibers are positioned in a row between two guide holes. The ferrules are aligned by the two guide pins and two guide holes and then held with the clamp spring to achieve a low connection loss for multifiber ribbon connections. Refractive index matching material is used between the ferrule endfaces to reduce the

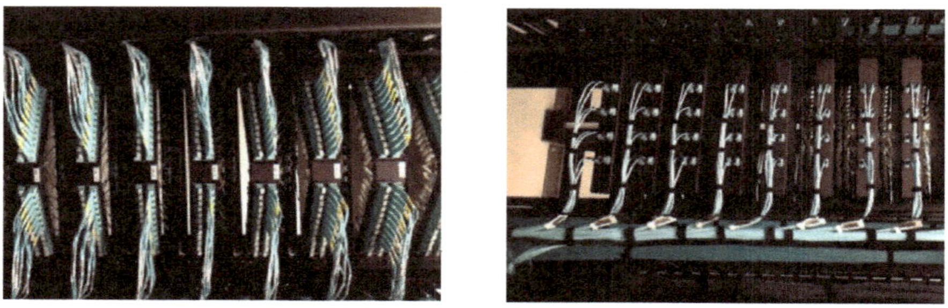

(a) (b)

Figure 2. Example of optical fiber wiring (a) with single-fiber connectors and (b) with multifiber connectors.

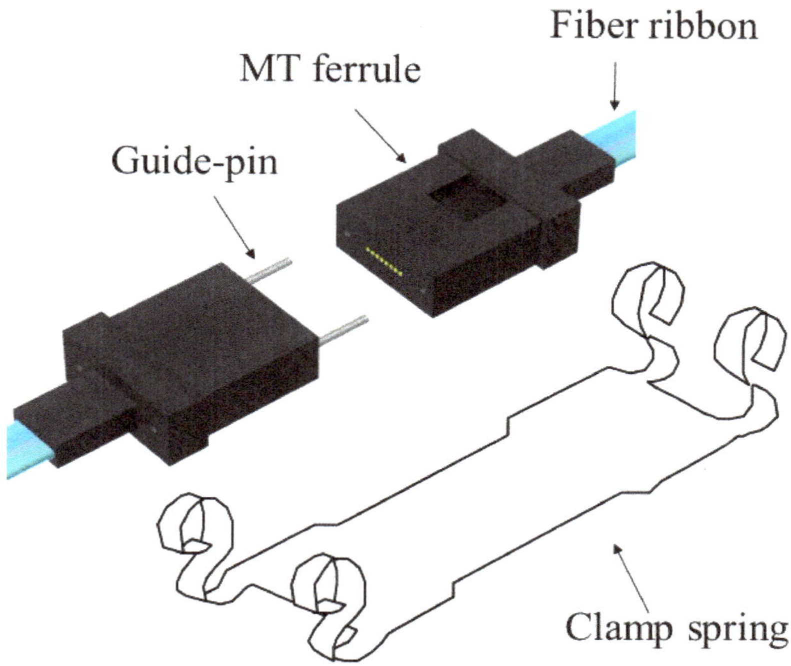

Figure 3. MT multifiber connector.

Fresnel reflection caused by an air-gap [10, 11]. The device is disconnected simply by removing the clamp spring and guide pins. The 4-, 8-, and 16-fiber MT connectors are currently used in the FTTH services in Japan.

Figure 4(a) shows the endface of an MT ferrule. In the MT ferrule endface, several fiber holes are positioned between two guide holes. The origin O is designated as the middle point of two guide hole centers. Each fiber hole is designed to be arranged with the designated fiber pitch on the basis of the origin. **Figure 4(b)** shows the fiber hole eccentricity of the MT ferrule. The actual fiber hole positions on the fabricated MT ferrule endface are different from the designated ideal positions because the molds are not perfectly accurate and/or plastic might deform during the MT ferrule fabrication. This large fiber hole eccentricity might result in large insertion loss of an MT connector, so it must be minimized for low insertion-loss MT connectors. Therefore, a technique has been developed to inspect for fiber eccentricities in MT ferrules, which is described in Section 4.

One application of the MT connector is as a multifiber push-on (MPO) connector [4–6, 12]. **Figure 5** shows the structure of the MPO connector, which is composed of two plugs and an adaptor. The plug contains an MT ferrule that has an endface that is obliquely polished with a slight fiber protrusion to enable physical contact. It is important for physical contact type connectors to eliminate an air gap between fiber ends [13–16]. Two guide pins are fitted into two guide holes in one ferrule to align the ferrules. The plug and adaptor are engaged by fitting a pair of elastic hooks into corresponding grooves. This connector provides easy push-pull reconnections without the need for refractive index matching material.

Figure 6 shows the fabrication process of an MPO plug. First, the mold is fabricated by using high precision mechanical engineering. A high precision mold is used, and an MT ferrule is

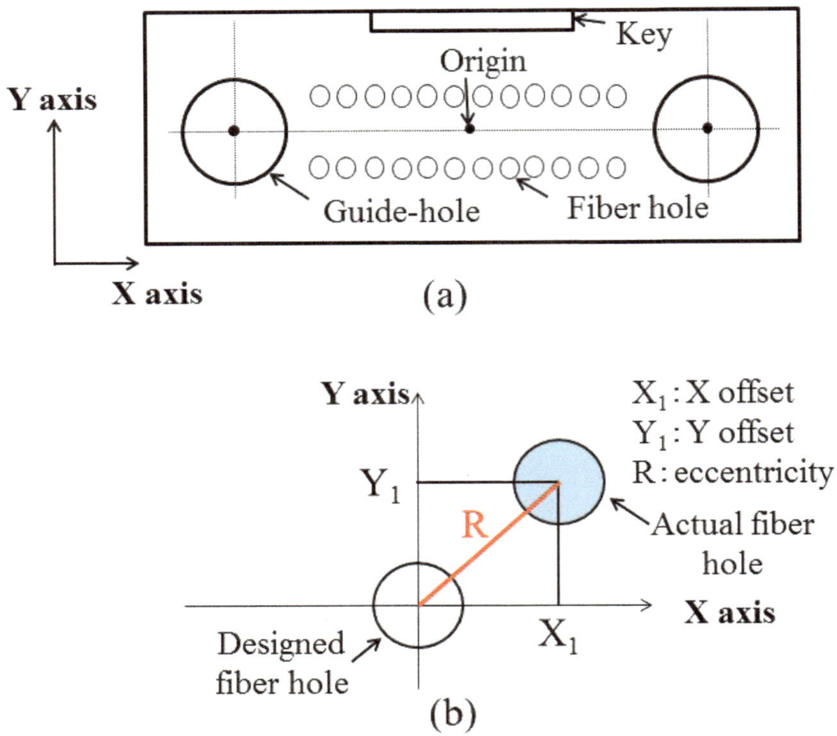

Figure 4. (a) Endface and (b) fiber hole eccentricity of MT ferrule.

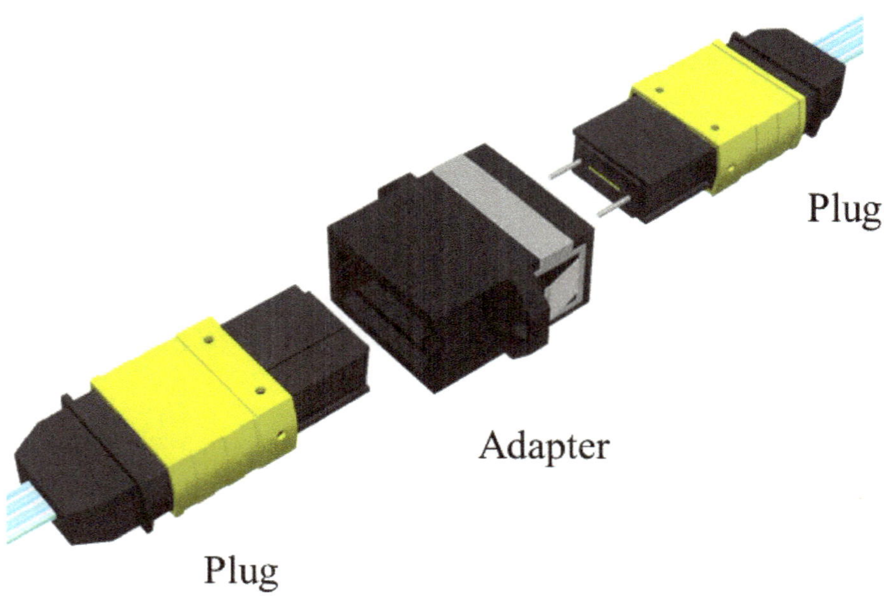

Figure 5. MPO multifiber connector.

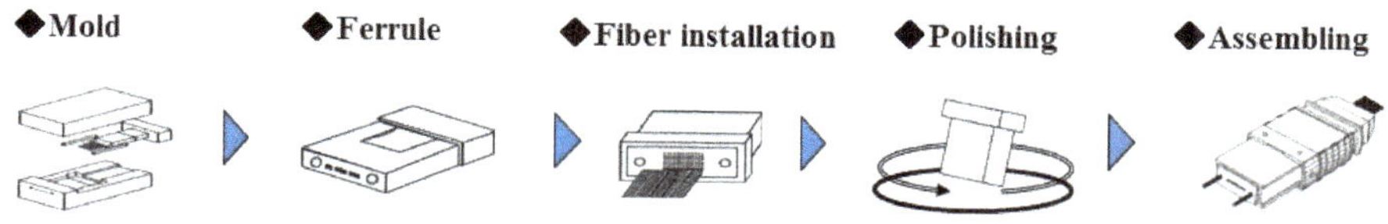

Figure 6. Fabrication process of MPO plug.

fabricated from plastic resin. This MT ferrule is a key device for MT/MPO connectors. Next, the multifiber ribbon is installed and fixed with adhesive in the MT ferrule. The MT ferrule containing the multifiber ribbon is then obliquely polished. Finally, an MPO plug houses the obliquely ended MT ferrule with multifibers. This MPO fabrication process contains two steps: ferrule making and connector assembling. Currently, fewer than 10 companies make ferrules, but more than 100 companies assemble connectors. More companies will make ferrules and assemble connectors in the future.

3. Innovative multifiber connectors based on MT/MPO connectors

This section reports our innovative multifiber connectors based on MT/MPO connector technologies. **Figure 7** shows three target single-mode multifiber connectors. The first is an extremely lower insertion-loss multifiber connector, which is detailed in Section 3.1. The second is a higher density multifiber connector that enables two high-density optical fiber cables to be connected as soon as possible, which is explained in Section 3.2. The third is a functional multifiber connector that enables multifibers to be switched, which is described in Section 3.3.

3.1. Lower insertion-loss multifiber connectors

We have investigated extremely great performances of multifiber connectors and developed angled physical contact (APC) MPO connectors that have an insertion loss lower than 0.2 dB and return loss higher than 60 dB. This section explains these APC-MPO connectors. To create lower insertion-loss connectors, fiber core offsets from the ideal positions must be decreased. APC-MPO connectors have two types of fiber core offset. One is due to deviation of components and the other is due to shear force caused by compressing two obliquely ended MT ferrules. These two fiber core offsets are detailed below.

Optical loss in a fiber connection has four causes: fiber-core offset, fiber-axis tilt, fiber-end separation, and mode field mismatch [17, 18]. With MT connectors, fiber-core offset is the dominant factor generating optical loss, and so we have limited our considerations to this factor. **Figure 8** shows four deviations: d_1, d_2, d_3, and d_4. d_1 is the fiber hole position deviation. d_2 is the deviation due to the clearance between a fiber and a fiber hole. d_3 is the deviation due to the clearance between a guide pin and a guide hole. d_4 is the fiber-core eccentricity [19]. The four deviations dependently affect total fiber core deviation. To realize an insertion loss of lower than 0.2 dB, the total deviation must be less than 0.5 μm.

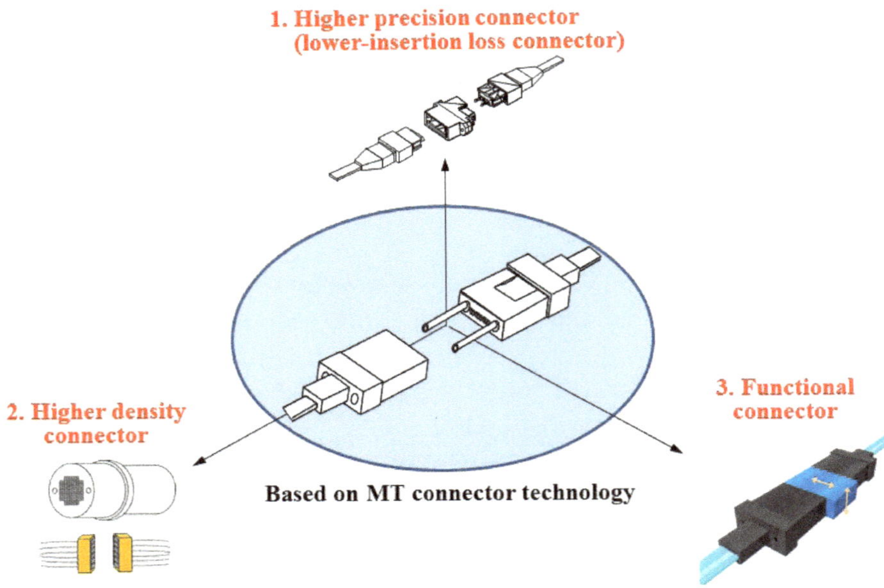

Figure 7. Three targets of single-mode multifiber connector technologies.

Figure 9(a) and **(b)** shows side views of connected MT ferrules and an MT ferrule endface in an MPO plug. The MPO connector uses MT ferrules obliquely polished at 8 degrees and can provide high return-loss characteristics. In contrast, the shear force occurs in the connected obliquely ended MT ferrules when two MT ferrules are compressed. This shear force leads to the Y offset of connected fibers in the MT ferrule. By considering the Y offset, fiber positions are designed to be previously offset, as shown in **Figure 9(b)**. This offset quantity is dependent on clearance between a guide hole and a guide pin and deformed material of an MT ferrule. The fiber offset must be controlled for low insertion-loss MPO connectors.

On the basis of the above design, we fabricated three pairs of high precision single-mode 12-fiber APC-MPO connectors by using three different fabrication molds. The fabricated APC-MPO connectors used high precision ferrules with d_1 of less than 0.5 μm and severely

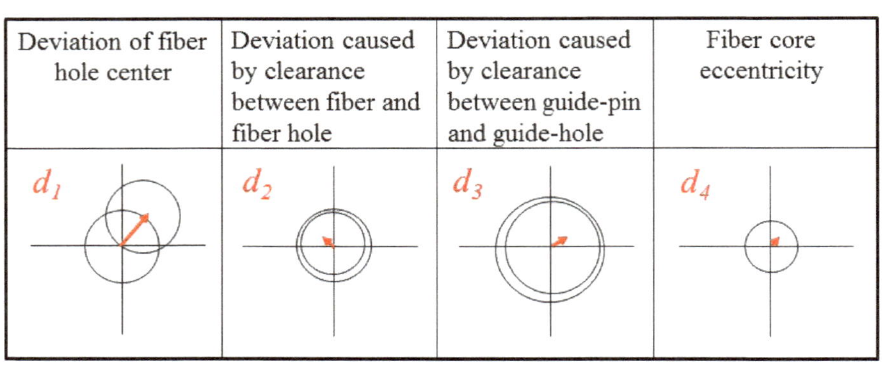

Figure 8. Four deviations resulting in fiber core offset.

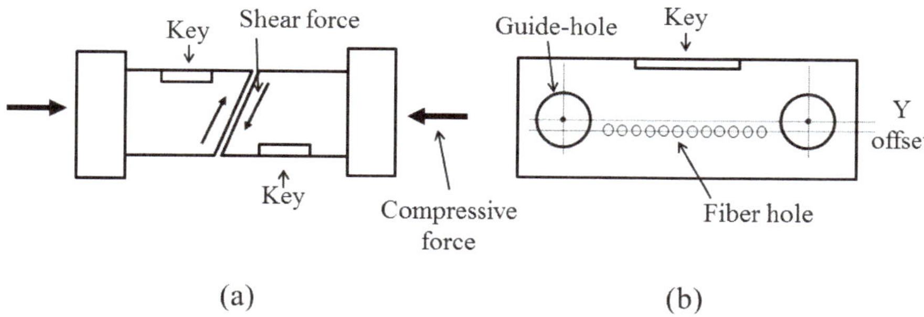

Figure 9. (a) Side view of connected MT ferrules and (b) MT ferrule endface in MPO plug.

controlled guide pins and optical fibers with lower d_2, d_3, and d_4. The fabricated APC-MPO connectors also used the controlled Y offset quantity. We measured the insertion losses and return losses using 1.31 μm LD, a power meter, and a backreflection meter. **Figure 10** shows histograms of the insertion losses and return losses of the connectors. The average insertion loss was 0.08 dB with a maximum of 0.17 dB. The return losses were greater than 64 dB. Consequently, we have revealed that the fabricated APC-MPO has an excellent performance.

3.2. Higher density multifiber connectors

We have developed a higher density multifiber connector to enable two high-density optical fiber cables to be connected as soon as possible [20–23]. **Figure 11(a)** and **(b)** shows the novel high-density preconnectorized cable joint and endface of high-count fiber MT connectors, respectively. The cable joint consists of an arrayed high-count fiber MT connector and a cable joint housing and can connect high-density optical fiber cables such as 2000-fiber cables at one time. When it is connected, each MT connector is aligned with two guide pins and pushed toward the opposite ferrule by a spring. The MT connectors are aligned and connected individually. This structure is the same as those of a conventional MT connector and MPO connector. The connector endface is polished at a right angle. The connectors need an index matching material to eliminate the Fresnel reflection caused by an air gap. We used a conventionally sized 2.5-mm-thick MT ferrule that has guide holes of 4.6 mm in pitch and 0.7 mm in diameter.

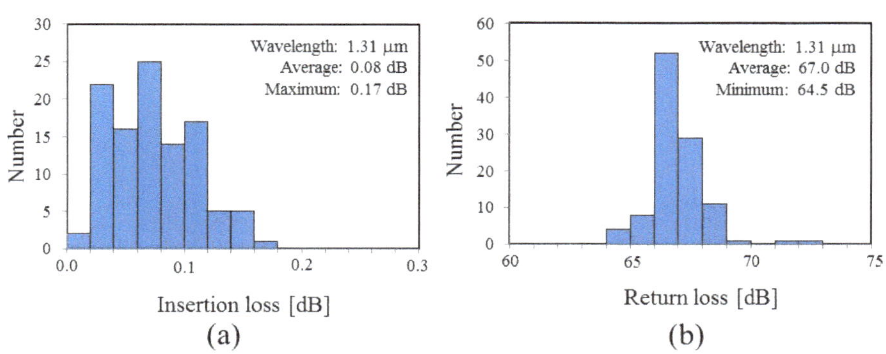

Figure 10. (a) Insertion loss and (b) return loss of fabricated high precision APC-MPO connectors.

This is because this type of ferrule is utilized widely in optical cable components such as MPO connectors. The optical fiber ribbons are generally installed into fiber holes of a ferrule. We used a conventional 12-fiber ribbon with a 0.3 thickness and 0.25 fiber pitch to assemble our fabricated connector. By considering the thickness of the fiber ribbon and that of the MT connector ferrule, we figured that seven is the maximum number of fiber ribbons installable in an MT ferrule. Therefore, we decided that 84 (7 rows of 12) is the target number of fibers installed in our high-count fiber MT connector as shown in **Figure 11(b)**.

On the basis of the above design, we fabricated an 84-fiber MT connector and high-density preconnectorized cable joint for 2000 fiber cables. **Figure 12(a)** and **(b)** shows endface photographs of the fabricated high-count fiber MT connector and high-density preconnectorized cable joint, respectively. The high-count fiber MT connector ferrule consists seven rows of 12 fiber ends between the two guide holes and the high-density preconnectorized cable joint has a 4 × 6 array of a high-count 84-fiber MT connector. Consequently, the fabricated high-density preconnectorized cable joint can potentially connect 2016 fiber cables at one time.

The fabricated high-count 84-fiber MT connector was measured at a wavelength of 1.31 μm. We connected six pairs of high-count fiber MT connectors using the same cramp spring as the conventional MT connectors. The average and maximum insertion losses were 0.40 and 1.8 dB, which are slightly larger than the average (0.3 dB) and maximum (1.0 dB) of the conventional 8-fiber MT connectors. This is because the fabricated fiber hole positions of 84-fiber MT ferrules are thought to have a large offset from the designed positions. However, we think that these offsets can be decreased with the same precision as with the conventional MT connector. The assembly of the high-density preconnectorized cable joint must not generate any excess loss. We demonstrate the insertion-loss changes of the 84-fiber MT connectors in the center and the corner of the installed cable joint. **Figure 13(a)** and **(b)** shows the measured insertion-loss change of the high-count 84-fiber MT connectors located at the center and corner of the cable joint, respectively. In **Figure 13**, there is no loss change of the high-count fiber MT connector before and after assembly. Consequently, there turned out to be no excess loss in the cable joint. Although it takes about 20 hours to connect two sets of

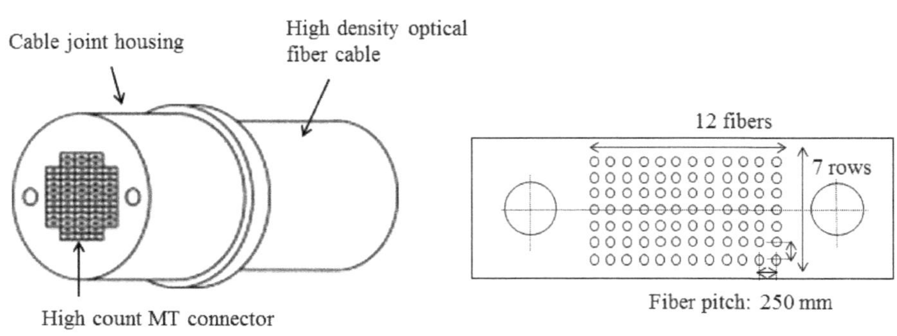

(a) (b)

Figure 11. (a) High-density preconnectorized cable joint and (b) high-count fiber MT connector.

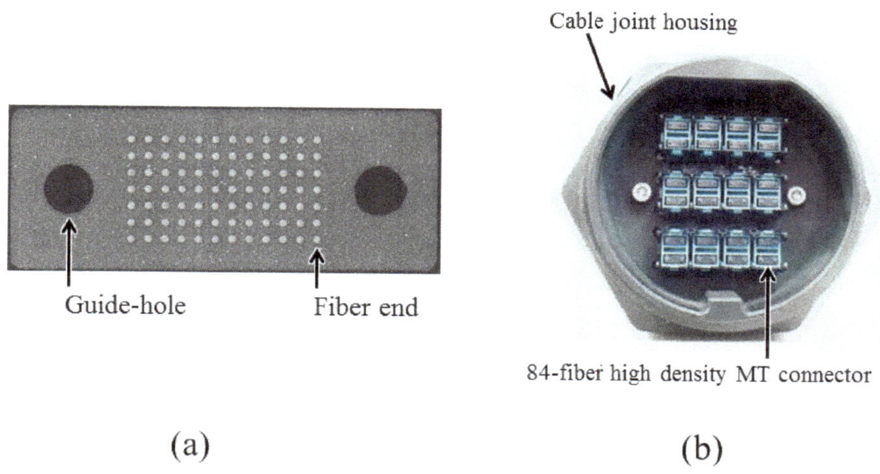

Figure 12. Endface pictures of (a) fabricated high-count fiber MT connector and (b) high-density preconnectorized cable joint.

2000 single-mode optical fiber cables using a fusion splice, it took about 5 minutes to connect two sets of 2000 single-mode optical fiber cables using the fabricated cable joint. The cross-sectional area of the cable joint was less than one-fifth that of a conventional optical closure. We have found that the fabricated cable joint can significantly reduce the time and space needed to join high-density optical fiber cables.

3.3. Functional multifiber connector for switching fibers

We have designed new kinds of functional multifiber connectors on the basis of MT connector technologies [24]. One is a device for switching fibers [25], which is described in this section. Figure 14 shows the basic structure and the switching mechanism of the proposed

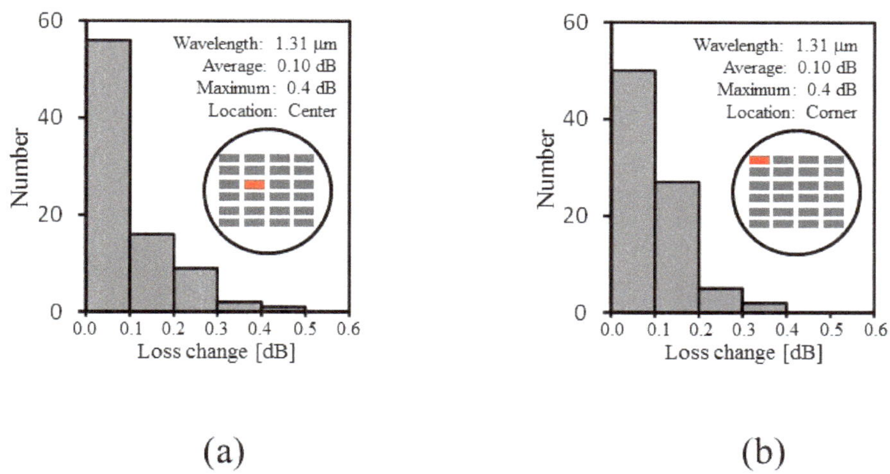

Figure 13. Measured insertion-loss change of the high-count 84-fiber MT connectors located at (a) center and (b) corner of cable joint.

switch. **Figure 14(a)–(c)**, respectively, shows the fiber connection states before, during, and after switching from one fiber to another. First, two guide pins are fixed at the bottom of the guide holes with the elastic materials in the multifiber array, as shown in **Figure 14(a)**. Next, when the conventional MT ferrule is pushed up with a certain force in a perpendicular direction, the two guide pins begin to slide in the same direction because the elastic materials change to the horizontal direction, as shown in **Figure 14(b)**. Finally, two guide pins are fixed at the top of the guide holes in the multifiber array by the restored elastic materials again after two guide pins have finished sliding, as shown in **Figure 14(c)**. Consequently, a certain optical fiber in the conventional MT ferrule can be switched from one optical fiber to another in the multifiber array.

We discuss the various applications of the proposed optical switch. **Figure 15** shows four types of switches. Types a and b are based on the conventional MT ferrule and multifiber array. In type a, two guide pins slide parallel to the multifiber arrangement between two guide holes. With this switch, a fiber from a row in the MT ferrule can be switched from one optical fiber to another in the same row in the multifiber array. In type b, two guide pins slide vertically in relation to the multifiber arrangement with two rows. In this switch, the optical fibers can be switched by the unit of the row. Type c can be inserted and removed between the two endfaces of conventional MT ferrules. With this switch, two guide pins can slide either parallel or vertical to the multifiber arrangement. This switch can be applied to conventional MT connectors and is useful both for installing MT connectors and for existing MT connectors in current optical fiber systems. Type d is an insertion and removal switch in which two guide pins slide vertically in relation to the multifiber arrangement. Therefore, the optical fibers are switched by the unit of the MT ferrule. Consequently, by using the proposed switching method where two guide pins slide in special guide holes, various types of switches with multifiber arrays can be realized for various uses.

We fabricated types a and c for an optical switch. The structure and performances of the type-a optical fiber switch are described elsewhere [25]. This section explains the type c optical fiber

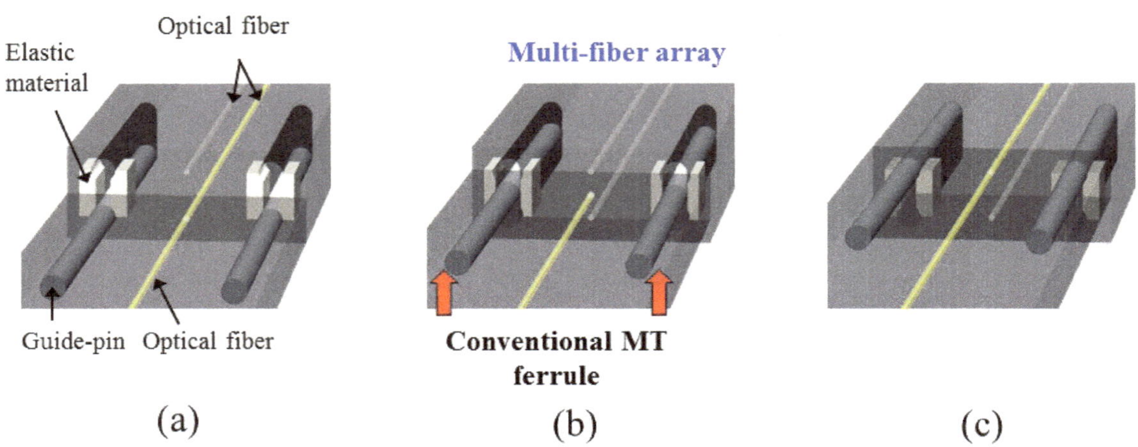

Figure 14. Basic switch structure and the switching mechanism (a) before, (b) during, and (c) after switching.

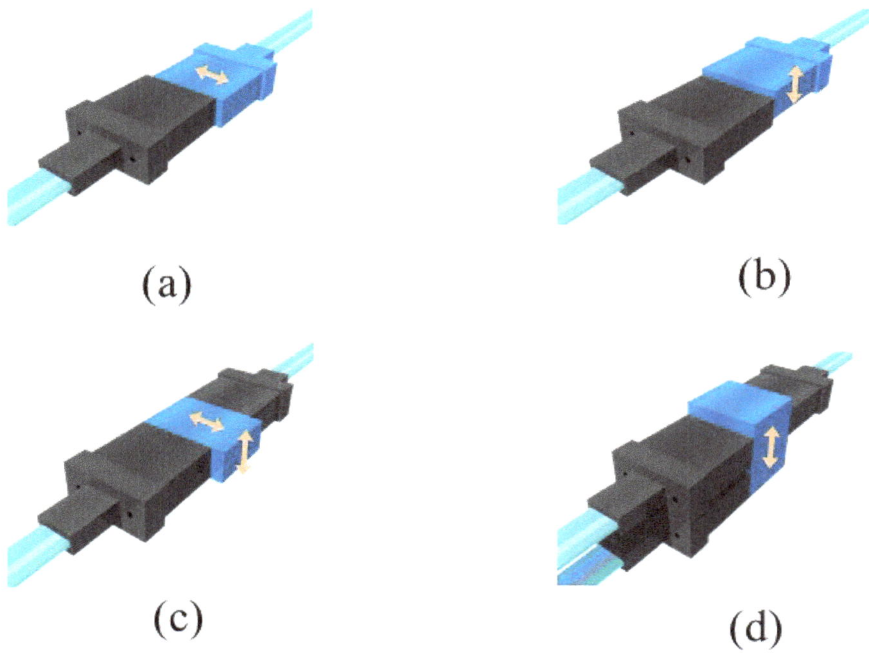

Figure 15. Various types of switches. (a) Fiber switch in a row, (b) fiber row switch, (c) insertable and removable switch, and (d) MT ferrule switch.

switch. **Figure 16(a)** and **(b)** shows the structure and a photograph of the fabricated multifiber-ribbon switch, which is composed of two MT ferrules and a multifiber array. The multifiber array has a novel guide hole structure in which two guide pins slide vertically to the four-fiber arrangement. The MT ferrules have two rows of four fibers. **Figure 17** shows the switching mechanism. In condition A before switching (a), the upper fiber ribbon in MT ferrule 1 is connected to optical fibers in the multifiber array and is also connected to the upper fiber ribbon in MT ferrule 2. When MT ferrule 1 is pushed up manually with a certain force in a perpendicular direction, the two guide pins slide in the opposite direction. In condition B after switching (b), where two guide pins in the multifiber array slide to the 0.25 mm length vertically to the arrangement direction, the lower fiber ribbon in MT ferrule 1 is connected to optical fibers in the multifiber array and is also connected to the upper fiber ribbon in MT ferrule 2. Therefore, the upper fiber ribbon in MT ferrule 1 is switched to the lower fiber ribbon in MT ferrule 1.

We measured the connection losses of the fabricated multifiber-ribbon switch at a wavelength of 1.3 μm using an LD and an optical power meter. The conventional two MT ferrules and multifiber array are connected with all four guide pins and a special clamp spring. Refractive index matching material is used between the ferrule and the multifiber array endfaces (both connections 1 and 2). **Figure 18** shows the measured connection losses. The average connection loss is 1.1 dB. These connection losses are higher than those of the conventional MT connectors because the fabricated fiber positions are thought to have large misalignments of the offset from the designed positions. However, we think that these offset misalignments can be minimized with the same precision as with the conventional MT connector. We also measured

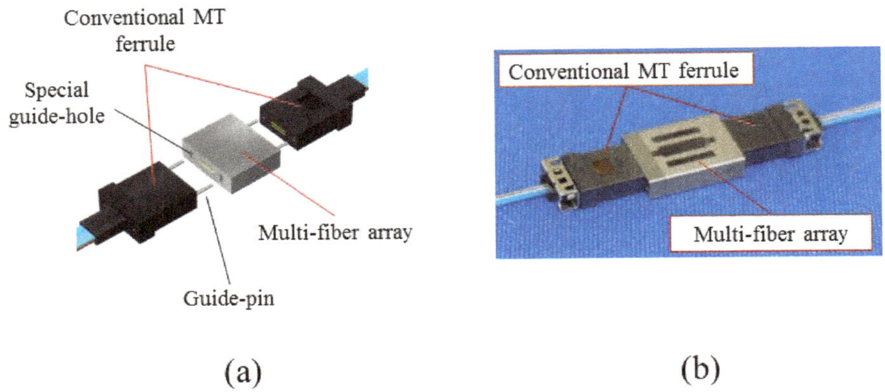

Figure 16. Fabricated fiber-ribbon switch. (a) Structure and (b) photograph of insertable and removable switch.

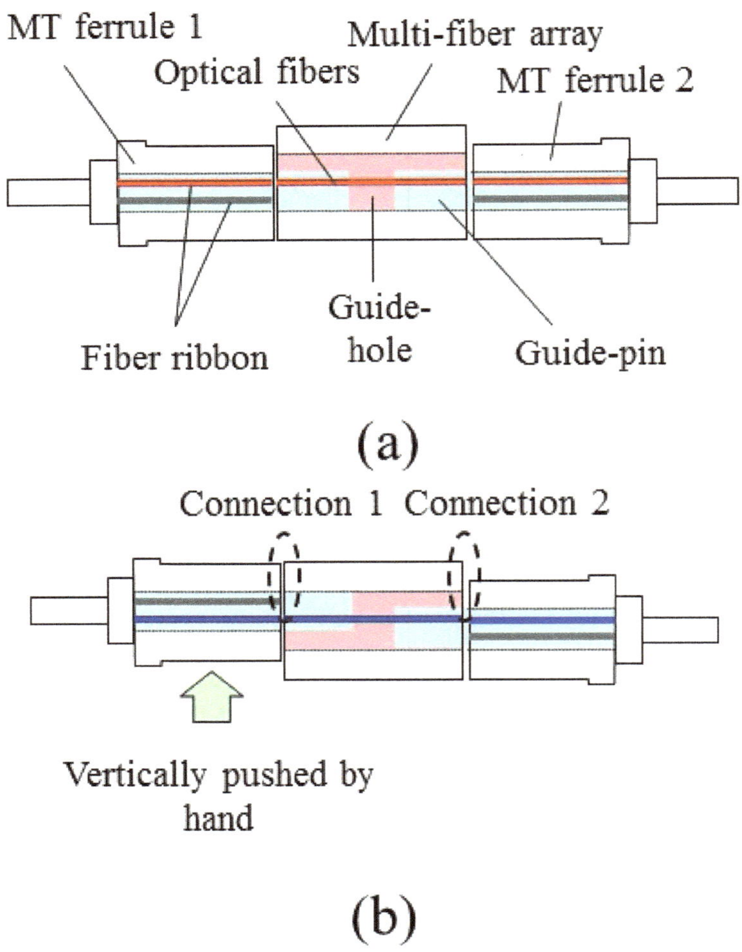

Figure 17. Switching mechanism of fiber-ribbon switch. (a) Before switching (condition A) and (b) after switching (condition B).

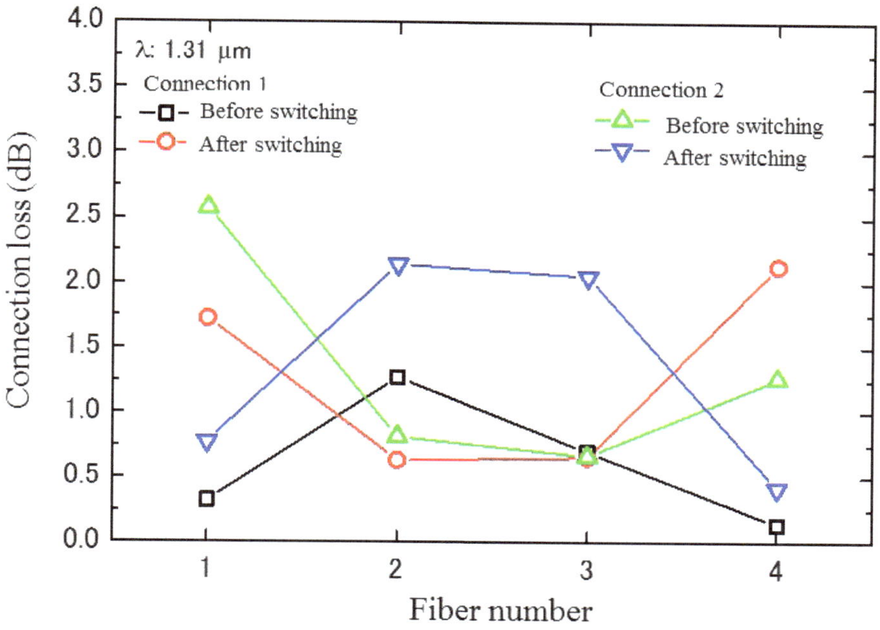

Figure 18. Connection losses of fabricated fiber-ribbon switch.

the force needed to push the conventional MT ferrule or multifiber array by hand in order to switch fibers. The measured forces were about 9.5 N. Consequently, we revealed that the fabricated switch can provide ease of handling.

4. Novel inspection technique for MT ferrule

To fabricate a high precision multifiber connector, MT ferrules need to be inspected. To shorten the time it takes to measure MT ferrules using conventional inspection equipment, we have created a novel inspection technique that has both fast measurement and low cost features. This section describes the new inspection technique.

Figure 19 shows the **(a)** structure and **(b)** observation area of conventional inspection equipment for MT ferrules [26]. The conventional equipment consists of a light source, high precision sample stage, an objective lens with 20 times, a charge-coupled device (CCD) camera with 0.4 M pixels, an image processor, and a computer. As shown in **Figure 19 (b)**, the observation area of conventional equipment is 0.2 × 0.17 mm, which corresponds to covering area of a fiber hole. To inspect whole fiber holes and two guide holes on an MT ferrule, the MT ferrule sample must be scanned on the high precision sample stage. Therefore, the conventional inspection equipment takes a long time to measure MT ferrules. We have thus designed and fabricated new inspection equipment to shorten the measurement time.

Figure 20 shows the **(a)** structure and **(b)** observation area of new inspection equipment for MT ferrules [27]. The novel equipment consists of a light source, an objective lens with 1 time, a CCD camera with 10 M pixels, an image processor, and computer. To reduce measurement

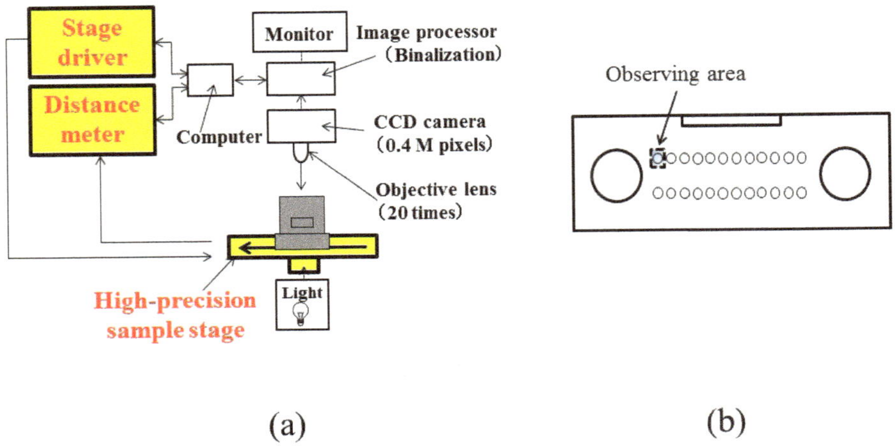

Figure 19. (a) Structure and (b) observation area of conventional inspection equipment for MT ferrules.

time, the scanning measurement must be eliminated. Therefore, the objective lens is changed from the conventional 20-time lens to a 1-time lens. Consequently, the observation area is expanded to be covered by the whole endface of an MT ferrule: 6.4 × 4.6 mm. In addition, the new inspection equipment is given a simpler structure and made more cost effective by eliminating the conventional high precision sample stage.

Our goal is to give the new inspection equipment the same measurement accuracy as the conventional equipment: less than 0.1 μm. The measurement accuracy of the new inspection equipment generally worsens because the objective lens is changed. To improve this worsened measurement accuracy, two methods were used. One is to change the pixels of the CCD camera and the other is to use multivalue processing with an image processor.

Figure 21 shows a photograph of the fabricated inspection equipment for MT ferrules. The fabricated inspection equipment is simpler and more cost-effective because it does not need a high precision sample stage, a stage driver, or a laser distance meter. We have experimentally demonstrated the multivalue processing using an image processor. **Figure 22** shows

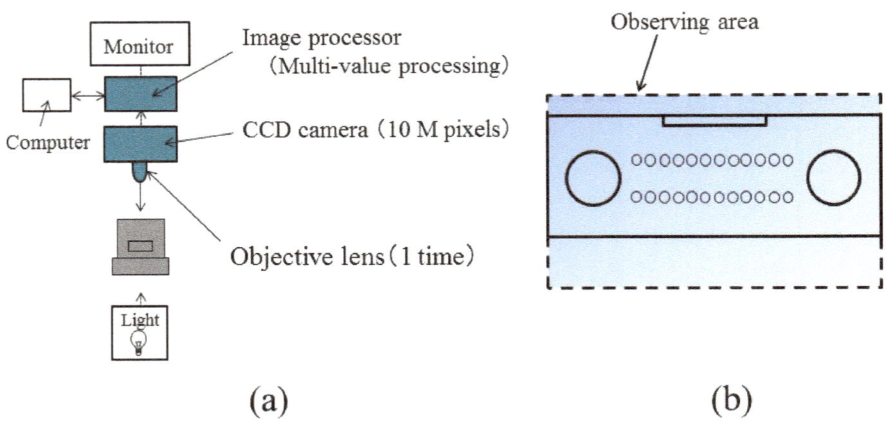

Figure 20. (a) Structure and (b) observation area of novel inspection equipment for MT ferrules.

experimental results for measurement time and accuracy for a 24-fiber MT ferrule. The measurement accuracy is the standard deviation of nine measurements. As the multivalue with the image processor becomes larger, the normalized standard deviation decreases whereas the normalized measurement time increases. These results indicate that there are optimal conditions for both high speed and high accuracy measurement. **Figure 23** shows measured fiber hole offsets for the same 24-fiber MT ferrule using the conventional and the fabricated inspection equipment. The offset results for the conventional and fabricated equipment are almost the same. The measured accuracy with the fabricated inspection equipment is less

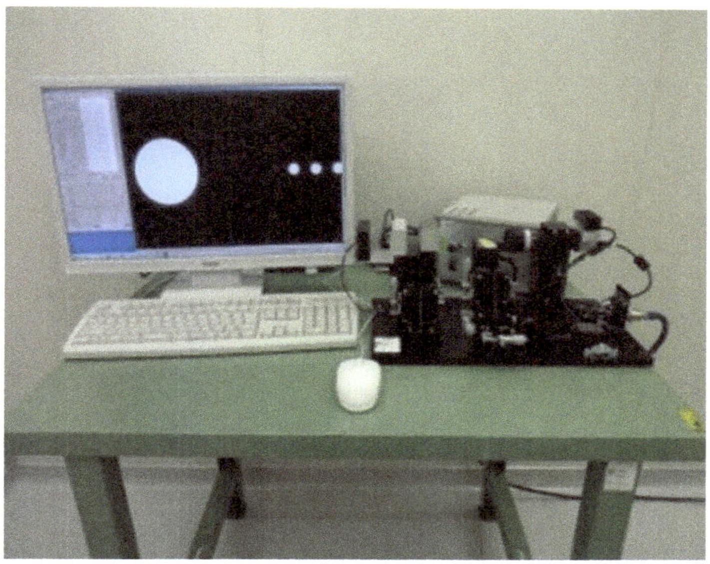

Figure 21. Photograph of the fabricated inspection equipment for MT ferrules.

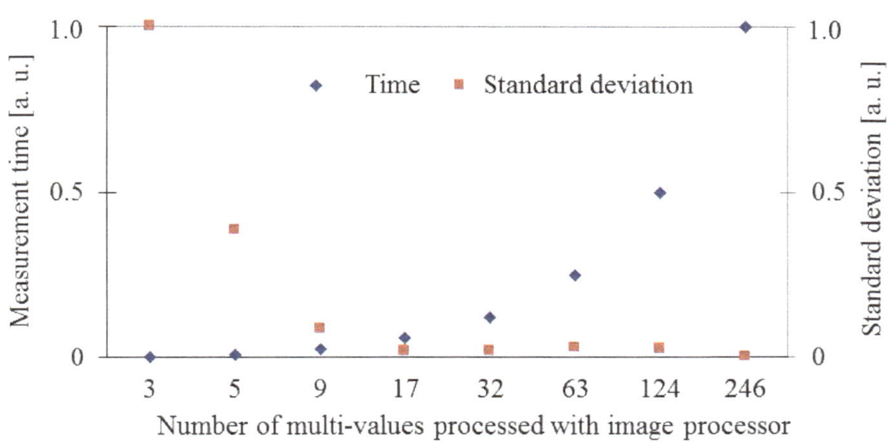

Figure 22. Experimental results of measurement time and accuracy to 24-fiber MT ferrule.

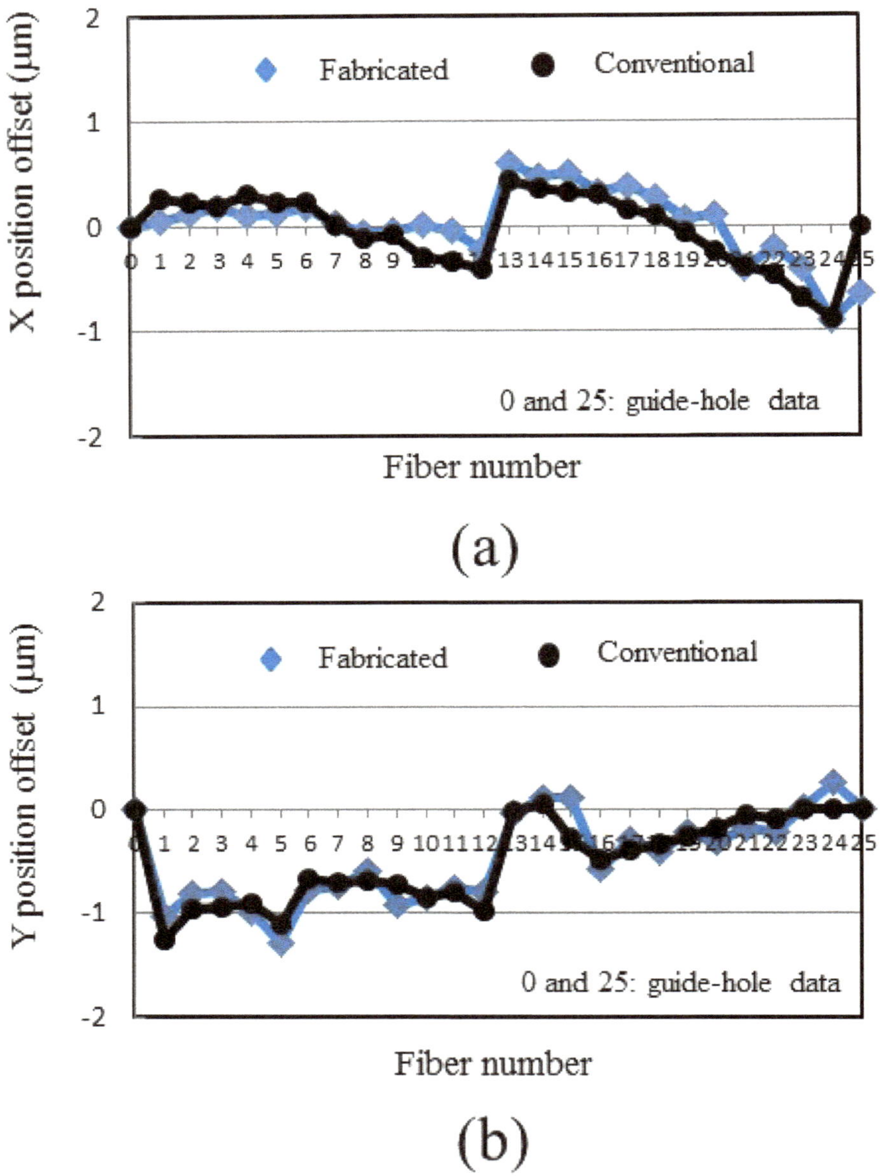

Figure 23. Measured fiber hole offsets of same 24-fiber MT ferrule using the conventional and fabricated inspection equipment. (a) X position offset and (b) Y position offset.

than 0.1 μm, which is as good as that of the conventional inspection equipment. It takes 3 minutes 40 seconds and 11 minutes 30 seconds to measure 24-fiber and 84-fiber MT ferrules using the conventional inspection equipment, respectively. In contrast, it takes 10 seconds each to measure 24-fiber and 84-fiber MT ferrules using the fabricated inspection equipment. Consequently, we reveal that the fabricated inspection technique can measure MT ferrules with high speed, cost effectiveness, and high accuracy.

5. Conclusion

This chapter reported the latest mechanically transferable (MT) and multifiber push-on (MPO) multifiber connector technologies.

Low insertion-loss and high return-loss angled physical contact (APC)-MPO single-mode multifiber connectors were developed. For these multifiber connectors to achieve a high performance, it is necessary to reduce deviations of all components and to control fiber offset due to shear force caused by compressing two obliquely ended MT ferrules.

Next, MT single-mode 84-fiber connectors were developed to realize higher density multifiber connectors. We have developed a novel high-density optical cable joint that has a 24-connector array consisting of new 84-fiber MT connectors. The new cable joining technique significantly shortens the time needed to join the optical fiber cables.

Novel optical fiber switches were developed on the basis of MT multifiber connector technology. The switches use a manual guide-pin slide method in special guide holes in a novel multifiber array. The fabricated fiber switches can be easily pushed up at a force of less than 10 N by hand.

Because high precision multifiber connectors need to be measured and inspected, a new inspection technique for MT ferrules was developed and equipment for it was fabricated. The fabricated inspection equipment can measure MT ferrules with high speed, cost effectiveness, and high accuracy.

These single-mode multifiber connectors can be used as key technologies for advanced optical fiber communication systems.

Acknowledgements

The author thanks Drs. K. Shiraki and T. Kurashima at NTT Access Service Systems Laboratories (ANSL) for their encouragement and helpful suggestions. The author is deeply grateful to R. Koyama, K. Saito, Y. Yamaguchi, and Y. Machida at ANSL for discussing the low insertion APC-MPO connectors and high-density MT connectors; T. Shimizu and C. Fukai at ANSL for discussing the novel optical switch based on MT connector technology; K. Noda and K. Nakashoji for helpful discussions regarding the inspection technique for MT ferrules; and Dr. T Satake at USConec Ltd. for providing the pictures of **Figure 2**.

Author details

Mitsuru Kihara

Address all correspondence to: kihara@osakac.ac.jp

NTT Access Network Service Systems Laboratories, Nippon Telegraph and Telephone Corporation, Hanabatake, Tsukuba, Ibaraki, Japan (He is currently with Osaka Electro-Communication University)

References

[1] Ministry of Internal Affairs and Communications. Subscriber Number and Share of Telecommunications [Internet]. 2016. Available from: http://www.soumu.go.jp/main_content/000454424.pdf [Accessed: February 12, 2017]

[2] Satake T, Nagasawa S, Arioka R. A new type of demountable plastic-molded single-mode multifiber connector. Journal of Lightwave Technology. 1986;**LT-4**(8):1232–1236. DOI: 10.1109/JLT.1986.1074860

[3] Satake T, Nagasawa S, Hughes M, Lutz S. MPO-type single-mode multi-fiber connector: Low-loss and high-return-loss intermateability of APC-MPO connectors. Optical Fiber Technology. 2011;**17**:17–30. DOI: 10.1016/j.yofte.2010.09.004

[4] Nagasawa S, Yokoyama Y, Ashiya F, Satake T. A high-performance single-mode multifiber connector using oblique and direct endface contact between multiple fibers in a plastic ferrule. IEEE Photonics Technology Letters. 1991;**3**:937–939. DOI: 10.1109/68.93269

[5] Nagasawa S, Ashiya F, Satake T. Multifiber optical connector plug with low reflection and low insertion loss. 1993. US Patent 5214730

[6] Kihara M, Nagasawa S, Tanifuji T. Design and performance of an angled physical contact type multifiber connector. Journal of Lightwave Technology. 1996;**14**(4):542–547. DOI: 10.1109/50.491391

[7] Nakajima T, Terakawa K, Toyonaga M, Kama M. Development of optical connector to achieve large-scale optical network construction. In: Proceedings of the International Wire and Cable Symposium (IWCS). 2006. pp. 439–443

[8] Kihara M, Koyama R, Abe Y, Kobayashi, Tomita S. Novel field installable optical fiber splice and connector for optical access networks. Optical Fiber Technology. 2013;**19**:269–278. DOI: 10.1016/j.yofte.2013.02.007

[9] Kihara M, Okada M, Hosoda M, Iwata T, Yajima Y, Toyonaga M. Tool for inspecting faults from incorrectly cleaved fiber ends and contaminated optical fiber connector end surfaces. Optical Fiber Technology. 2012;**18**:470–447. DOI: 10.1016/j.yofte.2012.07.004

[10] Kihara M, Nagasawa S, Tanifuji T. Temperature dependence of return loss for optical fiber connector. IEEE Photonics Technology Letters. 1995;**7**:795–797. DOI: 10.1109/68.393209

[11] Kihara M, Nagasawa S, Tanifuji T. Return loss characteristics of optical fiber connectors. Journal of Lightwave Technology. 1996;**14**:1986–1991. DOI: 10.1109/50.536966

[12] Takaya M, Kihara M, Nagasawa S. Design and performance of a multifiber backpanel type connector. IEEE Photonics Technology Letters. 1996;**8**:655–657. DOI: 10.1109/68.491570

[13] Kihara M, Tomita S, Haibara T. Influence of wavelength and temperature changes on optical performance of fiber connections with small gap. IEEE Photonics Technology Letters. 2006;**18**:2120–2122. DOI: 10.1109/LPT.2006.883256

[14] Kihara M, Uchino M, Omachi X, Watanabe H. Fault analysis: Optical performance investigation of fiber connections with imperfect physical contact. OSA Optics Letters. 2011;**36**:4791–4793. DOI: 10.1364/OL.36.004791

[15] Kihara M, Uchino M, Omachi M, Watanabe H. Investigation into optical performance of fiber connections with imperfect physical contact. Journal of Lightwave Technology. 2013;**31**:967–974. DOI: 10.1109/JLT.2013.2240256

[16] Kihara M, Watanabe H. Corrections to "Investigation into optical performance of fiber connections with imperfect physical contact". Journal of Lightwave Technology. 2016;**34**:3592–3595. DOI: 10.1109/JLT.2016.2579650

[17] Marcuse D. Loss analysis of single-mode fiber splices. Bell System Technical Journal. 1997;**56**(5):703–718. DOI: 10.1002/j.1538-7305.1977.tb00534.x

[18] Kihara M. Optical performance analysis of single-mode fiber connections. In: Yasin M, Harun SW, Arof H, editors. Optical Fiber Communications and Devices. InTech; 2011. pp. 239–256. ISBN: 978-953-307-954-7

[19] Shibata K, Takaya M, Nagasawa S. Design and performance of high-precision MT-type connector for 1.55-μm zero-dispersion-shifted fiber-ribbon cables. IEEE Photonics Technology Letters. 2001;**13**:136–138. DOI: 10.1109/68.910513

[20] Haibara T, Tomita S, Matsumoto M, Yabuta T. High-speed, low-loss connection techniques for high-count pre-connectorized cables. In: Proceedings of the International Wire and Cable Symposium (IWCS). 1991. pp. 296–302

[21] Tomita S, Matsumoto M, Haibara T, Nakashima T, Kihara M. Optical fiber cable and connector technology for FTTH network. The Institute of Electronics, Information and Communication Engineers. 1992;**E75-B**:862–870. ISSN: 0916–8516

[22] Takaya M, Shibata K. Design and performance of very high-density 60-fiber ferrules. Journal of Lightwave Technology. 2003;**21**:1549–1556. DOI: 10.1109/JLT.2003.812460

[23] Koyama R, Saito K, Fukai C, Kihara M, Kurashima T. A novel high-density pre-connectorized cable joining technique for single-mode optical fiber cables using new 84-fiber MT connectors. In: Proceedings of the International Wire & Cable, Connectivity Symposium (IWCS). 2016. pp. 572–575. Available from: http://www.iwcs.org/polopoly_fs/1.3285734.1474652248!/fileserver/file/687359/filename/099_0779-000072.pdf [Accessed: March 26, 2017].

[24] Kihara M, Nagasawa S, Takaya M, Tanifuji T. A new type of demountable optical device for coupling single-mode planar waveguides to a multifiber array using passive alignment. IEEE Photonics Technology Letters. 1995;**7**:1453–1455. DOI: 10.1109/68.477280

[25] Shimizu T, Kihara M, Izumita H, Kurashima T. Novel optical switch using manual guide-pin slide method based on MT multifiber connector technique. In: Proceedings of the OFC/NFOEC. 2013. pp. JTh2A.24. DOI: 10.1364/NFOEC.2013.JTh2A.24

[26] Measurement method for the hole location of a multiway connector plug. IEC 61300-3-27. JIS C5961 6.10

[27] Kihara M, Koyama R, Fukai C, Katayama K, Kurashima T, Novel equipment for inspecting fiber eccentricities in MT/MPO connector to enable high speed measurement. In: Technical Reports of the OFT (Japanese); 2017. OFT2016-176. Print edition: ISSN 0913-5685. Online edition: ISSN 2432–6380

7

Network Virtualization Over Elastic Optical Networks: A Survey of Allocation Algorithms

Danilo Bórquez-Paredes, Alejandra Beghelli and Ariel Leiva

Additional information is available at the end of the chapter

Abstract

Network virtualization has emerged as a paradigm for cloud computing services by providing key functionalities such as abstraction of network resources kept hidden to the cloud service user, isolation of different cloud computing applications, flexibility in terms of resources granularity, and on-demand setup/teardown of service. In parallel, flex-grid (also known as elastic) optical networks have become an alternative to deal with the constant traffic growth. These advances have triggered research on network virtualization over flex-grid optical networks. Effort has been focused on the design of flexible and virtualized devices, on the definition of network architectures and on virtual network allocation algorithms. In this chapter, a survey on the virtual network allocation algorithms over flexible-grid networks is presented. Proposals are classified according to a taxonomy made of three main categories: performance metrics, operation conditions and the type of service offered to users. Based on such classification, this work also identifies open research areas as multi-objective optimization approaches, distributed architectures, meta-heuristics, reconfiguration and protection mechanisms for virtual networks over elastic optical networks.

Keywords: optical fibre networks, flexible-grid/elastic networks, network virtualization, resource allocation algorithms

1. Introduction

Cloud computing has emerged as a new network paradigm [1]. Built on the success of grid computing applications, cloud computing implements the idea of 'computing as a utility' in a more commercially-oriented vision. Thus, the customer pays per use of computing facilities under the conditions stated in a service level agreement (SLA), having dynamic scaling of resources and transparent access to network services, unaware of the location and hardware/software characteristics of the required resources [2]. Apart from high bandwidth, cloud computing applications require the following functionalities from the underlying physical network [1]:

- Abstraction: The technology/implementation specific details of the physical network resources are hidden to the users, due to the "computing as a utility" philosophy.
- Isolation: Different cloud computing applications should not interfere with each other in the access to common physical resources.
- Flexible resource granularity: The amount of resources (storage, processing power and bandwidth) required by different cloud computing applications might vary significantly.
- On-demand setup/tear down: For efficiency, network resources should be set-up/torn down with a highly dynamic, rapidly reconfigurable and programmable network environment, something not possible with the current status of Internet [3].
- Resiliency: Grid/cloud computing applications should continue running in spite of failures affecting the optical network.

Network virtualization, which extends the well-known concepts of server and storage virtualization to networks, is envisaged as a key enabling technology for cloud computing services. As such, the benefits of running cloud applications on top of virtual networks (as opposed to on top of virtual servers alone, as usually done [4, 5]), was evidenced by several preliminary studies on network virtualization for cloud computing. In Ref. [6], resource allocation of cloud-based data centres services was proposed by abstracting the service requests as virtual network requests. In Ref. [7], a network virtualization platform that acts as a mediator between the cloud user requirements and the physical resources was proposed. In Ref. [8], a new network architecture based on network virtualization was proposed for cloud computing applications where the geographic location of servers is relevant. In Ref. [9], a network operator perspective was given about the convenience of network virtualization as an enabler for cloud computing. Nowadays, the benefits of network virtualization for cloud services are well identified in terms of cost, agility, resilience and multi-tenancy [10–12].

The underlying network over which network virtualization takes place is of fundamental importance to guarantee a good service. Arguably, the two most important requirements regarding the underlying network are the bandwidth capacity and the variety in the bandwidth granularity of connections, to allow for a high number of cloud computing applications with different bandwidth requirements. Both requirements would be naturally provided by flexible-grid optical networks [13, 14]. By overcoming the rigid spectrum allocation of current fixed-grid wavelength-division multiplexing (WDM) networks, elastic optical networks would make better use of the band C by allocating each connection the bandwidth just required. Depending on the bit rate and the modulation format, a gain in bandwidth usage between 33 and 100% could be achieved by using flexible-grid networks instead of a fixed one operating with a spectral width of 50 GHz [15]. Finally, flex grid would allow a wide bandwidth granularity of connections: bit rates from 10 Gbps to 1 Tbps are envisaged [13].

Given the impact that network virtualization is expected to have on the ever-increasing cloud computing area and the potential for significant bandwidth increase and bandwidth granularity offered by flexible-grid optical networks, in this survey, we review the efforts on network virtualization over optical flexible-grid networks.

The remaining chapter is as follows: Section 2 reviews the fundamental concepts of network virtualization and flexible-grid optical networks; Section 3 discusses the main challenges in the area of network virtualization over flexible-grid optical networks; Section 4 presents a taxonomy of the proposals found in the literature to allocate virtual networks over a flexible-grid underlying transport network; and Section 5 concludes the chapter highlighting the open research lines in the area.

2. Fundamental concepts

2.1. Network virtualization

Network virtualization refers to the creation of different isolated virtual networks on top of a common physical substrate. The isolation feature means that the information transmitted through a particular virtual network cannot be retrieved or affected by other existing virtual networks and the operation of the different virtual networks cannot affect the operation of the physical substrate [16].

Among the main features of network virtualization environments, we found several of the requirements imposed by cloud computing applications, namely, coexistence of different virtual networks, isolation between coexisting virtual networks, programmability, dynamicity, flexibility and heterogeneity [17].

By implementing cloud applications on virtual networks (i.e. one virtual network for each different cloud computing application), several benefits can be identified:

- Resource allocation based on maximum load could be avoided, leading to a more cost-effective operation, as the virtual network associated to the cloud application would request just the resources needed for proper operation. Some virtual network environments have even considered the possibility of reconfiguring the virtual network during operation (e.g. exploiting the feature of on-line virtual server migration) to adapt to time-variant requirements from the applications [2, 18].
- Isolation between different cloud applications for access to common physical resources
- Resiliency against node/server failures, due to the server-migration feature of virtualization environments
- Implementation of proprietary non-standard protocols for specific cloud applications requirements

Network virtualization has been envisaged as a very useful tool in network research and industry. In research, the test of new routing algorithms, network protocols or network controllers can be done by establishing a virtual network, without interrupting the normal operation of a physical network or deploying a physical network for tests. Thus, the production network may become the testbed [19]. An early example of this type of use was PlanetLab [20–22], established in 2002 for distributed systems and network research. Other efforts have been GENI in USA [23], FEDERICA and OneLab2 in Europe [24, 25], Akari in Japan [26] and FIBRE in a joint effort between Brazil and Europe [27]. For a review of several precursor experimental initiatives, see Ref. [17]. In an industry, network virtualization can offer separate networks for different units in a company, differentiation of services based on bandwidth usage (e.g. voice and video) or a rapid and flexible creation of sub-networks for different projects [28, 29]. For example, in a data centre each client can have its own topology and control its traffic flows. Finally, different service providers can share the same network infrastructure being unaware of the others.

As a way of illustration, **Figure 1** shows a schematic of a network virtualization system. The lower part shows the physical substrate, made of five nodes and six bidirectional links. The available capacity of physical links, measured in capacity units (c.u.), is shown next to each link. The upper part shows two of the virtual networks (three-node rings with three bidirectional links each) that have been established on the physical network. The capacity unit required by the virtual links are shown near to each link. Dotted lines represent the mapping between virtual and physical links. Both virtual networks can have virtual links established over the same physical link and a virtual link can require more than one physical link to be established. For the sake of clarity, the mapping of the nodes is not shown but it can be deduced by identifying the physical nodes at the extreme of the physical links associated to the virtual links. The decision about whether establishing a new virtual network is possible or not and what virtual link/node is established in what physical link/node is made by a virtual network allocation algorithm.

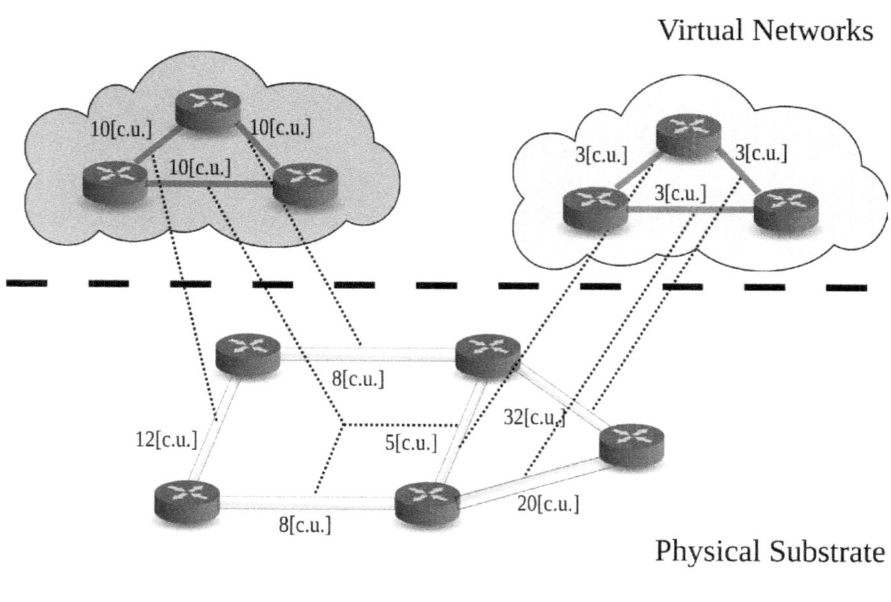

Figure 1. Two virtual networks established on the same physical substrate.

2.2. Mathematical modelling for network virtualization

The physical network is modelled by a directed graph $\mathcal{P} = (Np, Lp, Rp^t, Cp)$, where Np and Lp are the sets of physical nodes and links, respectively; Rp^t is the set of resources of type t in the physical nodes (for example, storage and processing resources; $t \in \mathbb{N}$) and Cp the set of resources at the physical links (optical bandwidth).

Analogously, the i-th virtual network can be modelled by a directed graph $\mathcal{V}_i = (Nv_i, Lv_i, Rv_i^t, Cv_i)$, where Nv_i is the set of virtual nodes and Lv_i the set of virtual links; Rv_i^t is the set of resources of type t required by each virtual node of the virtual network $\mathcal{V}_i$ (e.g. storage and processing resources) and Cv_i is the set of resources required by the virtual links (optical bandwidth).

The information required to execute the resource allocation algorithm is as follows:

1. $\mathcal{A} = \{\mathcal{V}_i\}$: Set containing the identification of all virtual networks $\mathcal{V}_i$ already established over the physical network $\mathcal{P}$.
2. $\mathcal{S}nv_k$: Set of all virtual nodes already established on the physical node $k \in Np$.
3. $\mathcal{S}lv_m$: Set of virtual links already established on physical link $m \in Lp$.

Every time the resource allocation algorithm must process a new virtual network request, at least the following two constraints must be met to be able to accept such request:

$$r_k^t \geq \sum_{\forall nv \in \mathcal{S}nv_k} r_{nv}^t \; ; \forall t \tag{1}$$

$$c_m \geq \sum_{\forall lv \in \mathcal{S}lv_m} c_{lv} \tag{2}$$

where r_k^t is the total number of resources of type t in physical node k, r_{nv}^t is the number of resources of type t allocated to virtual node nv, c_m is the total number of resources in physical link m and c_{lv} is the number of resources allocated to virtual link lv.

Eqs. (1) and (2) forbid that the number of resources allocated to the virtual nodes/links established in a particular physical node/link exceed the capacity of that node/link.

Additionally, depending on the type of physical network, extra constraints might appear on the allocation of resources to the virtual links. In the case of an optical network, fixed and flexible-grid networks impose different constraints. We review these two types of optical networks and their associated constraints in the following.

2.3. Fixed-grid optical network

In a circuit-switched optical network, each circuit is carried by an optical channel/carrier, based on the wavelength division multiplexing (WDM) technique. Currently, such optical channels operate in the range 1530–1565 nm, known as band C.

In a fixed-grid optical network, the optical carriers are determined by their central frequency and use a fixed amount of spectrum. According to the specification ITU-T G.694.1 [30], the selectable spectrum widths are 12.5 GHz, 25 GHz, 50 GHz and 100 GHz. Once a spectrum width is selected, all optical channels in a link are established with such spectral width. Depending on the selected spectral width, the central frequency used by the n-th optical channel is given by the following equation:

$$193.1 + n \times W \, THz \tag{3}$$

where $W \in \{0.0125; 0.025; 0.05; 0.1\}$ denotes the spectral width selected and n is an integer number whose range depends on the spectral width as follows: $n \in [-123, 227]$ for $W = 0.0125$; $n \in [-61, 113]$ for $W = 0.025$; $n \in [-30, 56]$ for $W = 0.05$; $n \in [-15, 28]$ for $W = 0.1$.

Figure 2 shows an example of the spectral usage of a fixed-grid link where six optical channels have been established: two optical channels at 10 Gbps using the on-off keying (OOK) modulation format, three channels at 40 Gbps modulated with dual polarization-quadrature phase shift keying (DP-QPSK) and one channel at 100 Gbps, also modulated with DP-QPSK. The spectral width of each channel is equal to 50 GHz and the central frequencies are determined by Eq. (3). It is common practice to identify the channels by their equivalent wavelength as well. Thus, in **Figure 2**, the corresponding wavelength of each channel has been written between brackets under the central frequency.

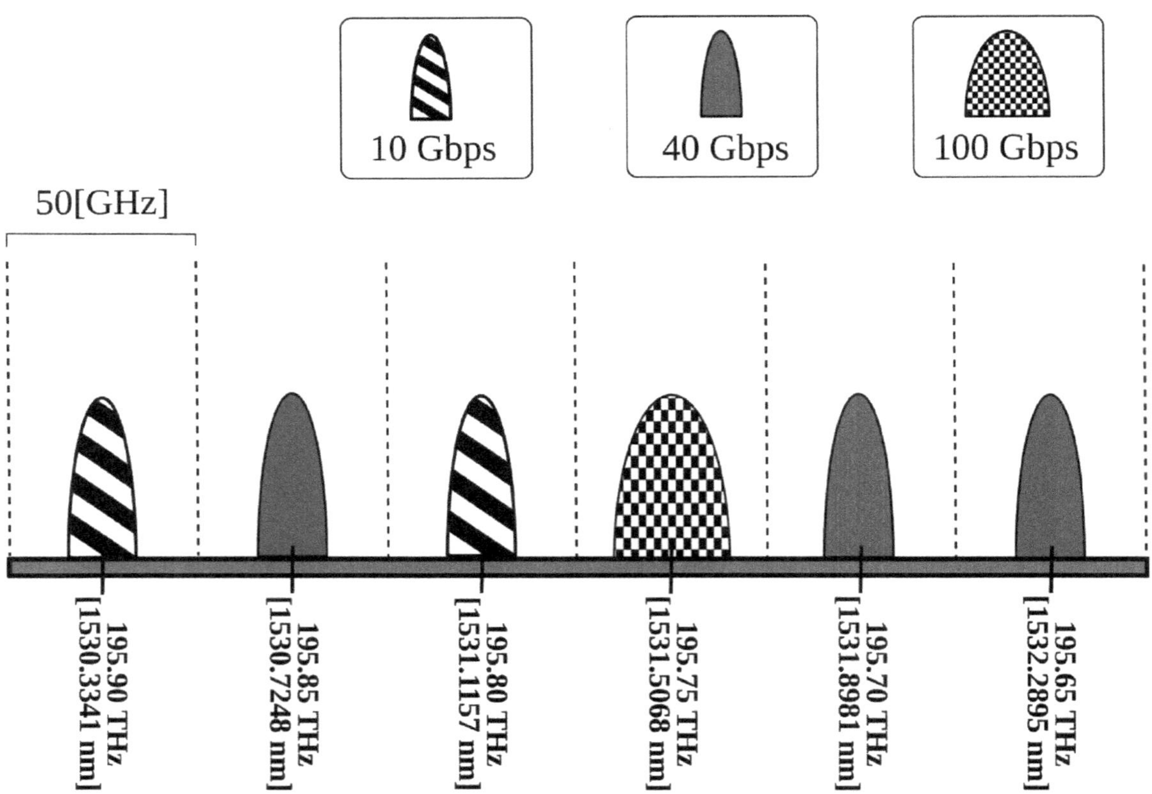

Figure 2. Frequency allocation to different transmission rate optical channels in a fixed-grid link.

In fixed-grid optical networks (in the absence of wavelength converters), the wavelength continuity constraint must be met. That is, the optical channel used by the virtual link must use the same central frequency and spectral width in all the physical links used. In networks operating with multiple transmission rates (as shown in **Figure 3**), additional constraints to deal with the signal degradation of higher bit rates channels mainly due to cross-phase modulation [31–33] may be required: for example, some channels should be left unused as guard bands or an optical reach (the maximum distance an optical signal can travel without exceeding a threshold on the bit error rate) be established.

The main drawback of fixed-grid optical networks is the inefficient spectrum usage [34], as observed in **Figure 2**, where channels are allocated more spectrum than effectively required: both a 10 Gbps OOK-modulated channel and a 40 Gbps channel modulated with DP-QPSK require a bandwidth equal to 25 GHz [34, 35], whereas a 100 Gbps channel modulated with DP-QPSK requires just 37.5 GHz [34]. To increase the spectrum usage, the flexible allocation of it has been proposed [14, 34]. This type of networks is known as flexible-grid or elastic optical networks.

2.4. Flexible-grid optical networks

In a flexible-grid optical network, the spectral width of a channel can be varied depending on the data transmission requirements [36]. Thus, the spectrum is divided in small units, typically of 12.5 GHz, known as frequency slot units (FSU) [34]. By using a different number of contiguous FSUs, different spectral widths can be achieved [37, 38] depending on the transmission requirements of the signal, such as the modulation format and the bit rate.

As a way of illustration, **Figure 3** shows the same six channels of **Figure 2**, now operating in a flexible-grid system. The numbers of 12.5 GHz FSUs required are 2, 2 and 3 for the 10, 40 and 100 Gbps channels, respectively. Thus, the flexible-grid allocation uses just 54.2% of the spectrum originally required (162.5 GHz instead of 300 GHz).

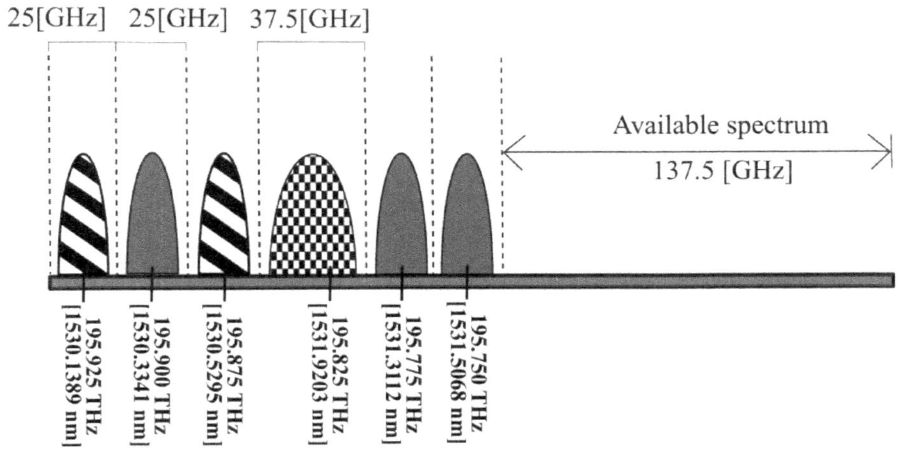

Figure 3. Frequency allocation to different transmission rate optical channels in a flexible-grid link.

Single-carrier and multi-carrier (super-channel) can be used to create an optical connection. In the latter, the overall bit rate is achieved through lower-rate sub-carriers. Examples of these systems are Co-WDM, Nyquist-WDM and time frequency packing [34, 39, 40]. In general, multi-carrier systems require a lower number of FSUs and exhibit a longer optical reach than single-carrier systems with the same total bit rate and modulation format [41, 42].

Regarding the modulation formats, there are bi-level and multi-level types. In a bi-level modulation format, as OOK and binary phase shift keying (BPSK) [42], the symbol rate equals the bit rate. In a multi-level modulation format, as QPSK and x-quadrature amplitude modulation (x-QAM) [41, 42], the symbol rate is lower than the bit rate of the bi-level type, leading to a lower requirement of FSUs. However, the optical reach of multi-level modulation formats is lower than that of bi-level [34, 36], highlighting a trade-off between number of FSUs and optical reach [34, 43, 44].

Once the number of FSUs required by a virtual link has been determined, the establishment of such link must meet at least two additional constraints: FSU continuity and FSU contiguity constraints. The FSU continuity constraint is analogous to the wavelength continuity constraint (exactly the same FSUs must be used in every physical link selected to establish a virtual link). The FSU contiguity constraint imposes that, if more than one FSU is required to establish a virtual link, then these FSU must be contiguous in the spectrum [45].

The sequence of physical links used to establish a virtual link meeting the FSU continuity and contiguity constraints is known as a spectrum path.

3. Research challenges in virvtual network allocation over flexible-grid optical networks

In the following, the main challenges in the research area of network virtualization over flexible-grid optical networks are discussed.

3.1. Performance metrics

A performance metric allows defining the quality of an algorithm to carry out its task. Thus, usually the (single) objective of an algorithm is the maximization or minimization of a performance metric. However, for a complex algorithm such as a virtual network allocation algorithm, there are several performance metrics that could be optimized.

Most published results have focused on minimizing the virtual network request rejection rate [46–56, 64]. The main advantage of using the performance metric is that it allows evaluating the ability of the algorithm to accommodate new virtual networks on the physical substrate. However, given that the blocking depends on many parameters (the physical and virtual network topologies, the capacity availability in physical nodes and links, the capacity requirements of virtual nodes and links [50–52]), to identify the best algorithm is necessary knowing exactly the network configuration where the algorithm will operate (something difficult to achieve in dynamic scenarios) or running extensive simulation experiments with different network configurations (a time-consuming task).

Instead of registering the blocking ratio, a computationally simpler metric consists on registering the number of virtual network establishment requests received when the first blocking (rejection) occurs [48, 49]. A good algorithm would aim at registering such event at the latest possible instant. If used in conjunction with the blocking ratio, the first blocking metric can give information about the instant when the network starts saturating (when the first blocking occurs) and the dynamic of the system once such saturation state is reached.

Maximizing the traffic carried by the physical network due to the established virtual networks has also been the objective of some algorithms [54, 57–59]. As with the blocking ratio, the value of this performance metric depends on the topologies of the physical and virtual networks as well as the capacity of physical nodes/links and capacity requirements of virtual nodes/links, which makes difficult drawing general conclusions about the quality of different algorithms. Additionally, the lack of information about the number of virtual networks rejected does not allow measuring the quality of the service offered to the users. Thus, it should be used in conjunction with the blocking ratio.

Guaranteeing a given level of availability (e.g. 0.99999) to a virtual network has not been addressed by the proposed virtual network allocation algorithms to date, although availability (the fraction of time that a service is in operative state) is one of the most important quality of service metrics in a service level agreement (SLA). However, some efforts have been carried out in guaranteeing operation under specific failure conditions [49, 53, 59, 61, 62].

All previous performance metrics somehow aim to evaluate the capacity of the algorithm to offer a good quality of service. However, the main challenge in evaluating the performance of complex algorithms is selecting a performance metric that can capture the quality of the service offered to the user as well as the cost in achieving such quality.

To offer physical resources to a virtual network, the service provider incurs expenditure and operational costs due to the acquisition and maintenance of transponders, regenerators, optical cables, optical amplifiers and ROADMs (reconfigurable optical add drop multiplexer) [60]. Thus, algorithms aiming at minimizing the cost have also been studied. This metric has been mostly used in static scenarios [46, 56, 61, 62], and it is useful for the network planning stage. In dynamic scenarios, it can be used to determine the cost per virtual network, the total cost of providing the network virtualization service during a period of time or the cost incurred to achieve a given performance in terms of blocking ratio or traffic carried.

To date, quality-of-service-related metrics and cost have been studied separately. The algorithm is designed to minimize/maximize one of them whilst the other one is just measured. Thus, a multi-objective optimization approach that evaluates quality (as blocking or availability) and the cost incurred to achieve the required quality would deliver more realistic information about the best algorithm alternative from a network operator perspective.

3.2. Network virtualization dynamics characterization

To date there are no commercial network virtualization systems over flexible-grid optical networks. In Ref. [63], an experimental system is reported, but traffic is artificially generated. Therefore there are no empirical statistics that help to model the structure (virtual topologies and their capacity requirements) and dynamic of such system. In terms of structure, it would be useful knowing how to model the virtual topologies and their capacity requirements. Such knowledge would facilitate the evaluation of allocation algorithms in terms of simulation, the only technique used so far to evaluate performance of dynamic systems.

In terms of structure, different works make different assumptions regarding the topologies of the virtual networks and their capacity requirements. **Table 1** summarises the main models used to characterize the virtual topologies. In it, the name of each physical and virtual topology is given along with its number of nodes ($|Np|$, $|Nv|$) and links ($|Lp|$, $|Lv|$). When a number lower than one is provided for $|Lv|$, it means that the probability interconnection between a node pair is given. The column 'Node/Link requirement' corresponds to the percentage of usage of the physical node and link by any virtual node and link, respectively. The symbol '-' implies that such information is not found in the chapter.

As most works (15 of 17) use a medium-sized physical network (NSFNet or DTNet) for evaluation, future works should consider at least one of these topologies as the physical substrate to facilitate comparison among different proposals. No pattern can be observed in terms of the virtual topologies, with most works using mesh topologies with different degrees of connectivity. Regarding resource requirements, all proposals require no more than 10% of the physical node/link resources. The rest uses percentages of a few units.

Regarding dynamism, the most used distribution to model the virtual network request inter-arrival time is the exponential [46, 50–56, 64]. The holding time is usually modelled by an exponential distribution [52, 55], a deterministic value [64] or infinite (to model incremental traffic) [48, 49].

3.3. Physical impairments

It is expected that flexible-grid optical networks can accommodate channels (used to implement virtual links) at rates from 10 Gbps to 1 Tbps. Such channels, in the same way as fixed-grid channels, will be affected by several physical impairments that degrade the quality of the signal transmission. Additionally to typical physical impairments, as attenuation, chromatic dispersion, four-wave mixing (FWM) and amplified spontaneous emission (ASE) noise [65], in elastic optical networks the non-linear effect of cross phase modulation (XPM) takes relevance because of the existence of channels with different modulation formats in the same link. Due to the XPM effect, channels using intensity-based modulation formats (e.g. OOK typically used in 10 Gbps channels) interfere negatively in the quality of the signal of phase-modulated channels (e.g. BPSK and QPSK, used for higher bit rate channels) [66].

Most previous works have not considered this situation, with some of them assuming an ideal physical substrate [50, 54] whereas others have resorted to simplified models. For instance, in Refs. [48, 49, 51, 53, 57, 61, 62], the degradation is summarized in the figure of the maximum optical reach of signals, in Refs. [46, 56, 58, 59], the use of guard bands to all channels is used to simulate an ideal substrate, whereas in Refs. [47, 52, 55, 64], guard bands (to all channels or selectively added to channels most affected by the XPM degradation) are added to the limitation of the optical reach.

Physical topology (|Np|, |Lp|)	Virtual topology (|Nv|, |Lv|)	Node/link requirement	Work
NSFNet (14, 21)	Mesh (2–6, 0.5	-/0.35–3.5%	[46]
	Mesh (-, E%)	-/-	[47]
	- (2–5, -)	0.32–2.5%/0.31–2.5%	[48]
	Mesh (5–7, -)	1%/0.62–0.93%	[52]
	Ring (5–7, 5–7)	10%/0.62–0.93%	
	- (-, -)	-/2–16%	[53]
	Mesh (2–5, 0.5)	0.29–0.86%/0.67–3.33%	[54]
	Ring (3–7, 3–7)	1%/0.63–15.94%	[55]
	Mesh (3–5, 3–10)	1%/0.625–3.125%	[56]
	Mesh (4–7, 4–14)	1%/1.25–5%	[58]
	Ring (2–4,2–4)	-/-	[59]
	Mesh (5,7)	2–4%/-	[62]
6-node (6, 8–10)	Mesh (2–3, 0.5)	2–6%/2–6%	[51]
	Ring (3, 3)	10%/1.25–5%	[58]
	- (4–5, 3.6–3.8)	-/-	[61]
	Mesh (3–4,0.5)	1–4%/-	[64]
DTNet (14, 23)	Mesh (3–4. 0.5)	0.5–5%/0.5–5%	[51]
	Mesh (3–10, 0.5)	1–4%/-	[58]
	Ring (2–4,2–4)	-/-	[59]
	Mesh (2–5, 1–15)	-/0.85–3.12%	[61]
	Mesh (3–4, 0.5)	0.5–5%/0.5–5%	[64]
US network (24, -)	- (2–5, -)	0.06–2%/0.31–2.5%	[50]
Random (50, 141)	Mesh (2–10, 0.5)	0.5–10%/0.5–10%	[51, 52]
CORONET (75, 99)	- (-,11.4)	-/-	[54]
ARPANET (20, 32)	Mesh (3–7, 0.5)	0.2–1.2%/0.5–5%	[58]

Table 1. Characteristics of virtual network requests used in the literature.

3.4. Resource allocation to virtual networks

The selection of the physical nodes and links to be allocated to a virtual network is a $\mathcal{NP}$-Hard problem [67]. Thus, most proposals solving this problem over flexible-grid optical networks have resorted to heuristics [46–59, 61, 62, 64] and a few of them have proposed integer linear models [51, 58, 59, 61, 64], but mostly in the context of a static scenario where the random nature of the virtual network requests is not a problem.

Much work is still needed in identifying the features of good performing heuristics to allocate virtual networks as well as evaluating the performance of meta-heuristics.

3.5. Spectrum fragmentation

Under dynamic operation, as a result of the resource release from virtual network that depart from the network, voids in the spectrum are generated. A void is a set of contiguous available FSUs between portions of allocated FSUs (or between a portion of allocated FSUs and the beginning/end of the band), as shown in **Figure 4**.

Due to the FSU contiguity constraint, the existence of these voids is problematic, as they fragment the spectrum. As a result, a virtual link could not be implemented due to the lack of enough contiguous FSUs, leading to a higher blocking ratio. For example, in the situation depicted in **Figure 4**, although three FSUs are available, a virtual link requiring three FSUs could not be established because of the contiguity constraint.

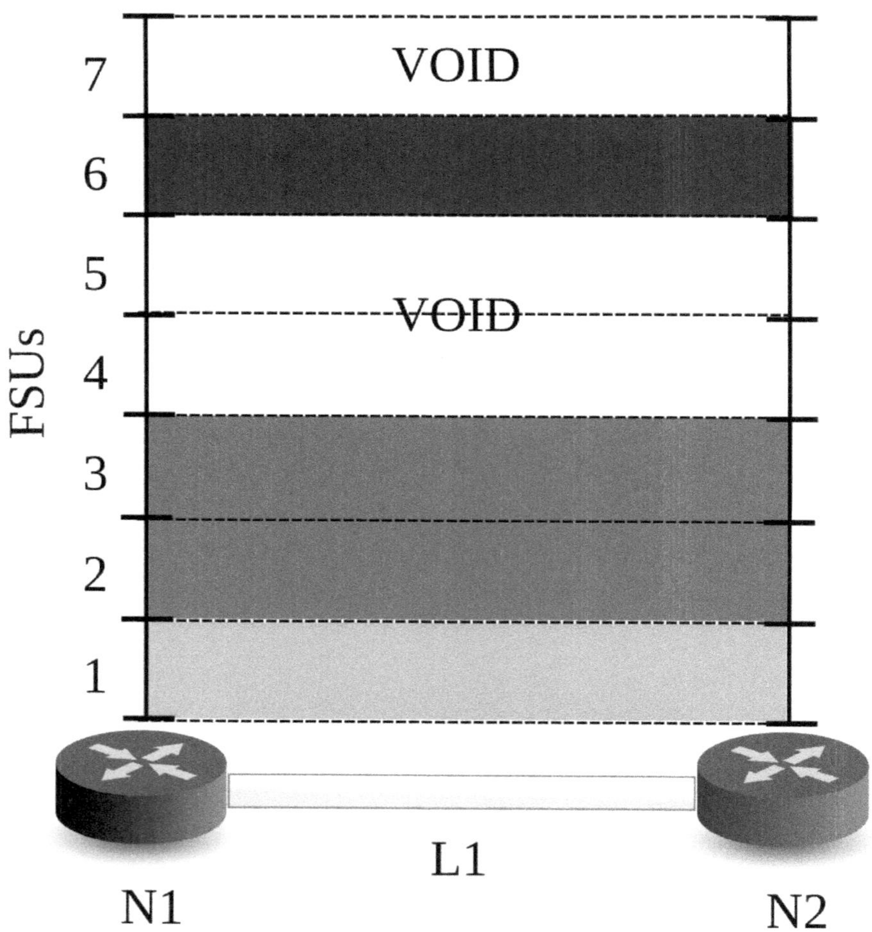

Figure 4. Spectrum fragmentation.

To decrease the spectrum fragmentation, the re-allocation of FSUs to the different channels in a link has been proposed in the area of flexible-grid networks by Refs. [68–72]. In Ref. [73], the impact of avoiding fragmentation on the blocking ratio can be seen.

In Ref. [54], a technique of spectrum defragmentation in the area of virtual networks over flexible-grid optical networks is reported, showing that the blocking ratio decreases with respect to an algorithm without defragmentation. However, defragmentation is costly as computation time and additional resources must be used to apply it. This highlights a trade-off between the blocking ratio decrease and the frequency of defragmentation. Further research on the interplay of allocation algorithms and defragmentation techniques is required.

4. Taxonomy

Figure 5 shows a comprehensive classification of the resource allocation algorithms in the area of network virtualization over flexible-grid optical networks. The taxonomy includes current proposals, but it is generic enough as to include algorithms not studied yet.

In the taxonomy, each possible algorithm is defined by three main dimensions: its performance metric, its operation conditions and the type of service offered to the user. In the following each of these dimensions are described as well as the different choices available in each one of them.

4.1. Performance metric

The most commonly used performance metric in the literature is the blocking ratio [46–56, 64]. Although the use of the same metric would facilitate comparison, due to the different assumptions made on the physical and virtual topologies, a direct comparison is not always possible.

The variant of blocking, first blocking, has been used in Refs. [48, 49]. Remaining metrics used in reported works are the traffic carried by the physical networks [54, 57–59] and cost-related metrics [46, 56, 61, 62].

Although published works do not explicitly mention the performance metric of availability, few works make assumptions on the operation conditions of the network that allow guaranteeing 100% availability. In Refs. [53, 59, 61, 62], only single link failures are assumed. Thus, the allocation of two link-disjoint spectrum paths to implement each virtual link is enough to ensure the operation of every virtual network. In Ref. [49], single link/node failures are assumed and, then by allocating two node/link-disjoint spectrum paths to each virtual link, a 100% availability is provided. Note that if the system violates the assumptions on the type of failure that can occur (e.g. a double link failure occurs in a system designed to tolerate single link failure), 100% availability cannot be guaranteed anymore.

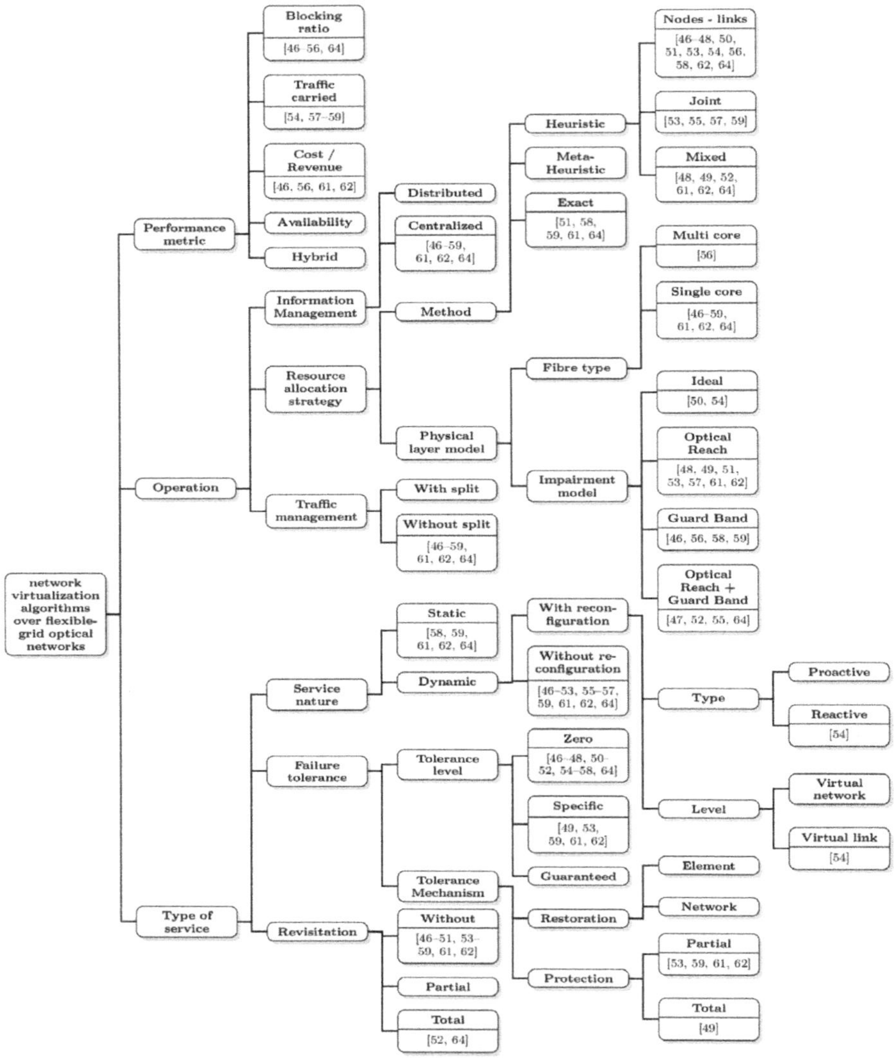

Figure 5. Taxonomy of network virtualization algorithms over optical flexible-grid networks.

Guaranteeing availability under any type of failure has not been researched in the area of network virtualization over flexible-grid networks neither the combination of quality and cost performance metrics.

4.2. Operation

4.2.1. Information management

To date, all proposals implicitly assume centralized information management [46–59, 61, 62, 64]. That is, a central entity has global knowledge of the network status and the resource allocation algorithm is executed every time a new virtual network request is generated. In fact, the first proposals for architecture with a virtual network controller are based on a centralized scheme, as the one proposed in Ref. [74]. Centralized systems are suitable when the time between successive requests is long enough for the central controller to execute the resource allocation algorithm.

In Ref. [75], different distributed virtual network allocation approaches are discussed in the context of packet networks. Results show that a distributed operation reduces the delay in mapping a virtual network and the number of messages required to be exchange to coordinate the allocation. In Ref. [76], the impact of a distributed virtual network reconfiguration approach on the interruption time of the service is studied in the context of fixed-grid networks. Although the distributed operation has advantages in terms of resilience against failures, lower computation times and network congestion due to message exchange, it has increased complexity in terms of control plane network (more controllers), synchronization of messages and a potential decreased performance due to the obsolescence of information. These aspects are yet to be studied in network virtualization systems over flexible-grid networks.

4.2.2. Resource allocation strategy

The virtual network allocation strategy must consider two aspects: the method used to solve the problem of embedding the virtual network on the physical network and the model used to characterize the constraints of the physical substrate.

There are three general methods to solve the problem of the virtual network embedding:

a. Exact methods: These are the techniques that find the global optimal solution to a problem. However, they are computationally complex and thus, they are usually applied only to small instances of the problems with slow dynamics. In real dynamic systems, where a solution must be found in short-time scales, this type of method is not feasible. However, in simulation environments, an integer linear programming (ILP) model can be solved for each virtual network request to be used as a benchmark, as done in Ref. [51]. In the area of virtual network allocation over flexible-grid networks, most ILP models have been used to solve the problem in a static scenario (virtual networks permanently established, not allocated on-demand). Works in Refs. [51, 58, 59, 61, 64] apply ILP to allocate a set of predefined virtual networks on a small physical network (six nodes) with the objective of minimizing cost.

b. Meta-heuristics: These are the generic algorithms capable to adapt to different problems by adjusting their parameters and configurations. Usually, they find very good quality solutions, but cannot guarantee the optimum solution as exact methods do. Work in Refs. [77, 78] proposed the use of genetic algorithms and ant colony to solve the problem of virtual network embedding in conventional networks, respectively. No works have been reported on flexible-grid networks as the physical substrate, neither in static nor in dynamic scenarios.

c. Heuristics: These are ad-hoc algorithms that do not guarantee a global optimum solution, designed for a specific problem. However, they are computationally simpler than the previous techniques. Most works in the area of network virtualization over flexible-grid networks resort to heuristics [46–59, 61, 62, 64], mainly focused on dynamic scenarios.

Normally, heuristics designed to solve complex problems, divide the original problem in sub-problems easier to solve separately. This approach is applied in this area as well. The original problem of mapping a virtual network is divided in node mapping (allocation of a physical node to a virtual node) and link mapping (allocation of a spectrum path to a virtual link). Most proposals map nodes first to then establish the virtual links connecting them [46–48, 50, 51, 53, 54, 56, 58, 62, 64].

To map the nodes and links, the heuristic must define the order in which the virtual and physical nodes/links are processed. To do so, a ranking is elaborated for each set of physical/virtual nodes/links and the first element in the ranking of virtual nodes/links is attempted to be mapped in the first element of the ranking of the physical nodes/links. The most common criterion to build the physical node ranking is the amount of available resources [48, 49, 58]. A function of the computing capacity and the nodal degree [50], a function of the number of sub-carriers of each transponder in the physical node and the slice capability of the physical node [46] and the node index [64] have also been used. Criteria to rank the virtual nodes are the amount of resources required [48, 58], the nodal degree [50] or the node index. The case where the virtual nodes must be established in specific physical nodes (defined in the virtual network establishment request), as in Ref. [47], is a particular case of a node/link mapping, as all virtual nodes are established in the specified physical nodes (if enough resources are available) before establishing the virtual links.

Physical links can be ranked in terms of their distance [48, 50, 58], cost [64] or number of available FSUs. Finally, virtual links are ranked in terms of their FSU requirements [47, 53, 58, 64].

Given that the solution found by solving the node/link mapping sub-problems sequentially is expected to be of lower quality than solving the original problem, an attempt to solve both problems jointly was proposed in Refs. [53, 55, 57, 59]. In these works, a sub-set of all possible mapping patterns for the nodes of a virtual network are evaluated and the one using the lowest slot layer (slot layer of a mapping pattern is the highest FSU used) [57], lowest cost [53] or best Hamming-inspired distance [55] is selected.

Finally, the approach of alternating the allocation of virtual nodes and links (mixed) has also been studied in Refs. [48, 49, 52, 61, 62, 64]. For example, in Ref. [61], the virtual nodes at the ends of each virtual link are mapped to then map the virtual link, showing results close to the ILP approach in a static scenario.

Apart from the FSU continuity and contiguity constraints, the solution methods can use one of several models to characterize additional constraints of the physical substrate. To date, the following models have been used:

a. Ideal, where no signal degradation is assumed [50, 54].

b. Optical-reach-based, this is the simplest model where the maximum distance covered by a spectrum path is determined solely by the modulation format and the bit rate, as in Refs. [48, 49, 51, 53, 57, 61, 62].

c. Guard-band-based, where a given number of FSUs might be left unused between channels of different bit rate, as in Refs. [46, 56, 58, 59].

d. Optical reach and guard band, where the optical reach is determined by the modulation format and the bit rate. Since the optical reach can decrease due to effect of neighbouring signals, by adding (selectively or not) guard bands between channels [47, 52, 55, 64] such detrimental effect can be mitigated.

4.2.3. Traffic management

In the context of packet networks, the split of traffic of a virtual link into several paths in the physical substrate has been proposed as a way of increasing the probability of accepting a virtual network establishment request [79]. In a flexible-grid optical network where a virtual link requiring M contiguous FSUs must be established but no path has more than $x < M$ contiguous FSUs, such situation could be solved by establishing the virtual link along several spectrum paths in such a way that the total number of FSUs used along all the paths equal M. Such mechanism could be enabled by recently introduced sliceable or multi-flow transponders [80, 81]. This approach has not been explored in the area of network virtualization over flexible-grid networks.

4.3. Type of service

4.3.1. Service nature

The service provider can offer a static or dynamic service. In the former case, the virtual network demands are known *a priori* and they are established permanently, whether they are used to transmit information or not [58, 59, 61, 62, 64]. In the latter case, virtual networks are established and released on demand.

In a dynamic service, spectrum experiences fragmentation. As a result, even when there is an enough number of FSUs to accommodate a new virtual network, these FSUs might not meet the contiguity constraint, leading to the rejection of requests. To decrease spectrum fragmentation, some dynamic systems reconfigure the established connections. Several works have evaluated the impact of reconfiguration on point-to-point connections on flexible-grid optical networks [68–72]. As expected, reconfiguration decreases blocking [54] at the expense of higher complexity of the control plane.

There are two types of reconfiguration techniques: proactive or reactive [82]. The former re-allocate resources before a blocking condition occurs, either in a synchronous or asynchronous way. In Refs. [69–71], pro-active reconfiguration algorithms are presented for point-to-point connections over flexible-grid optical networks. Reconfiguration may take place every time a given number of virtual networks request has been received. No proactive systems have been reported in network virtualization over flexible-grid networks. Reactive reconfiguration techniques re-allocate resources only when a new request cannot be accepted. In Ref. [54], a reactive reconfiguration method to re-allocate virtual networks over fixed-grid networks is presented, getting lower rejection rates than not reconfiguring at low-medium loads.

Reconfiguration can be applied at two different levels: re-allocation of complete virtual networks or re-allocation of a sub-set of virtual links/nodes, as in Ref. [54] in flexible-grid networks or [83] in fixed-grid networks. None of these cases has been studied in network virtualization systems over flexible-grid optical networks.

4.3.2. Fault tolerance

A network virtualization service can offer different levels of fault tolerance: zero, specific or guaranteed. Most works reported to date have studied systems without fault tolerance at all [46–48, 50–52, 54–58, 64]. In that case, the occurrence of any type of failure interrupts the operation of the virtual networks operating over the physical component affected by the failure. A specific survivability system is capable of continuing operation in spite of the occurrence of specific types of failures. Normally, these systems are designed to survive the most common failure events (e.g. a cable cut) and remain unprepared for unlikely events (as a node failure). In the area of network virtualization over flexible-grid networks, the algorithm proposed in Refs. [53, 59, 61, 62] can survive only to single link failures, whereas Ref. [49] can survive single link or node failure. Finally, a guaranteed survivability system ensures that limits on downtime are not exceed, no matter what the type of failure, as done in Refs. [84, 85] in a context different from network virtualization. If such condition is violated, the service provider is enforced to pay an economic compensation to the user. Such approach has not been explored in the area of network virtualization over flexible-grid networks.

Fault tolerance mechanisms can also be classified as proactive (protected systems) or reactive (restored systems). Protected systems allocate backup resources when the primary resources for the virtual network are allocated [49, 53, 59, 61, 62]. Therefore, upon failure occurrence, the time to recover from failure is shorter than reactive systems. Protected systems can allocate a complete backup virtual network (total protection) [49] or backup to some components (partial protection, e.g. only virtual links have backup resources) [53, 59, 61, 62]. Protected systems can also be classified as dedicated or shared. In the former, backup resources are dedicated to the corresponding primary resource. In the latter, a backup resource is shared among several primary resources. No research has been reported on the area of shared protection for virtual networks over flexible-grid networks. Restored systems allocate resources to the virtual networks affected by a failure only once the failure has occurred; as a result, the recovery time is longer, but a lower amount of backup resources are required. Restoration can be carried out for complete virtual networks or only for the part of them affected by the failure. Restoration on virtual networks over flexible-grid networks has not been researched yet.

4.3.3. Revisitation

Revisitation allows the establishment of two virtual nodes from the same virtual network in the same physical node [16]. Revisitation has been proposed in the context of overlay networks [86] as a way of emulating larger networks on small testbeds. In virtual network systems over flexible-grid networks, revisitation has been used in Ref. [64] and the impact of it on blocking was studied in Ref. [52] showing a decrease of blocking ratio of two orders of magnitude with respect to the same algorithm without revisitation.

Revisitation has been little researched in the literature, probably because a real application for it has not been found yet. For example, for research on new Internet protocols, delay and bandwidth utilization are two key metrics that could not be measured if two virtual nodes are hosted in the same physical node. For cloud replication services would not be useful either, as the replicas must be allocated to geographically different sites. However, it is mentioned as one of the four key architectural principles of network virtualization in Ref. [16], where it would be useful to help the service providers to manage highly complex tasks and facilitate virtual networks management.

In **Table 2**, a summary of the virtual network resource allocation proposed to date is presented. For each algorithm, all the dimensions presented in the taxonomy of **Figure 5** are specified.

	Performance metric	Operation			Type of service		
		Information management	Resource allocation strategy	Traffic management	Service nature	Failure tolerance	Revisitation
[46]	Blocking ratio cost/revenue	Centralized	Nodes-links Guard band	Without split	Dynamic without reconfiguration	Zero	Without
[47]	Blocking ratio	Centralized	Nodes-links Optical reach + Guard band	Without split	Dynamic without reconfiguration	Zero	Without
[48]	Blocking ratio	Centralized	Nodes-links Mixed Optical reach	Without split	Dynamic without reconfiguration	Zero	Without
[49]	Blocking ratio	Centralized	Nodes-links Optical reach	Without split	Dynamic without reconfiguration	Specific, total protection	Without
[50]	Blocking ratio	Centralized	Nodes-links Ideal	Without split	Dynamic without reconfiguration	Zero	Without
[51]	Blocking ratio	Centralized	Exact Nodes-links Optical reach	Without split	Dynamic without reconfiguration	Zero	Without
[52]	Blocking ratio	Centralized	Mixed Optical reach + Selective guard band	Without split	Dynamic without reconfiguration	Zero	Total
[53]	Blocking ratio	Centralized	Nodes-links Joint Optical reach	Without split	Dynamic without reconfiguration	Specific, partial protection	Without
[54]	Blocking ratio traffic carried	Centralized	Nodes-links Ideal	Without split	Dynamic with reconfiguration in virtual links	Zero	Without

	Performance metric	Operation			Type of service		
		Information management	Resource allocation strategy	Traffic management	Service nature	Failure tolerance	Revisitation
[55]	Blocking ratio	Centralized	Mixed Optical reach + Selective guard band	Without split	Dynamic without reconfiguration	Zero	Without
[56]	Blocking ratio cost/revenue	Centralized	Nodes-links Guard band	Without split	Dynamic without reconfiguration	Zero	Without
[57]	Traffic carried	Centralized	Joint Optical Reach	Without split	Dynamic without reconfiguration	Zero	Without
[58]	Traffic carried	Centralized	Exact Nodes-links Guard band	Without split	static	Zero	Without
[59]	Traffic carried	Centralized	Exact Joint Guard band	Without split	Dynamic without reconfiguration	Specific	Without
[61]	Cost/revenue	Centralized	Exact Mixed Optical reach	Without split	Static	Specific, partial protection	Without
[62]	Cost/revenue	Centralized	Nodes-links Mixed Optical reach	Without split	Dynamic without reconfiguration	Specific	Without
[64]	Blocking ratio	Centralized	Exact Nodes-links Mixed Optical reach +Guard band	Without split	Static Dynamic without reconfiguration	Zero	Total

Table 2. Summary of the characteristics of the algorithms reviewed.

5. Conclusions

Network virtualization has emerged as an enabling technology for cloud computing services. Such services would push even further the limits on bandwidth utilization, where flexible-grid optical networks will be the key to increase the network capacity of actually deployed optical networks. As a result, a new area of research focused on network virtualization over flexible-grid networks has emerged.

On such area, the research efforts focus on three main lines: design of flexible and virtualized devices, definition of network architectures and virtual network allocation algorithms.

In this chapter, a survey on the virtual network allocation algorithms over flexible-grid networks has been presented along with a classification of all possible proposals of algorithms by means of taxonomy. Such classification allowed the identification of several aspects that must be further investigated in the area:

- Multi-objective optimization approaches that allow to select resource allocation algorithms with a good compromise between quality and cost.
- The design and evaluation of distributed virtual network allocation algorithms.
- The application of meta-heuristics (as genetic algorithms, ant colony, etc.) to solve the virtual network allocation problem over flexible-grid networks.
- The study of the impact of traffic split on the performance of virtual network allocation algorithms.
- The effect and complexity of reconfiguration on the performance of network virtualization systems.
- The design and evaluation of shared protection mechanisms.
- The design and evaluation of shared protection and restored fault tolerance mechanisms.

Acknowledgements

Financial support from CONICYT through the Fondecyt Project 1130388 and 11140386 and the PhD Scholarship,CONICYT-PCHA/Doctorado Nacional/2015-21150525 is gratefully acknowledged.

Author details

Danilo Bórquez-Paredes[1*], Alejandra Beghelli[1] and Ariel Leiva[2]

*Address all correspondence to: danilo.borquez@uai.cl

1 Science and Engineering Faculty, Universidad Adolfo Ibañez, Chile

2 Electrical Engineering School, Pontificia Universidad Católica de Valparaíso, Chile

References

[1] Develder C, De Leenheer M, Dhoedt B, Pickavet M, Colle D, De Turck F, Demeester P. Optical networks for grid and cloud computing applications. Proceedings of the IEEE. 2012;**100**(5):1149–1167. DOI: 10.1109/JPROC.2011.2179629

[2] Gharbaoui M, Martini B, Castoldi P. Anycast-based optimizations for inter-data-center interconnections [Invited]. IEEE/OSA Journal of Optical Communications and Networking. 2012;**4**(11):B168–B178. DOI: 10.1364/JOCN.4.00B168

[3] GIGAOM. Why the Cloud Needs New Forms of Networking [Internet]. 2011. Available from: https://gigaom.com/2011/06/23/structure-network-virtualization-openflow/[Accessed: 28 December 2016]

[4] Alhazmi K, Abu Sharkh M, Ban D, Shami A. A map of the clouds: Virtual network mapping in cloud computing data centers. In: IEEE 27th Canadian Conference on Electrical and Computer Engineering; 4-7 May 2014; Toronto, Canada. IEEE; 2014. pp. 1–6. DOI: 10.1109/CCECE.2014.6901053

[5] Xiao L, Jiang W, Chen F, Guo L, Hu Y. A survey of cloud computing data virtualization service. Applied Mechanics and Materials. 2013;**441**:1016–1019. DOI: 10.4028/www.scientific.net/AMM.441.1016

[6] Sun G., Yu H., Anand V., Li L, Di H. Optimal provisioning for virtual network request in cloud-based data centers. Photonic Network Communications. 2012;**24**(2):118—131. DOI: 10.1007/s11107-012-0372-0

[7] Baroncelli F., Martini B., Castoldi P. Network virtualization for cloud computing. Annals of telecommunications—annales des télécommunications. 2010;**65**(11-12):713–721. DOI: 10.1007/s12243-010-0194-y

[8] Hao F, Lakshman T, Mukherjee S, Song H. Enhancing dynamic cloud-based services using network virtualization. SIGCOMM Computer Communication Review 2010;**40**(1):67–74. DOI: 10.1145/1672308.1672322

[9] Dang Tran F. Network Virtualization as an Enabler for Cloud Computing: A Telco Perspective [Internet]. 2009. Available from: http://www.itcspecialistseminar.com/presentations/itcss09_DangTran_pres.pdf [Accessed: 24 February 2017]

[10] I. Morita. Network control and virtualization for cloud and mobile services from carrier's point of view. In: Optical Fiber Communication Conference; 22–26 Mar 2015; Los Angeles, California, USA. OSA; 2015. p. W1G.3. DOI: 10.1364/OFC.2015.W1G.3

[11] Harter I, Schupke D, Hoffmann M, Carle G. Network virtualization for disaster resilience of cloud services. IEEE Communications Magazine. 2014;**52**(12):88–95. DOI: 10.1109/MCOM.2014.6979957

[12] Alshaer H. An overview of network virtualization and cloud network as a service. International Journal of Network Management. 2015;**25**(1):1–30. DOI: 10.1002/nem.1882

[13] Tomkos I, Palkopoulou E, Angelou M. A survey of recent developments on flexible/elastic optical networking. In: 14th International Conference on Transparent Optical Networks; 2-5 July 2012; England: Coventry; 2012. pp. 1–6. DOI: 10.1109/ICTON.2012.6254409

[14] Jinno M, Takara H, Kozicki B, Tsukishima Y, Yoshimatsu T, Kobayashi T, Miyamoto Y., Yonenaga K, Takada A, Ishida O, Matsuoka S. Demonstration of novel spectrum-efficient elastic optical path network with per&whyphen;channel variable capacity of 40 Gb/s to over 400 Gb/s. In: 34th European Conference on Optical Communication; 21-25 September 2008; Brussels, Belgium. IEEE; 2008. pp. 1–2. DOI: 10.1109/ECOC.2008.4729581

[15] Mayoral A, Gonzalez de Dios O, Lopez V, Fernandez-Palacios J. Migration steps towards flexi-grid networks. In: Future Network and Mobile Summit; 3-5 July 2013; Lisboa, Portugal. IEEE; 2013. pp. 1–9. DOI: 10.1364/JOCN.6.000988

[16] Chowdhury N, Boutaba R. A survey of network virtualization. Computer Networks. 2010;**54**(5):862–876. DOI: 10.1016/j.comnet.2009.10.017

[17] Chowdhury N, Boutaba R. Network virtualization: State of the art and research challenges. IEEE Communications Magazine. 2009;**47**(7):20–26. DOI: 10.1109/MCOM.2009.5183468

[18] Houidi I, Louati W, Zeghlache D, Papadimitriou P, Mathy L. Adaptive virtual network provisioning. In: 2nd ACM SIGCOMM Workshop on Virtualized Infrastructure Systems and Architectures; 30 August-03 September 2010; New Delhi, India. ACM; 2010. pp. 41–48. DOI: 10.1145/1851399.1851407

[19] Sherwood R, Gibb G, Yap K, Appenzeller G, Casado M, McKeown N, Parulkar G. Can the production network be the testbed? In: 9th USENIX Conference on Operating Systems Design and Implementation; 4-6 October 2010; Vancouver, Canada. USENIX Association; 2010. pp. 1–6.

[20] Bavier A, Bowman M, Chun B, Culler D, Karlin S, Muir S, Peterson L, Roscoe T, Spalink T, Wawrzoniak M. Operating system support for planetary-scale network services. In: 1st Conference on Symposium on Networked Systems Design and Implementation; 29-31 March 2004; San Francisco, California, USA. pp. 19–19. DOI: 10.1145/1113361.1113362

[21] PlanetLab. An Open Platform for Developing, Deploying and Accessing Planetary-Scale Services [Internet]. 2007. Available from: http://www.planet-lab.org/ [Accessed: 28 December 2016]

[22] Anderson T, Peterson L, Shenker S, Turner J. Overcoming the Internet impasse through virtualization. The Computer Journal. 2005;**38**(4):34–41. DOI: 10.1109/MC.2005.136

[23] GENI. Global Environment for Network Innovations [Internet]. 2006. Available from: http://www.geni.net [Accessed: 28 December 2016]

[24] FEDERICA. Federated E-Infrastructure Dedicated to European Researchers Innovating in Computing Network Architechtures [Internet]. 2008. Available from: http://www.fp7-federica.eu/ [Accessed: 28 December 2016]

[25] OneLab2. An Open Federated Laboratory Supporting Network Research for the Future Internet [Closed project] [Internet]. 2008. Available from: http://cordis.europa.eu/project/rcn/87273_en.html [Accessed: 28 December 2016]

[26] NICT. "AKARI" Architecture Design Project for New Generation Network [Internet]. 2006. Available from: https://web.archive.org/web/20130210083738/http://akari-project.nict.go.jp/eng/index2.htm [Accessed: 28 December 2016]

[27] Future Internet Brazilian Environment for Experimentation [Internet]. 2011. Available from: http://fibre.org.br/ [Accessed: 28 December 2016]

[28] CISCO. Easy Virtual Network—Simplifying Layer 3 Network Virtualization [Internet]. 2013. Available from: http://www.cisco.com/c/en/us/products/collateral/ios-nx-os-software/layer-3-vpns-l3vpn/whitepaper_c11-638769.html [Accessed: 28 December 2016]

[29] VMWARE. NSX for vSphere Network Virtualization Design Guide ver 3.0 [Internet]. 2014 [Updated: 2015]. Available from: https://communities.vmware.com/docs/DOC-27683 [Accessed: 28 December 2016]

[30] ITU-T. Spectral grids for WDM applications: DWDM frequency grid [Internet]. 2002 [Updated: 2012]. Available from: https://www.itu.int/rec/T-REC-G.694.1/en [Accessed: 24 February 2017]

[31] Sambo N, Secondini M, Cugini F, Bottari G, Iovanna P, Cavaliere F, Castoldi P. Modeling and distributed provisioning in 10-40-100-Gb/s multirate wavelength switched optical networks. Journal of Lightwave Technology. 2011;**29**(9):1248–1257. DOI: 10.1109/JLT.2011.2122245

[32] Chowdhury P, Tornatore M, Nag A, Ip E, Wang T, Mukherjee B. On the design of energy-efficient mixed-line-rate (MLR) optical networks. Journal of Lightwave Technology. 2012;**30**(1):130–139. DOI: 10.1109/JLT.2011.2177441

[33] Peng S, Nejabati R, Simeonidou D. Impairment-aware optical network virtualization in single-line-rate and mixed-line-rate WDM networks. IEEE/OSA Journal of Optical Communications and Networking. 2013;**5**(4):283–293. DOI: 10.1364/JOCN.5.000283

[34] Gerstel, M. Jinno, A. Lord, S. Yoo. Elastic optical networking: A new dawn for the optical layer? IEEE Communications Magazine. 2012;**50**(2):s12–s20. DOI: 10.1109/MCOM.2012.6146481

[35] Amaya N, Irfan M, Zervas G, Banias K, Garrich M, Henning I, Simeonidou D, Zhou YR, Lord A, Smith K, Rancano VJF, Liu S, Petropoulos P, Richardson DJ. Gridless optical networking field trial: Flexible spectrum switching, defragmentation and transport of 10G/40G/100G/555G over 620-km field fiber. In: 37th European Conference and Exhibition on Optical Communication; 18-22 September 2011; Geneva, Switzerland. pp. 1–3. DOI: 10.1364/OE.19.00B277

[36] Jinno M, Takara H, Kozicki B, Tsukishima Y, Sone Y, Matsuoka S. Spectrum-efficient and scalable elastic optical path network: architecture, benefits, and enabling technologies. IEEE Communications Magazine. 2009;**47**(11):66–73. DOI: 10.1109/MCOM.2009.5307468

[37] Christodoulopoulos K, Tomkos I, Varvarigos E. Dynamic bandwidth allocation in flexible OFDM-based networks. In: Optical Fiber Communication Conference; 6-10 March 2011; Los Angeles, California, USA. OSA; 2011. p. OTuI5. DOI: 10.1364/OFC.2011.OTuI5

[38] Patel A, Ji P, Wang T, Jue J. Optical-layer traffic grooming in flexible grid WDM networks. In: IEEE Global Communications Conference; 5-9 December 2011; Houston, Texas, USA. IEEE; 2011. pp. 1–6. DOI: 10.1109/GLOCOM.2011.6133618

[39] Sambo N, Meloni G, Paolucci F., Cugini F, Secondini M, Fresi F, Poti L, Castoldi P. Programmable transponder, code and differentiated filter configuration in elastic optical networks. Journal of Lightwave Technology. 2014;**32**(11):2079–2086. DOI: 10.1109/JLT.2014.2319859

[40] Cerutti I, Martinelli F, Sambo N, Cugini F, Castoldi P. Trading regeneration and spectrum utilization in code-rate adaptive flexi-grid networks. Journal of Lightwave Technology. 2014;**32**(23):4496–4503. DOI: 10.1109/JLT.2014.2359179

[41] Jinno M, Kozicki B, Takara H, Watanabe A, Sone Y, Tanaka T, Hirano A. Distance-adaptive spectrum resource allocation in spectrum-sliced elastic optical path network [Topics in Optical Communications]. IEEE Communications Magazine. 2010;**48**(8):138–145. DOI: 10.1109/MCOM.2010.5534599

[42] Lach E, Idler W. Modulation formats for 100G and beyond. Optical Fiber Technology. 2011;**17**(5):377–386. DOI: 10.1016/j.yofte.2011.07.012

[43] Eira A, Santos J, Pedro J, Pires J. Design of survivable flexible-grid DWDM networks with joint minimization of transponder cost and spectrum usage. In: 38th European Conference on Optical Communications; 16-20 September 2012; Amsterdam, Netherlands. IEEE; 2012. pp. 1–3. DOI: 10.1364/ECEOC.2012.P5.16

[44] Gringeri S, Basch E, Xia T. Technical considerations for supporting data rates beyond 100 Gb/s. IEEE Communications Magazine. 2012;**50**(2):s21–s30. DOI: 10.1109/MCOM.2012.6146482

[45] Rosa A, Cavdar C, Carvalho S, Costa J, Wosinska L. Spectrum allocation policy modeling for elastic optical networks. In: 9th International Conference on High Capacity Optical Networks and Enabling Technologies; 12-14 December 2012; Istanbul, Turkey. IEEE; 2012. pp. 242–246. DOI: 10.1109/HONET.2012.6421472

[46] J. Zhang, B. Mukherjee, J. Zhang, Y. Zhao. Dynamic virtual network embedding scheme based on network element slicing for elastic optical networks. In: 39th European Conference on Optical Communication; 22-26 September 2013; IET; 2013. London, England. pp. 1–3. DOI: 10.1049/cp.2013.1323

[47] Patel A, Ji P, Huang Y, Wang T. Distance-adaptive virtual network embedding in software-defined optical networks. In: OptoElectronics and Communications Conference held Jointly with the 18th International Conference on Photonics in Switching; 30 June-4 July 2013; Kyoto, Japan. IEEE; 2013. pp. 1–2.

[48] Patel A, Ye Z, Ji P. Cloud service embedding in software-defined flexible grid optical transport networks. In: Optical Fiber Communication Conference; 9-14 March 2014; San Francisco, California, USA. IEEE; 2014.p. M2H.2. DOI: 10.1364/OFC.2014.M2H.2

[49] Ye Z, Patel A, Ji P, Qiao C. Survivable virtual infrastructure mapping over transport software-defined networks (T-SDN). In: Optical Fiber Communication Conference; 9-14 March 2014; San Francisco, California, USA. IEEE; 2014. p. M3H.1. DOI: 10.1364/OFC.2014.M3H.1

[50] Gong L, Zhao W, Wen Y, Zhu Z. Dynamic transparent virtual network embedding over elastic optical infrastructures. In: IEEE International Conference on Communications; 9-13 June 2013; Budapest, Hungary. IEEE; 2013. pp. 3466–3470. DOI: 10.1109/ICC.2013.6655086

[51] Gong L, Zhu Z. Virtual optical network embedding (VONE) over elastic optical networks. Journal of Lightwave Technology. 2014;**32**(3):450–460. DOI: 10.1109/JLT.2013.2294389

[52] Barra E, Salinas R, Mora F, Tarifeno M, Beghelli A, Sambo N, Castoldi P. Virtual network provisioning over multi-line rate networks with fixed or flexible grid. In: 16th International Conference on Transparent Optical Networks; 6-10 July 2014; Graz, Austria. IEEE; 2014. pp. 1–4. DOI: 10.1109/ICTON.2014.6876709

[53] Khandaker F, Xie W, Jue J, Wang X, Zhang Q., She Q, Cankaya H, Palacharla P, Sekiya M. Survivable Virtual Optical Network mapping in spectrum and modulation format convertible flexible grid optical networks. In: Optical Fiber Communications Conference; 22-26 March 2015; Los Angeles, California, USA. pp. Th3J.1. DOI: 10.1364/OFC.2015.Th3J.1

[54] Shakya S, Pradhan N, Cao X, Ye Z, Qiao C. Virtual network embedding and reconfiguration in elastic optical networks. In: IEEE Global Communications Conference; 8-12 December 2014; Austin, Texas, USA. IEEE; 2014. pp. 2160–2165. DOI: 10.1109/GLOCOM.2014.7037128

[55] Gonzalez A, Barra E, Beghelli A, Leiva A. A sub-graph mapping-based algorithm for virtual network allocation over flexible grid networks. In: 17th International Conference on Transparent Optical Networks; 5-9 July 2015; Budapest, Hungary. IEEE; 2015. pp. 1–4. DOI: 10.1109/ICTON.2015.7193484

[56] Zhu R, Zhao Y, Zhang J, Yang H, Tan Y, Zhang J, Jue J, Wang N. Multi-dimensional resource virtualization in spectral and spatial domains for inter-datacenter optical networks. In: Optical Fiber Communication Conference; 20-22 March 2016; Anaheim, California, USA. IEEE; 2016. p. W2A.53. DOI: 10.1364/OFC.2016.W2A.53

[57] Wang X, Zhang Q, Kim I, Palacharla P, Sekiya M. Flexible virtual network provisioning over distance-adaptive flex-grid optical networks. In: Optical Fiber Communication Conference; 9-14 Mar 2014; San Francisco, California, USA. OSA; 2014 p. W3H.3. DOI: 10.1364/OFC.2014.W3H.3

[58] Zhang S, Shi L, Vadrevu C, Mukherjee B. Network virtualization over WDM and flexible-grid optical networks. Optical Switching and Networking. 2013;**10**(4):291–300. DOI: 10.1016/j.osn.2013.03.005

[59] Assis K, Peng S, Almeida R, Waldman H, Hammad A, Santos A, Simeonidou D. Network virtualization over elastic optical networks with different protection schemes. IEEE/OSA Journal of Optical Communications and Networking. 2016;**8**(4):272–281. DOI: 10.1364/JOCN.8.000272

[60] Zhou X, Nelson L, Magill P. Rate-adaptable optics for next generation long-haul transport networks. IEEE Communications Magazine. 2013;**51**(3):41–49. DOI: 10.1109/MCOM.2013.6476864

[61] Xie W, Jue J, Zhang Q, Wang X, She Q, Palacharla P, Sekiya M. Survivable virtual optical network mapping in flexible-grid optical networks. In: International Conference on Computing, Networking and Communications; 3-6 February 2014; Honolulu, Hawai. IEEE; 2014. pp. 221–225. DOI: 10.1109/ICCNC.2014.6785335

[62] Chen B. Power-aware virtual optical network provisioning in flexible bandwidth optical networks [Invited]. Photonic Network Communications. 2016;**32**(2):300–309. DOI: 10.1007/s11107-016-0609-4

[63] Wang W, Lin Y, Zhao Y, Zhang G, Zhang J, Han J, Chen H, Hou B, Ji Y, Zong L. First demonstration of virtual transport network services with multi-layer protection schemes over flexi-grid optical networks. IEEE Communications Letters. 2016;**20**(2):260–263. DOI: 10.1109/LCOMM.2015.2509066

[64] Zhao J, Subramaniam S, Brandt-Pearce M. Virtual topology mapping in elastic optical networks. In: IEEE International Conference on Communications; 9-13 June 2013; Budapest, Hungary.IEEE; 2013. pp. 3904–3908. DOI: 10.1109/ICC.2013.6655167

[65] Morea A., Chong A., Rival O. Impact of transparent network constraints on capacity gain of elastic channel spacing. In: Optical Fiber Communication Conference; 6-10 March 2011; Los Angeles, California, USA. IEEE; 2011. pp. 1–3. DOI: 10.1364/NFOEC.2011.JWA062

[66] Sambo N, Castoldi P, Cugini F, Bottari G, Iovanna P. Toward high-rate and flexible optical networks. IEEE Communications Magazine. 2012;**50**(5):66–72. DOI: 10.1109/MCOM.2012.6194384

[67] Chowdhury N, Rahman M, Boutaba R. Virtual network embedding with coordinated node and link mapping. In: IEEE International Conference on Computer Communications; 19-25 April 2009; Rio de Janeiro, Brazil. IEEE; 2009. pp. 783–791. DOI: 10.1109/INFCOM.2009.5061987

[68] Patel A, Ji P, Jue J, Wang T. Defragmentation of transparent Flexible optical WDM (FWDM) networks. In: Optical Fiber Communication Conference; 6-10 March 2011; Los Angeles, California, USA. pp. 1–3. DOI: 10.1364/OFC.2011.OTuI8

[69] Chen C, Chen X, Zhang M, Ma S, Shao Y., Li S, Suleiman M, Zhu Z. Demonstrations of efficient online spectrum defragmentation in software-defined elastic optical networks. Journal of Lightwave Technology. 2014;**32**(24):4099–4109. DOI: 10.1109/JLT.2014.2364515

[70] Khodashenas P, Comellas J, Perelló J, Spadaro S. Correlation between traffic granularity and defragmentation periodicity in elastic optical networks. Transactions on Emerging Telecommunications Technologies. 2014;**26**(7):1011–1018. DOI: 10.1002/ett.2795

[71] Zhang M, Shi W, Gong L, Lu W, Zhu Z. Bandwidth defragmentation in dynamic elastic optical networks with minimum traffic disruptions. In: IEEE International Conference on Communications; 9-13 June 2013; Budapest, Hungary. IEEE; 2013. pp. 3894–3898. DOI: 10.1109/ICC.2013.6655165

[72] Cugini F, Paolucci F, Meloni G, Berrettini G, Secondini M, Fresi F, Sambo N, Poti L, Castoldi P. Push-pull defragmentation without traffic disruption in flexible grid optical networks. Journal of Lightwave Technology. 2013;**31**(1):125–133. DOI: 10.1109/JLT.2012.2225600

[73] Borquez D, Beghelli A, Leiva A. Deadlock-Avoiding vs. greedy spectrum allocation algorithms in dynamic flexible optical networks. In: 17th International Conference on

Transparent Optical Networks; 5-9 July 2015; Budapest, Hungary. pp. 1–4. DOI: 10.1109/ICTON.2015.7193486

[74] Donadio P, Fioccola G, Canonico R, Ventre G. A PCE-based architecture for the management of virtualized infrastructures. In: IEEE 3rd International Conference on Cloud Networking; 8-10 October 2014; Luxembourg. IEEE; 2014. pp. 223–228. DOI: 10.1109/CloudNet.2014.6968996

[75] Belbekkouche A, Hasan M, Karmouch A. Resource Discovery and Allocation in Network Virtualization. IEEE Communications Surveys Tutorials. 2012;**14**(4):1114–1128. DOI: 10.1109/SURV.2011.122811.00060

[76] Marquezan C, Nobre J, Granville L, Nunzi G, Dudkowski D, Brunner M. Distributed reallocation scheme for virtual network resources. In: IEEE International Conference on Communications; 14-18 June 2009; Dresden, Germany. IEEE; 2009. pp. 1–5. DOI: 10.1109/ICC.2009.5198934

[77] Sun G, Anand V, Yu H, Liao D, Li L. Optimal provisioning for elastic service oriented virtual network request in cloud computing. In: IEEE Global Communications Conference; 3-7 December 2012; Anaheim, California, USA. pp. 2517–2522. DOI: 10.1109/GLOCOM.2012.6503495

[78] Fajjari I, Aitsaadi N, Pujolle G, Zimmermann H. VNE-AC: Virtual network embedding algorithm based on ant colony metaheuristic. In: IEEE International Conference on Communications; 5-9 June 2011; Kyoto, Japan. IEEE; 2011. pp. 322–340. DOI: 10.1504/IJCNDS.2011.039538

[79] Zinner T, Tran-Gia P, Tutschku K, Nakao A. Performance evaluation of packet reordering on concurrent multipath transmissions for transport virtualisation. Journal of Communication Networks and Distributed Systems. 2011;**6**(3):322–340. DOI: 10.1504/IJCNDS.2011.039538

[80] Jinno M, Takara H, Sone Y, Yonenaga K, Hirano A. Multiflow optical transponder for efficient multilayer optical networking. IEEE Communications Magazine. 2012;**50**(5):56–65. DOI: 10.1109/MCOM.2012.6194383

[81] Sambo N, D'Errico A, Porzi C, Vercesi V, Imran M, Cugini F, Bogoni A, Poti L, Castoldi P. Sliceable transponder architecture including multiwavelength source. IEEE/OSA Journal of Optical Communications and Networking. 2014;**6**(7):590–600. DOI: 10.1109/JOCN.2014.6850200

[82] Ahmed J, Solano F, Monti P, Wosinska L. Traffic re-optimization strategies for dynamically provisioned WDM networks. In: 15th International Conference on Optical Network Design and Modeling; 8-10 February 2011; Bologna, Italy. IEEE; 2011. pp. 1–6.

[83] Bai H, Gu F, Liang K, Rahnamay-Naeini M, Khan S, Hayat M, Ghani N. Virtual network reconfiguration in optical cloud substrates. In: Optical Fiber Communications Conference; 9-14 March 2014; San Francisco, California, USA. IEEE; 2014. pp. 1–6. DOI: 10.1364/OFC.2014.M3H.7

[84] Mello D, Pelegrini J, Ribeiro R, Schupke D, Waldman H. Dynamic provisioning of shared-backup path protected connections with guaranteed availability requirements. In: 2nd International Conference on Broadband Networks; 7 October 2005; Boston, Massachusetts, USA. IEEE; 2005. pp. 1320–1327. DOI: 10.1109/ICBN.2005.1589761

[85] Tarifeño M, Beghelli A, Moreno E. Availability-driven optimal design of shared path protection in WDM networks. Networks. 2016; 68(3):224–237. DOI: 10.1002/net.21695

[86] Touch J. Dynamic Internet overlay deployment and management using the X-Bone. In: International Conference on Network Protocols; 14-17 November 2000; Osaka, Japan. IEEE; 2000. pp. 59–68. DOI: 10.1109/ICNP.2000.896292

8

Multi-Period Attack-Aware Optical Network Planning under Demand Uncertainty

Konstantinos Manousakis, Panayiotis Kolios and Georgios Ellinas

Additional information is available at the end of the chapter

Abstract

In this chapter, novel attack-aware routing and wavelength assignment (Aa-RWA) algorithms for multiperiod network planning are proposed. The considered physical layer attacks addressed in this chapter are high-power jamming attacks. These attacks are modeled as interactions among lightpaths as a result of intra-channel and/or inter-channel crosstalk. The proposed Aa-RWA algorithm first solves the problem for given traffic demands, and subsequently, the algorithm is enhanced in order to deal with demands under uncertainties. The demand uncertainty is considered in order to provide a solution for several periods, where the knowledge of demands for future periods can only be estimated. The objective of the Aa-RWA algorithm is to minimize the impact of possible physical layer attacks and at the same time minimize the investment cost (in terms of switching equipment deployed) during the network planning phase.

Keywords: physical layer attacks, routing and wavelength assignment, optical networks, multi-period planning, demand uncertainty

1. Introduction

In wavelength division multiplexed (WDM) optical networks, wavelength routing is used for establishing communication between source-destination pairs. In these networks, data are transmitted over all-optical WDM channels called lightpaths. A connection is established by utilizing a lightpath, which is determined by choosing a path between the source and the destination and allocating a wavelength on all the links of the path. The selection of the path and wavelength is an important optimization problem and is known as the routing and wavelength assignment (RWA) problem [1].

In WDM optical networks, transparent optical cross-connects (OXCs) are used in order to provide efficient space and wavelength switching functions [2]. An OXC takes as input signals at multiple wavelengths and some of these wavelengths can be dropped locally, while others pass through by switching them to the appropriate output ports. For the implementation of OXCs, wavelength selective switch (WSS) technology is used for the deployment of cost-effective and dynamic wavelength-switched networks [3].

In transparent optical networks, where data signals remain in the optical domain until they reach their destinations, connections are vulnerable to physical layer attacks. An attack is defined as an intentional action against the ideal and secure functioning of the network. One type of attack in optical networks is high-power jamming which can affect the signal through in-band jamming that is the result of intra-channel crosstalk or out-of-band jamming that is the result of inter-channel crosstalk and nonlinearities [4]. This type of attack propagates through the transparent network affecting several connections, and as a consequence, the localization of this kind of attack is a difficult problem. Due to the high bit rates of optical networks and the interaction of the connections, a jamming attack can potentially cause a huge amount of information loss. Therefore, the limitation of attack propagation is a crucial consideration in optical network planning. An overview of security challenges in communication networks can be found in Ref. [5].

Physical layer attacks in optical networks have been studied by several researchers [6–10]. In these works, the concept of attack-aware routing and wavelength assignment (Aa-RWA) is analyzed. Specifically, in Ref. [6], authors proposed an integer linear program (ILP) formulation and a tabu search heuristic algorithm for the routing sub-problem in optical networks in order to minimize the effect of out-of-band jamming and the gain competition caused in optical fibers and optical amplifiers, respectively. In Ref. [7], authors proposed ILP formulation and heuristic algorithms for the wavelength assignment sub-problem in optical networks in order to minimize the in-band jamming attack caused in optical nodes. In Ref. [8], authors proposed ILP and heuristic algorithms based on simulated annealing techniques in order to minimize the in-band and out-of-band jamming attacks. Moreover, in Ref. [9, 10], authors proposed a greedy randomized adaptive search procedure (GRASP) heuristic and an ILP formulation, respectively, for the placement of power equalizers in order to limit the jamming attack propagation in transparent optical networks.

Another important aspect in network planning that usually is not taken into account is the uncertainty of the connection requests. In most cases, the demands are considered to be known before network planning; however, in some cases, network planning must be performed for a period of time where the demand requests can only be forecasted with uncertainty. One approach to deal with demand uncertainty is by overprovisioning, essentially allocating many resources that can satisfy any traffic demand. However, this approach requires a high cost investment (capital expenditure—capex) from the network operators [11]. More sophisticated approaches to deal with demand uncertainty are necessary in order to achieve a cost-effective network investment strategy [12].

Stochastic programming (SP) [13] and robust optimization (RO) [14] are the main alternative techniques to deal with uncertain data both in a single period and in a multi-period decision making process. In SP, the probability distribution functions of the underlying stochastic parameters must be known. On the other hand, RO addresses the uncertain nature of the problem without making specific assumptions on probability distributions. The uncertain parameters are assumed to belong to a deterministic uncertainty set. RO adopts an approach that addresses uncertainty by guaranteeing the feasibility and optimality of the solution against all instances of the parameters within the uncertainty set.

In Ref. [15], authors apply robust optimization in order to incorporate the uncertainty of demands into the network upgrade problem. Under the robust network upgrade model, the network planning can be performed by tuning the trade-off between network cost and robustness level. Further, in Ref. [16], authors propose multi-period network planning approaches based on SP, where the demands are forecasted over periods of time and the network investments are performed based on these forecasts.

In this chapter, novel Aa-RWA algorithms are proposed to address the problem of multi-period network planning under demand uncertainty with the objective to minimize the impact of possible physical layer attacks and at the same time to minimize the network infrastructure investment cost. Physical layer attacks are modeled as interactions among connections through in-band and out-of-band channel crosstalk. Moreover, the investment cost is taken into account in this formulation via the number of WSSs required in order to minimize the impact of a possible physical layer attack.

The simulation results show that when the distribution of demands for all the time periods is taken into account in advance, better results can be obtained in terms of the number of WSSs required to be placed in the network nodes so as to minimize the impact of a jamming attack, compared to the case where the distribution is known only for the period under consideration.

The chapter is organized as follows. Section 2 describes the network architecture, while Section 3 describes the planning approaches for demand uncertainty. In Section 4, the physical layer attacks in optical networks are presented, and in Section 5, the problem of attack-aware RWA with given traffic demands is solved. This is followed in Section 6 by the attack-aware RWA under demand uncertainties. Performance results are presented in Section 7, while Section 8 presents some concluding remarks.

2. Network and node architecture

An optical network topology is represented by a connected graph $G = (V, E)$, where V denotes the set of optical cross-connects (nodes) and E denotes the set of (point-to-point) single-fiber links (edges). Each fiber link is able to support a common set $C= \{1,2,\ldots,W\}$ of, W, distinct wavelengths. Source-destination pairs are equipped with transmitter-receiver pairs, also known as transponders (TSP), in order to transmit/receive data. Optical nodes currently deployed in optical networks are based on two architectures. The first architecture utilizes a broadcast-and-select (BS) configuration and the second a route-and-select (RS) configuration. Both of these optical node architectures consist of two stages and can remotely configure all transit traffic and only differ in the implementation of their first stage. The building components of these node architectures are the WSSs. A WSS can steer each optical channel present on its input port toward one of its output ports according to the desired routing choice.

BS-based nodes (**Figure 1**) include a splitter first stage (1 × N) that implicitly provides a broadcast capability toward all outputs. In a BS-based architecture, the WSS functionality (second stage) resembles a multiplexer (it switches each individual wavelength to a certain output). Although this is a simple and popular architecture, the loss introduced by the power splitters limits its scalability and can only be utilized in network nodes with small degrees.

RS architecture nodes (**Figure 2**) on the other hand have a WSS first stage (1 × N) that provides on-demand routing to the required output. The basic advantage of the RS-based architecture with respect to the BS-based architecture is that the through loss is not dependent on the degree of the node. However, it requires additional WSSs at the input stage, which makes it more costly to be implemented.

Both implementations have a WSS second stage (N × 1) that provides the selection of the wavelengths at the output fibers, allowing full switching flexibility (any wavelength from any incoming fiber can pass through or any wavelength from the add/drop terminals can be added/dropped).

In order to deal with the losses introduced by the power splitters of the BS-based architecture and the high cost of the RS-based architecture, a hybrid architecture can also be used (**Figure 3**). This architecture contains either splitters (1 × N) or WSSs (1 × N) at the input ports as can be seen in **Figure 3**. In essence, hybrid nodes are constructed by replacing splitters with WSSs at the input stage of the BS-based nodes.

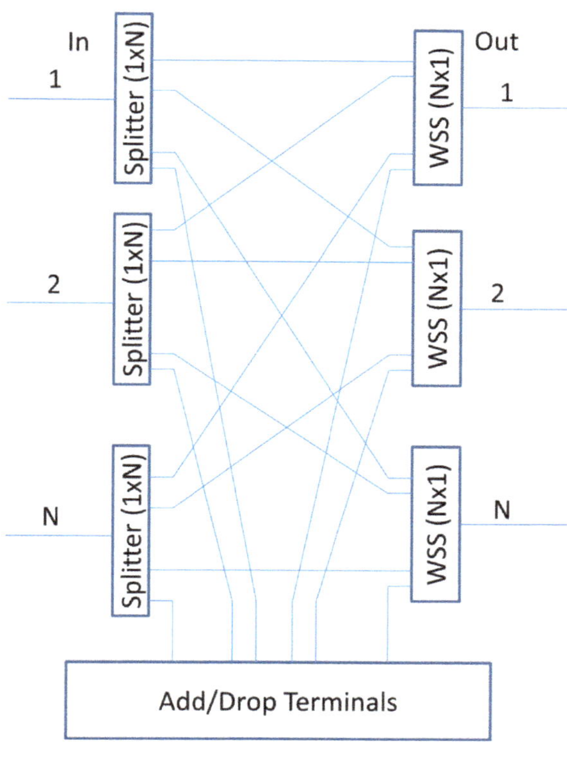

Figure 1. Broadcast-and-select-based node architecture.

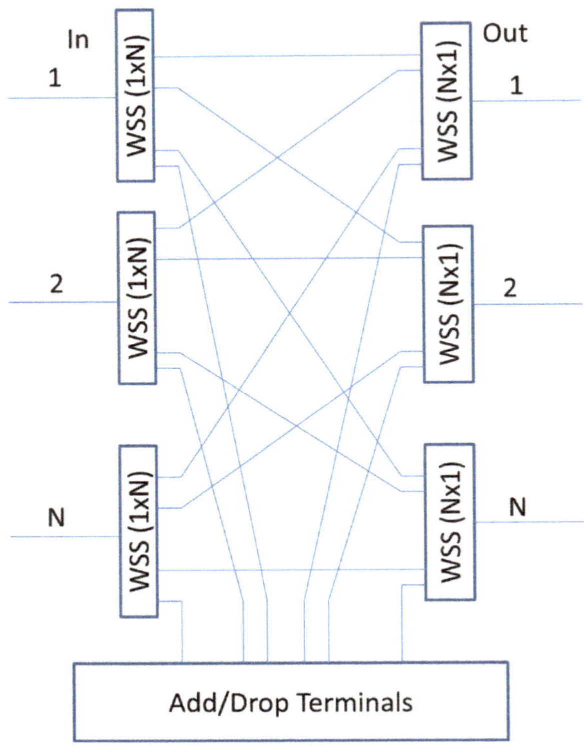

Figure 2. Route-and-select-based node architecture.

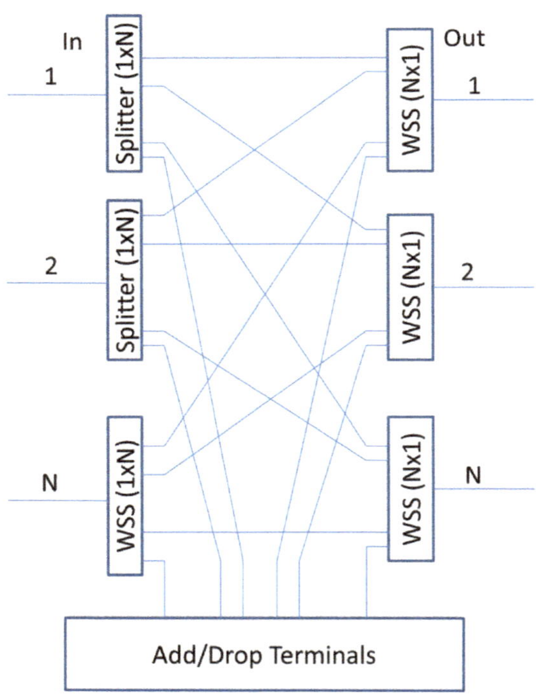

Figure 3. Hybrid node architecture.

Depending on the network traffic, it is envisioned that a fraction of the network nodes will be BS-based, other nodes will be RS-based and the rest will be hybrid nodes. The objective of the proposed algorithms of this chapter is to use hybrid nodes in order to minimize the lightpath interactions and at the same time to minimize the network cost. This means that WSSs are placed only in some of the input ports and specifically only at the locations that are necessary in order to allow only the necessary wavelengths to pass through the WSS and avoid all crosstalk interactions. Thus, by using hybrid nodes and not RS-based nodes, we can minimize the network cost while at the same time eliminating crosstalk interactions and consequently protecting the network against jamming attacks.

3. Planning approaches for demand uncertainty

In order to provide cost-efficient network solutions, it is necessary to plan optical networks over a long-time horizon. When dealing with optical networks, where the cost to build the network is high and the investment that takes place should last for a long time, sophisticated planning decisions must take place to ensure that the network infrastructure will not require any major upgrades over a predetermined amount of time. The problem becomes more involved in the case of future traffic demand forecasts that include uncertainty, as network planning decisions must be taken without the exact knowledge of future traffic demands. In this case, these decisions will be based on estimations. In the remaining of this chapter, the proposed multi-period network planning approaches with uncertain traffic demands are discussed. The planning approaches assume that for the first period, the demands follow a known distribution and for the periods that follow the demands are increased based on a multiplicative factor.

The multi-period network planning problem in this chapter will be investigated for two different period-planning types as detailed below.

3.1. Incremental network planning

This approach considers the demands of the next period and optimizes the investment cost in each period. Therefore, the solution is calculated sequentially for each period. The solution can be optimal for each period but not jointly for all the periods under consideration. Once the solution is provided for one period, then this solution affects the solution of the periods that follow. This is due to the fact that the solution of one period is assumed to be fixed and the solutions of the periods that follow are now based upon the previously found solutions.

3.2. Multi-period network planning

This approach considers the demands of all periods and optimizes the investment cost from the beginning of the planning period, that is the multi-period approach minimizes the network cost over all periods at once. Therefore, the demand distribution for every time period is necessary. This approach can calculate an optimal overall solution and provide decisions for the investment strategy of network operators.

4. Physical layer attacks

In general, the physical layer attacks in transparent optical networks can be grouped in two main categories: eavesdropping and service disruption.

In eavesdropping, the purpose of an attacker is to passively analyze the traffic in the network after gaining access to the information through an unauthorized observation method. To gain mid-span access to the fiber, the eavesdropper has to cut through and strip away the cable's outer jacket to access the individual fibers in its center.

Service disruption can be performed through high-power jamming attacks and can be classified into three sub-categories based on the effects it inflicts on the signal:

i. in-band jamming which is the result of intra-channel crosstalk,

ii. out-of-band jamming that is the result of inter-channel crosstalk and nonlinearities, and

iii. gain competition in optical amplifiers, where a high-power jamming signal can increase its own power, thus resulting in reduction in the gain of the rest of the co-propagating channels on the same fiber.

These types of attacks propagate through the transparent network affecting several connections, and as a consequence, the localization of an attack is a difficult problem. Due to the high bit rates of optical networks and the interaction of the connections, a jamming attack can cause a huge amount of information loss. Therefore, the limitation of attack propagation is a crucial consideration in designing transparent WDM optical networks.

The focus of this study is to deal with service disruption and especially with in-band and out-of-band jamming attacks.

4.1. In-band jamming attack

High-power in-band jamming attack is an attack that can be performed through the intra-channel crosstalk effect. Intra-channel crosstalk is the effect of power leakage between lightpaths crossing the same switch and using the same wavelength due to non-ideal isolation of the inputs/output ports of the switching fabric. Intra-channel crosstalk cannot be filtered out, since the interfering signal is on the same wavelength as the one affected. Thus, a high-power jamming signal can cause significant leakage inside the switches between lightpaths that are on the same wavelength as the attacking signal.

Figure 4 illustrates an example of a high-power jamming attack in node n_1 of the network through lightpath (p_1, w_i). In this figure, the attacker uses the lightpath (p_1, w_i) in order to attack the network. The attacking signal initially affects lightpath (p_0, w_i), through intra-channel

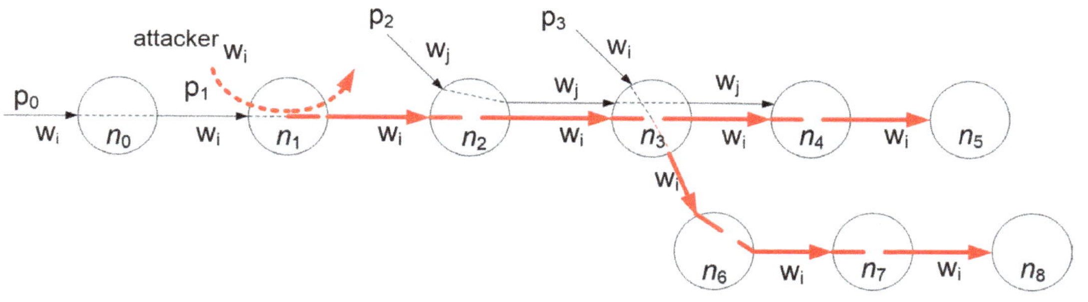

Figure 4. High-power in-band jamming attack propagation.

crosstalk because this lightpath uses the same wavelength and is crossing the same node as the attacking lightpath. In turn, lightpath (p_0, w_i) becomes an attacker too called "secondary attacker". Thus, lightpath (p_0, w_i) spreads the attack further to lightpath (p_3, w_i).

4.2. Out-of-band jamming attack

High power out-of-band jamming attack is an attack that can be performed through the inter-channel crosstalk effect. Inter-channel crosstalk results due to the power leakage between adjacent channels.

Figure 5 illustrates the high-power out-of-band signal propagation through the inter-channel crosstalk effect. In this case, lightpath (p_1, w_{i+1}) is used by an attacker in order to attack the network. Lightpath (p_1, w_{i+1}) then affects lightpath (p_0, w_i) as the two lightpaths co-propagate along the same fiber utilizing adjacent wavelengths. Then, the affected lightpath (p_0, w_i) becomes a "secondary attacker" and affects lightpath (p_3, w_{i-1}).

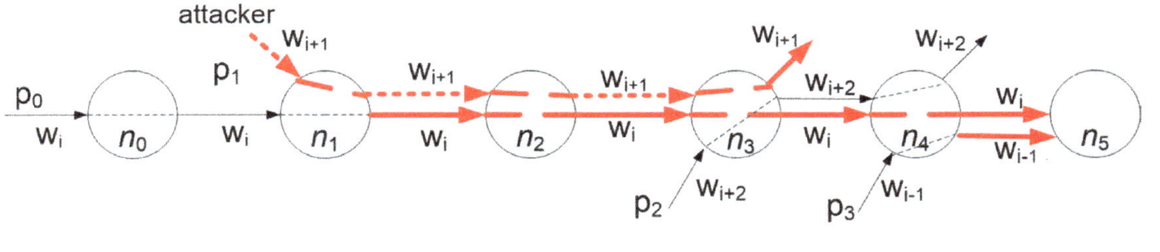

Figure 5. High-power out-of-band jamming attack propagation.

5. Attack-aware routing wavelength assignment

In this section, a heuristic algorithm is presented for the Aa-RWA with given demands in order to minimize the propagation of physical layer attacks. The algorithm aims at minimizing the interactions among lightpaths in order to avoid the propagation of high-power jamming attacks, in terms of affected lightpaths through intra- and inter-channel crosstalk. As discussed above, with these types of attacks, an affected lightpath can also affect other lightpaths, thus spreading the attack to other parts of the network. The goal of the Aa-RWA techniques is then to minimize as much as possible the spread of any attack that can occur in the network.

The proposed heuristic approach solves the problem by sequentially serving one-by-one the connections and consists of two phases. In the first phase, k candidate paths are calculated for each requested connection. In the second phase, the algorithm establishes the connections sequentially with the objective to minimize the number of in-band and out-of-band lightpath interactions.

5.1. Finding candidate paths

In the first phase, k candidate paths are identified for serving each requested connection. These paths are selected by employing a k-shortest path algorithm. The k-shortest path algorithm pre-calculates for each source-destination pair (s, d) a set of k candidate paths P_{sd} as follows: first, the shortest path is calculated using Dijkstra's algorithm, and then, the cost of the links which belong to the shortest path is doubled and Dijkstra's algorithm is executed again. This procedure is repeated until k paths are found. After a subset P_{sd} of candidate paths for each source-destination pair (s, d) is computed, the total set of computed paths is given as input to the next phase of the algorithm.

5.2. Attack-aware RWA

This section describes the heuristic algorithm for establishing the connections, one-by-one, in some particular order with the objective to minimize the lightpath interactions through the crosstalk effect.

5.2.1. Definitions

Each link l of the network is characterized by a Boolean wavelength availability vector $BWAV_l(i)$, $1 \le i \le W$, whose i^{th} element is equal to 0 if the ith wavelength of link l is utilized by a connection and is equal to 1, otherwise. W is the number of wavelengths that each fiber is able to support.

Each path p is characterized by a Boolean wavelength availability vector $BWAV_p(i)$, $1 \le i \le W$. The $BWAV_p$ consisting of links $l \in p$ is defined as the Boolean AND operation to the $BWAV_l$ of these links in each of the wavelengths of the $BWAV_l$ vectors.

$$BWAV_p = AND_{l \in p}\left(BWAV_l\right) \tag{1}$$

Thus, the element $BWAV_p(w)$ is equal to 1 if wavelength w is available over path p. The above equation enforces the wavelength continuity constraint among the links comprising a path. Each element $BWAV_p(i)$ represents a lightpath (p, w) between source-destination pairs (s, d).

5.2.2. Algorithm description

The aim of the heuristic algorithm is to establish Λ_{sd} lightpaths for (s, d) under the current utilization state of the network, given in the form of the wavelength availability vectors $BWAV_l$, for all l and the established lightpaths up to that point. The objective of the Aa-RWA heuristic algorithm is to minimize the number of lightpaths that interact with other lightpaths through intra- and inter- channel crosstalk and thus to minimize the propagation of high-power jamming signal attacks.

The wavelength utilization $BWAV_p$ of the candidate pre-calculated paths for the source-destination pair (s, d) is computed based on the $BWAV_l$ of the links. For each demand, the lightpath (p, w), from the set of candidate lightpaths with the smallest number of in-band and out-of-band channel interactions with the already established lightpaths, is chosen. To evaluate this, the wavelength availability vectors $BWAV_l$ are used to identify the interactions of established lightpaths. Then, the lightpath with the minimum sum of in-band and out-of-band channel interactions is established.

After establishing the lightpath (p, w), the corresponding $BWAV_l$ is updated. The algorithm at each step establishes a requested connection Λ_{sd}. If there are no available wavelengths, then the connection is blocked. Subsequently, the algorithm establishes lightpaths for all the connection requests in sequential order. The output of the algorithm is a set of established lightpaths in terms of paths and wavelengths. For each lightpath, the algorithm also returns two scalars that represent the number of inter-channel and the intra-channel interactions of this lightpath with the other established lightpaths.

6. Attack-aware routing and wavelength assignment under demand uncertainty for multi-period planning

As emphasized above, multi-period network planning is crucial in avoiding overprovisioning WSSs within hybrid nodes. As such, the aforementioned Aa-RWA algorithm is extended in this section to consider the demand forecasts of future time periods and in doing so to ensure that the WSS placement considers the changing network characteristics. In line with the most popular period-planning types available in the literature, the Aa-RWA algorithm is applied for both the incremental network planning case as well as the multi-period planning approach. In the former case, the Aa-RWA algorithm is applied in each step, the WSS placement for that step is decided, and the subsequent period considers the presence of those WSSs in the network when running the Aa-RWA algorithm for the next time period. In the multi-period approach on the other hand, the in-band and out-of-band interactions in each node are calculated for all time periods by the Aa-RWA algorithm and then statistical measures are used to assess the level of interaction and the extent to which a WSS is needed at a specific node.

In either case, the level of in-band and out-of-band interactions (and the subsequent decision on WSS placement) is strongly governed by the demand uncertainties and the assumptions made on growth year after year. The growth factor is assumed to be the mean value around a normally distributed random variable of the actual traffic growth between source destination pairs and thus Monte Carlo simulations are conducted to investigate the overall performance under independent trials. Details of the network setup and the exact values considered are detailed in Section 7.

6.1. Incremental Aa-RWA network planning

In incremental Aa-RWA network planning, there is knowledge for the demand distribution for only one period at a time (the period under consideration). For this reason, decisions are taken only for the current period. The flowchart of the proposed algorithm is given in **Figure 6**. The algorithm takes as input N independent sets of demands. For each one of the N sets, the algorithm solves the problem according to the deterministic Aa-RWA algorithm as presented in Section 5 and produces N outputs with metrics related to in-band and out-of-band interactions. These metrics associate two values for each input port of every network node. Specifically, these values count the number of lightpaths that interact though in-band and out-of-band crosstalk in the specific input port. Based on these values, the algorithm specifies the ports where WSSs should be placed. The assumption in this work is that in every period a maximum number of m WSSs can be placed due to budget constraints. The input ports where the WSSs are placed are chosen according to the maximum mean values of the in-band and out-of-band interactions. Subsequently, the output of each period contains the established lightpaths, and the next period takes as input the already established lightpaths and the placement of the WSSs from the previous period. The same procedure is followed for every period during the entire time horizon under consideration.

6.2. Multi-period Aa-RWA network planning

In multi-period Aa-RWA network planning, there is a priori knowledge for the demand distribution for all the time periods under consideration. Therefore, decisions are taken based on the traffic estimate for all time periods. The flowchart of the proposed algorithm is given in **Figure 7**. The algorithm takes as input N independent sets of demands for every one of the T periods (increasing over time based on a multiplicative factor as previously mentioned).

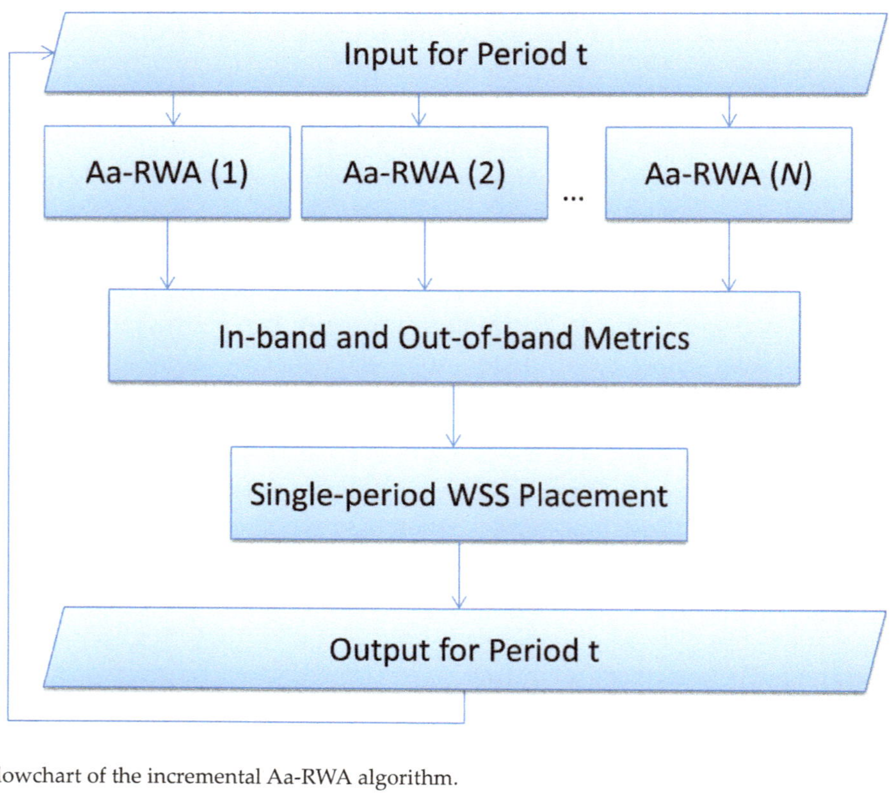

Figure 6. A flowchart of the incremental Aa-RWA algorithm.

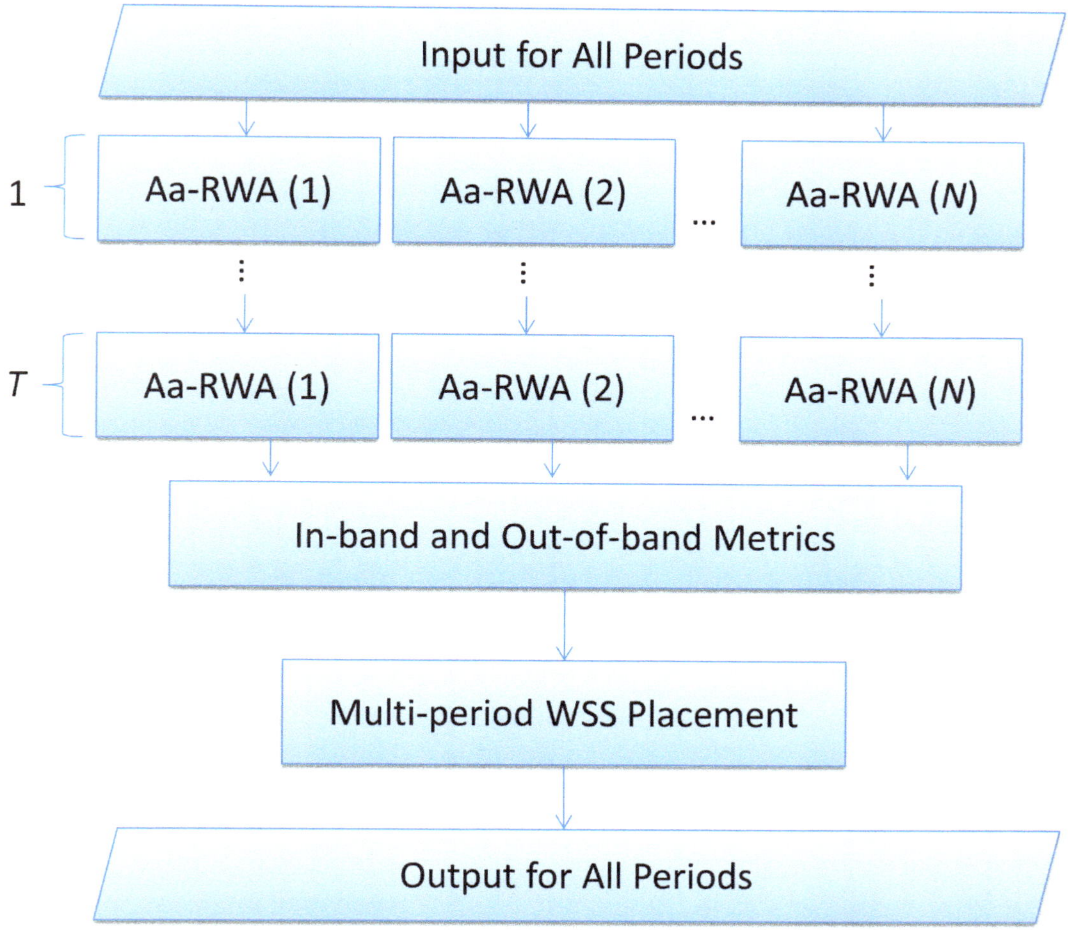

Figure 7. A flowchart of the multi-period Aa-RWA algorithm.

For each one of the N sets and for each time period, the algorithm solves the problem according to the deterministic Aa-RWA algorithm and produces $N*T$ outputs with metrics related to the in-band and out-of-band interactions. Based on these values, the "multi-period WSSs placement" module specifies the input ports and the time periods for the placement of the WSS. Again, the assumption is that in every period, a maximum number of m WSSs can be placed due to budget constraints. In this case, the placement of the WSSs is performed based on the maximum mean values of the in-band and out-of-band interactions over all instances and all periods.

7. Performance results

The network topology used in our simulations was the Geant-2 network topology [17] that has 34 nodes and 54 bidirectional links (108 fibers; shown in **Figure 8**). Each fiber is able to

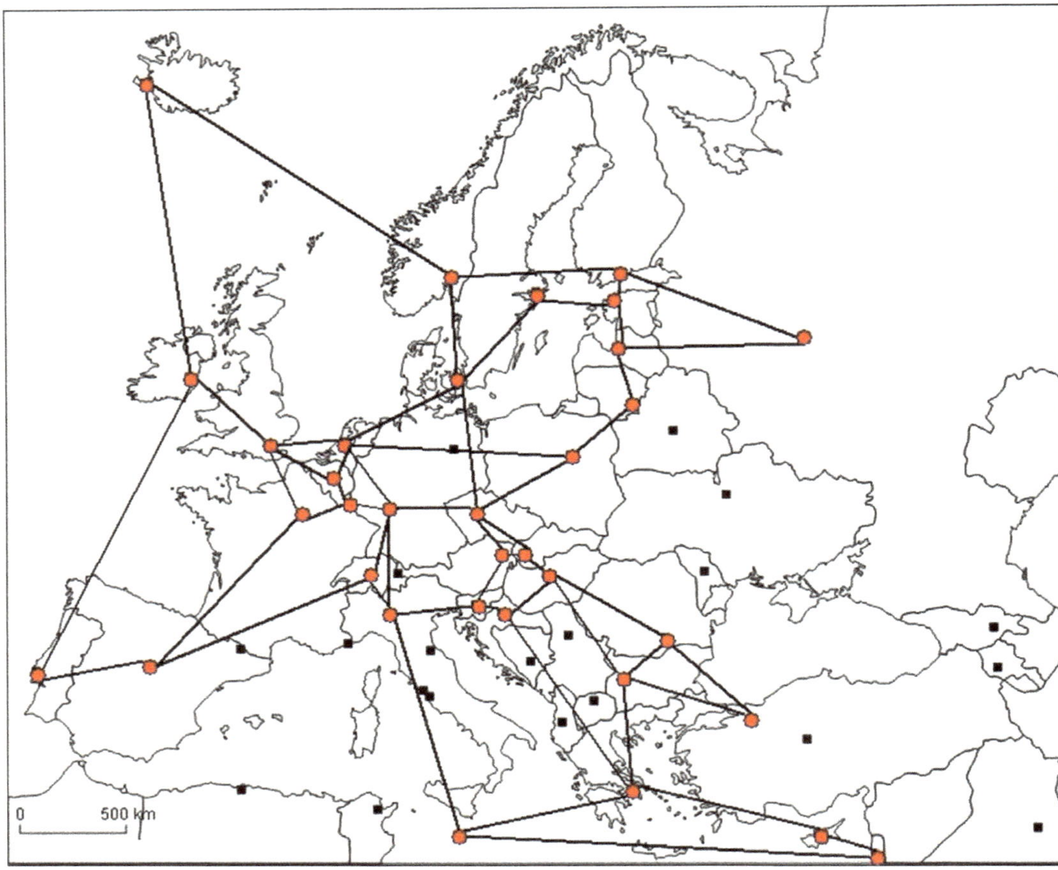

Figure 8. Geant-2 network topology: 34 nodes, 54 links.

support 80 wavelengths. The capacity of each wavelength was assumed equal to 10 Gbps. Initially, 50 different traffic matrices were produced with uniform distribution between source destination pairs and mean value equal to 1.35 Tbs of total requested capacity. Both algorithms (multi-period Aa-RWA and incremental Aa-RWA) were studied for five periods. The growth factor for each period was assumed to be equal to 1.5. The demand increase for each period applies for the source destination pairs that have a non-zero value at the initial traffic matrix. The algorithms for each source destination pair computed $k = 3$ alternative candidate paths.

In **Figure 9**, results for the multi-period Aa-RWA algorithm are depicted. Specifically, in **Figures 9(a)**, **(b)**, the mean values for inter-channel and intra-channel crosstalk for a horizon of five periods are presented, respectively. The mean values are the result of the 50 different traffic matrices. The inter-channel and intra-channel crosstalk per link (input port of a node) are the number of the interactions at this port. In **Figure 9**, the central mark of each box is the median, and the edges of the box are the 25th and 75th percentiles, the whiskers extend to the most extreme data points that are not considered outliers, and outliers are plotted individually.

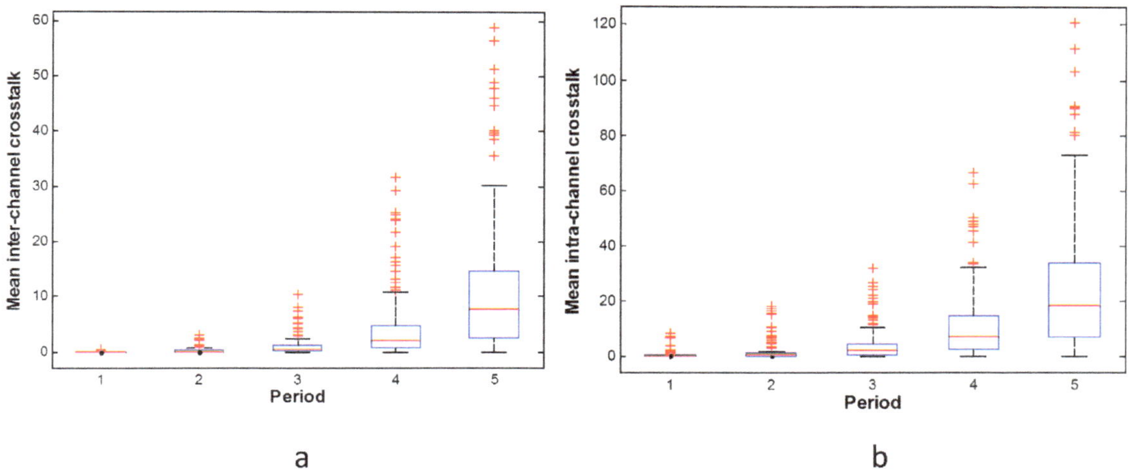

Figure 9. Mean values of multi-period Aa-RWA algorithm for (a) inter-channel crosstalk and (b) intra-channel crosstalk for a horizon of five periods.

Both inter- and intra-channel crosstalk increase exponentially with increasing traffic demands. However, as shown in **Figures 9(a), (b)**, specific links experience significantly higher crosstalk than others. Therefore, the required WSSs can be placed only at the input ports of the nodes that experience high crosstalk.

Incremental Aa-RWA algorithm follows the same trend as the multi-period Aa-RWA algorithm (as illustrated in **Figure 10**). Note that the trend would be completely different in the case where an attack-unaware RWA algorithm was used. In that case, all the periods would experience high values of crosstalk as can be found from the results of [8]. These results are not presented here, since the scope of this chapter is to plan an optical network in order to deal with physical layer attacks and therefore an attack-unaware RWA algorithm is out of the scope of this study.

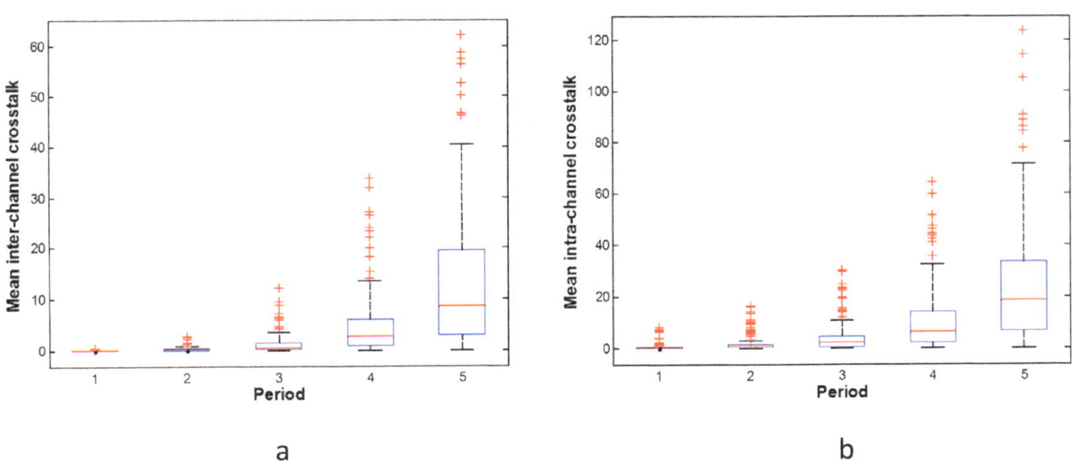

Figure 10. Mean values of incremental Aa-RWA algorithm for (a) inter-channel crosstalk and (b) intra-channel crosstalk for a horizon of five periods.

In **Figure 11**, the mean value of inter- and intra-channel crosstalk that the links experience during time period 5 is presented for the multi-period Aa-RWA algorithm. The results are presented in the form of histograms, where each column represents the number of links that have crosstalk between the ranges that are depicted in the x-axis of the histograms. From **Figure 11**, it is clear that a very small number of links have very high crosstalk, while the majority of links experience only a small crosstalk effect. This result offers a good indication that an addition of a small number of WSSs at the specific nodes where high crosstalk is experienced will significantly improve the performance of the network, thus minimizing the effect of a jamming attack. Note that the larger the number of links that appear in the leftmost bar, the smaller the crosstalk effect at the input ports of these nodes. Therefore, the best algorithms will be those where their histograms are more left shifted.

In **Figure 12**, the same histograms are presented for the case of the incremental Aa-RWA algorithm. Compared to the previous results of the multi-period case, the crosstalk effect of the incremental updating results to slightly increased inter-channel crosstalk and comparable intra-channel crosstalk. Nevertheless, the same crosstalk trends are observed here as well, where a small number of links experience significant crosstalk, while the rest of the links experience significantly lower crosstalk.

In **Figure 13**, the total number of required WSSs in order to minimize the impact of crosstalk effect per period is presented for the two proposed algorithms. For each period, the algorithms decide to place a WSS at the input port of a link when the mean values of the inter- and intra-channel crosstalk are above a certain threshold. Based on these decisions, the multi-period Aa-RWA algorithm requires less number of WSSs per period as compared to the incremental Aa-RWA algorithm. This is due to the fact the routing and wavelength assignment of the multi-period algorithm takes into account the future traffic demands, and the decisions are more appropriate. On the other hand, the incremental algorithm may decide to place a WSS in one period, and in future periods, there will be demands that would not be able to be established over already placed WSSs due to insufficient number of wavelengths. Thus, there would be not enough choices for efficient routing and wavelength assignment.

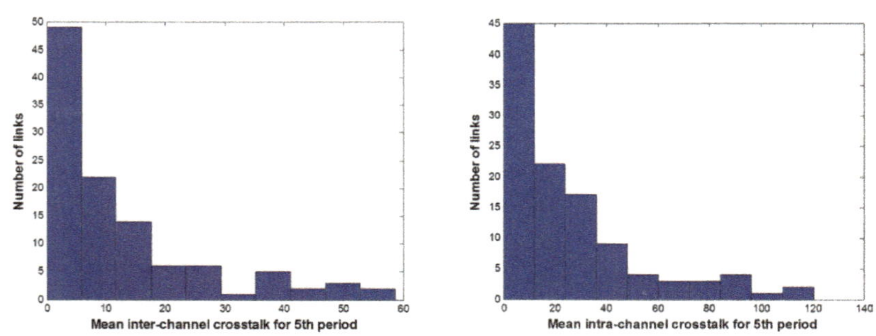

a b

Figure 11. Histogram for link (input ports of nodes) distribution related to (a) inter-channel and (b) intra-channel crosstalk interactions for multi-period Aa-RWA algorithm for the fifth period.

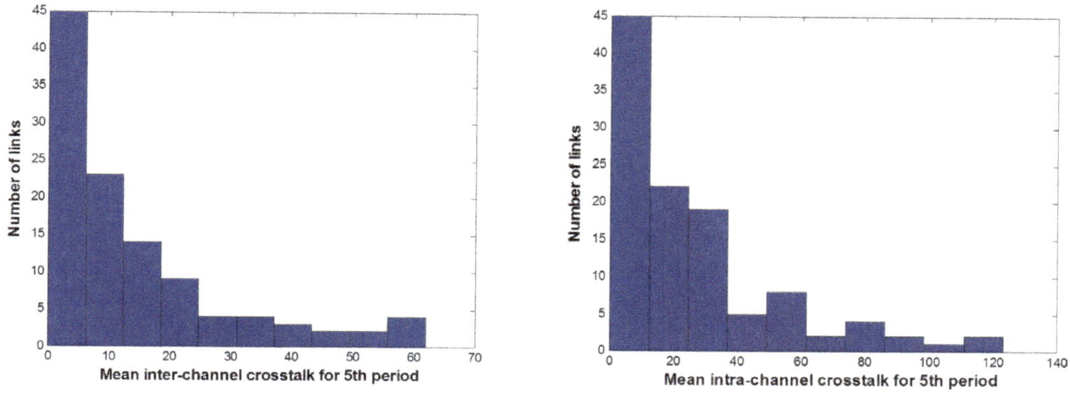

a b

Figure 12. Histogram for link (input ports of nodes) distribution related to (a) inter-channel and (b) intra-channel crosstalk interactions for incremental Aa-RWA algorithm for the fifth period.

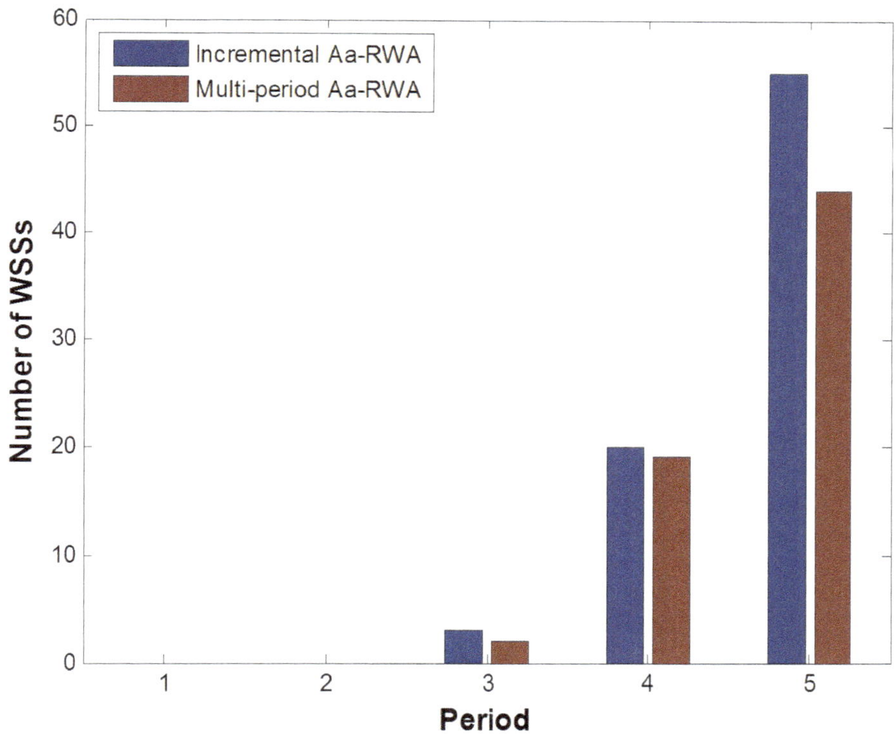

Figure 13. Number of required WSSs per period for the incremental Aa-RWA and the multi-period Aa-RWA algorithms.

8. Conclusions

This chapter proposed new attack-aware RWA algorithms for the multi-period planning of optical networks under demand uncertainty. These algorithms decide on the placement of wavelength selective switches at the input ports of network nodes and the period that the placement should be performed. The decisions are taken based on the distribution of the demands with the objective to minimize the impact of physical layer attacks over all periods. The algorithm that takes into account jointly all the time periods has a better performance than the algorithm that takes into account the periods in a sequential manner, resulting in a smaller number of required WSSs to be placed in the network so as to minimize the effect of a jamming attack.

Acknowledgements

This chapter is based upon work from COST Action CA15127 ("Resilient communication services protecting end user applications from disaster-based failures: RECODIS") supported by COST (European Cooperation in Science and Technology). This work has received funding from the People Program (Marie Curie Actions) of the European Union's Seventh Framework Program (FP7/2007–2013) under REA Grant Agreement n° 630853.

Author details

Konstantinos Manousakis*, Panayiotis Kolios and Georgios Ellinas

*Address all correspondence to: manousakis.konstantinos@ucy.ac.cy

Department of Electrical and Computer Engineering, KIOS Research Center for Intelligence Systems and Networks, University of Cyprus, Cyprus

References

[1] Ramaswami R, Sivarajan K. Routing and wavelength assignment in all-optical networks. IEEE/ACM Transactions on Networking. 1995;**3**(5):489–500. DOI: 10.1109/90.469957

[2] Ramaswami R, Sivarajan K, Sasaki G Optical Networks: A Practical Perspective. 3nd ed. Morgan Kaufmann; 2010. ISBN: 978-0-12-374092-2 Burlington, MA 01803, USA

[3] Kaman V, Helkey R, Bowers J. Multi-degree ROADM's with agile add-drop access. In: Proceedings of the IEEE Photonics in Switching; San Francisco, CA. 2007. DOI: 10.1109/PS.2007.4300729

[4] Mas C, Tomkos I, Tonguz OK. Failure location algorithm for transparent optical networks. IEEE Journal on Selected Areas in Communication. 2005;**23**(8):1508–1519. DOI: 10.1109/JSAC.2005.852182

[5] Furdek M, et al. An overview of security challenges in communication networks. In: Proceedings of the 8th IEEE International Workshop on Resilient Networks Design and Modeling (RNDM); Halmstad, Sweden. 2016. pp. 43–50. DOI: 10.1109/RNDM.2016.7608266

[6] Skorin-Kapov N, et al. A new approach to optical networks security: Attack-aware routing and wavelength assignment. IEEE/ACM Transactions on Networking. 2010;**18**(3): 750–760. DOI: 10.1109/TNET.2009.2031555

[7] Skorin-Kapov N, Furdek M, Pardo R, Mariño P. Wavelength assignment for reducing in-band crosstalk attack propagation in optical networks: ILP formulations and heuristic algorithms. European Journal of Operational Research. 2012;**222**(3):418–429. DOI: 10.1016/j.ejor.2012.05.022

[8] Manousakis K, Ellinas G. Attack-aware planning of transparent optical networks. Optical Switching and Networking. 2016;**19**(2):97–109. DOI: 10.1016/j.osn.2015.03.005

[9] Jirattigalachote A, Skorin-Kapov N, Furdek M, Chen J, Monti P, Wosinska L. Sparse power equalization placement for limiting jamming attack propagation in transparent optical networks. IEEE/OSA Journal of Optical Communications and Networking. 2011;**8**(4):249–258. DOI: 10.1016/j.osn.2011.06.008

[10] Manousakis K, Ellinas G. Equalizer placement and wavelength selective switch architecture for optical network security. In: Proceedings of the IEEE Symposium on Computers and Communication; Cyprus. 2015. DOI: 10.1109/ISCC.2015.7405631

[11] Verbrugge S, Colle D, Pickavet M, Demeester P. Common planning practices for network dimensioning under traffic uncertainty. In: Proceedings of the 4th IEEE International Workshop on the Design of Reliable Communication Networks (DRCN); October 2003; Banff, Alberta, Canada. pp. 317–324. DOI: 10.1109/DRCN.2003.1275372

[12] Leung D, Grover WD. Capacity planning of survivable mesh-based transport networks under demand uncertainty. Photonic Network Communications. 2005;**10**(2):123–140. DOI: 10.1007/s11107-005-2479-z

[13] Birge JR, Louveaux F. Introduction to Stochastic Programming. New York: Springer-Verlag; 2011. ISBN: 978-1-4614-0236-7

[14] Bertsimas D, Sim M. The price of robustness. Operations Research. 2004;**52**(1):35–53. DOI: 10.1287/opre.1030.0065

[15] Aparicio-Pardo R, Pavon-Marino P, Mukherjee B. Robust upgrade in optical networks under traffic uncertainty. In: Proceedings of the 16th IEEE International Conference on Optical Network Design and Modelling (ONDM); Colchester. 2012. pp. 1–6. DOI: 10.1109/ ONDM.2012.6210204

[16] Kronberger C, Schondienst T, Schupke DA. Impact and handling of demand uncertainty in multiperiod planned networks. In: Proceedings of the IEEE International Conference on Communications (ICC); Kyoto. 2011. pp. 1–6. DOI: 10.1109/icc.2011.5962890

[17] GÉANT2 Network. Available from: http://www.geant2.net/

9

The Future Electrical Multiplexing Technique for High Speed Optical Fibre

Rahmat Talib and Mohammad Faiz Liew Abdullah

Additional information is available at the end of the chapter

Abstract

Advancement in transmission technology based on fiber optic such as multiplexing technique is an attractive research area for future development of high capacity and high speed optical communication system. Typical electrical based multiplexing such as electrical time division multiplexing (ETDM) and duty cycle division multiplexing (DCDM) have difficulty to fulfil the requirements of modern fiber optic communication with practical solution. Multi slot amplitude coding (MSAC) is the latest multiplexing technique that has been proposed as an alternative to ETDM and DCDM. The results show that the spectral width is reduced by around 25%, not less than 55% improvement of chromatic dispersion (CD) tolerance, 0.6 dB better receiver sensitivity, and 1.5 dB better optical signal to noise ratio (OSNR) compared to DCDM for 30 Gbit/s transmission capacity. The spectral width for 3 × 10 Gbit/s, 4 × 10 Gbit/s and 5 × 10 Gbit/s MSAC is 60 GHz, which indicates improvement of spectral efficiency. This advantage is not possible to be achieved through ETDM technique. In addition, 10 GHz clock signal can be extracted from the MSAC signal which is important for recovery circuit at receiver since it is similar to symbol rate.

Keywords: optical fibre communication, multiplexing, optical modulation

1. Introduction

The current trend of internet use for social applications, such as facebook and twitter, is a strong indication of the substantial demand for information and communication technology. Therefore, behind the success story of these popular social applications is the capability of telecommunication infrastructure to handle huge amount of information transfer worldwide.

Optical communications have been widely used for this important task. Advancement in high capacity and high speed optical communication system has always become an important topic of discussion among the communications community. One of the important techniques in optical communication system is to realize high capacity data transportation through multiplexing technique. The technology is moving forward, where innovative alternative to the existing multiplexing method is indispensable to accomplish the future needs in optical communication implementation.

Generally, multiplexing is required to share the huge bandwidth of well-known fibre optic medium with many users, hence provides more cost efficient for practical implementation for high capacity data transformation. Typical implementation of multiplexing can be done in electrical or optical domain. Electrical-based multiplexing is very important technique due to the capability of electronic technology to switch the data with high speed, efficiency and reliability [1].

In this chapter, we report the investigation of system performance for recent electrical-based multiplexing technique known as multi-slot amplitude coding (MSAC) for high speed optical communication link. This work is very important in order to justify the benefit of MSAC as an innovative multiplexing compared to other electrical-based multiplexing such as electrical time division multiplexing (ETDM) and duty cycle division multiplexing (DCDM).

2. Electrical time division multiplexing (ETDM)

ETDM is a technique that adopting electronic circuit to execute multiplexing process in electrical domain. In EDTM concept, the bit from multiple input tributaries is arranged as a single output tributary by allocating all the bits with smaller time slot as compared to the time slot of input channel. As a result, the bit rate or speed of output tributary is higher than the input bit rate so that all the bits at the input can be transferred correctly to the output. The output tributary bit rate, R_T of ETDM is

$$R_T = NR \tag{1}$$

Where N is the number of input tributaries and R is the input tributary bit rate. The bit duration T_b of the input tributary is given by

$$T_b = \frac{1}{R} \tag{2}$$

The basic concept of EDTM implementation based on three input tributaries depicted in **Figure 1**. In this figure, ETDM operates with three input tributaries so the output bit rate of EDTM is $3R$ bits/s and the output bit duration becomes $T_b/3$. The output of ETDM can be used to modulate the light source for data transmission. Due to simple and efficient operation of ETDM technology, it is extensively adopted for commercial application for synchronous optical network (SONET) , synchronous digital hierarchy (SDH) and optical transport network (OTN). Currently, high speed electronic-based ETDM has been reported to support more than 40 Gbit/s serial data using advance material and state-of-the-art technology [2].

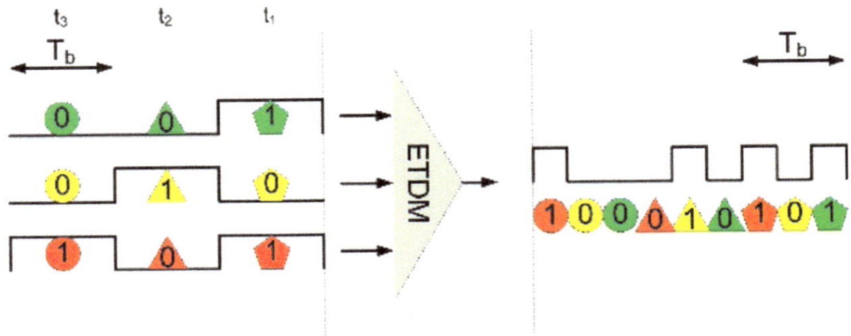

Figure 1. ETDM concept.

3. Duty cycle division multiplexing (DCDM)

Multiplexing process also can be realized using DCDM concept, and it has been proposed as an alternative technique to ETDM [3]. In order to implement DCDM concept, basic components can be used which are return-to-zero (RZ) convertor and electrical adder. Since, RZ convertor has the ability to adjust the duty cycle (DC) parameter, hence various DCs of binary signal can be obtained. In DCDM, each input tributary is applied to RZ convertor with the different predefined DC parameter. The outputs for all RZ convertors are combined by an electrical adder to generate the DCDM signal. This technique requires a synchronize bit and identical amplitude of non-return-to-zero (NRZ) format for all input tributary. **Figure 2** shows an example of DCDM signal generation for three tributaries. **Figure 2**a–c shows the 8 bits sequence (all possible combination) of input tributary 1, 2 and 3, respectively. **Figure 2d–f** shows the RZ format with 25% DC of tributary 1, 50% DC of tributary 2 and 75% DC of tributary 2, respectively, after NRZ to RZ conversion. Combination of signals in **Figure 2d–f** using electrical adder generates DCDM signal as shown in **Figure 2g**. Based on this multiplexing,

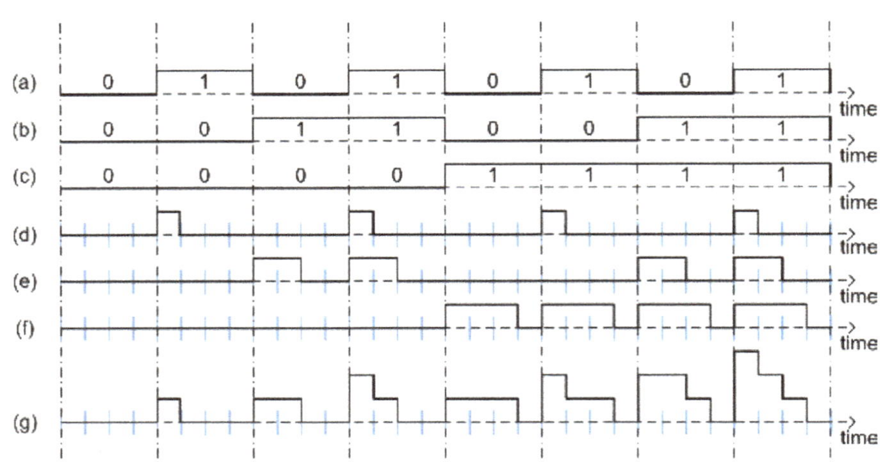

Figure 2. An example of DCDM signal.

a unique signal waveform or symbol can be obtained. There is a transition at the beginning of the DCDM symbol because RZ conversion turns the signal from high to low for bit 1 s.

4. Multi-slot amplitude coding (MSAC)

MSAC technique is a latest multiplexing concept to enhance the utilization of signal level for generating waveform or symbol compared to DCDM and ETDM [4]. In this technique, the multiplexer converts all possible combination bits of each tributary as MSAC symbol based on predefine translation rule. MSAC symbols can be obtained based on two parameters, which are the number of slots, *S* and number of signal levels, *M* as depicted in **Figure 3**. Assuming an equal duration of slot, the slot duration of MSAC symbol, *Td* is given by

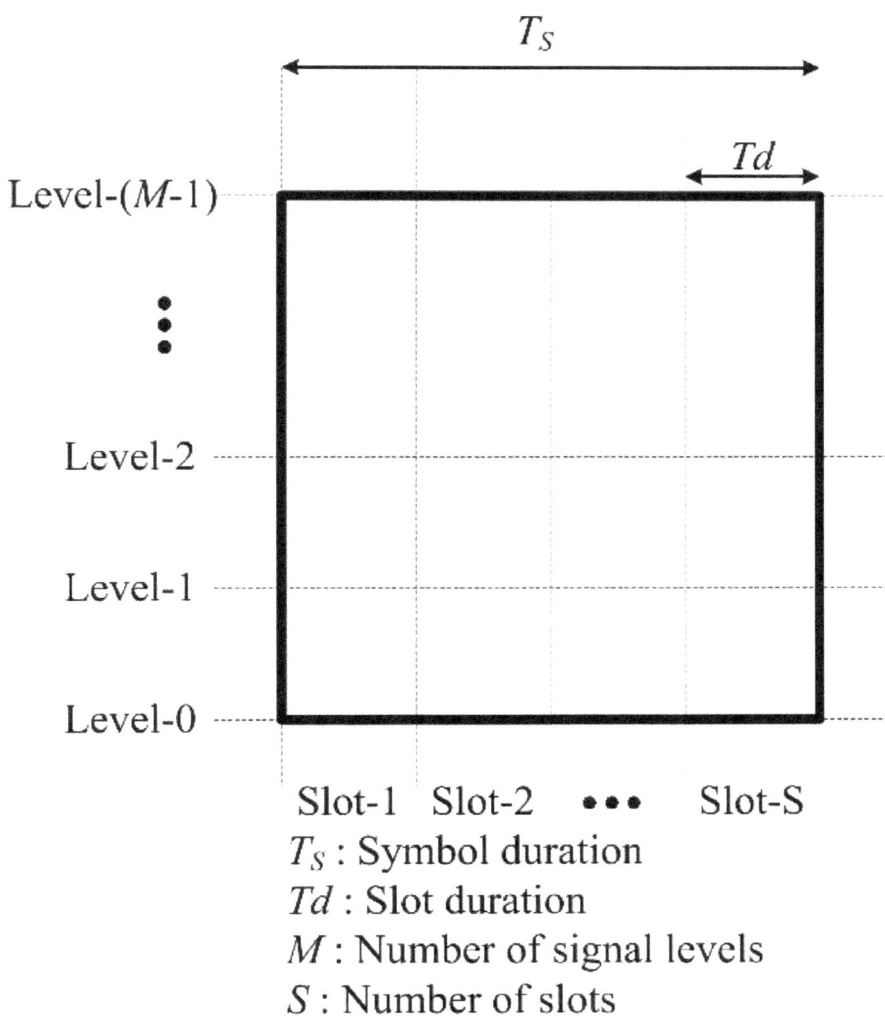

Figure 3. General format of symbol for MSAC.

$$Td = \frac{T_S}{S} \tag{3}$$

where T_S is the symbol duration. T_S is similar to the bit duration of the input tributary, T_b. Therefore,

$$T_S = T_b = \frac{1}{R} \tag{4}$$

where R is the input tributary bit rate.

For equal signal level spacing, the maximum amplitude of MSAC, A is

$$A = (M-1)\Delta \tag{5}$$

where Δ is amplitude spacing. **Figure 4** displays an example of MSAC symbol for three users. There are nine symbols ($x_1(t)$ to $x_9(t)$) for $S = 3$ and $M = 3$. Note that MSAC symbol allocates the first slot in symbol format as zero level.

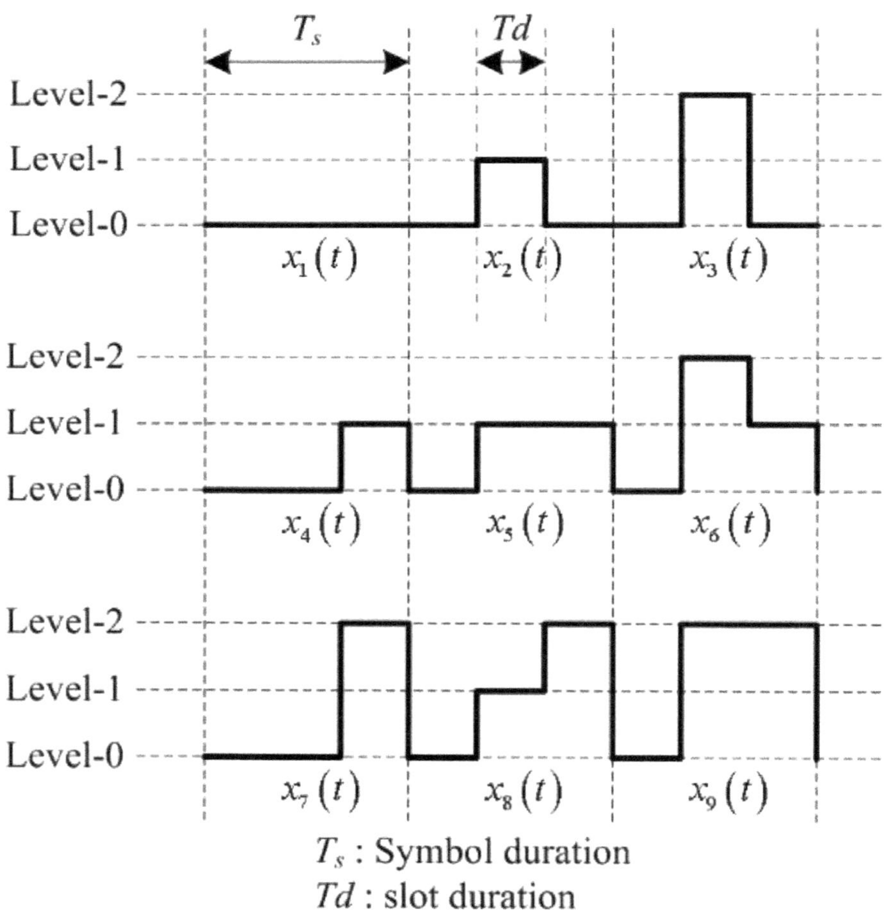

Figure 4. Example of symbol waveform for $M = 3$ and $S = 3$.

5. Optical communication system simulation using MSAC

Figure 5 shows the simulation setup for *N* tributaries MSAC in optical communication system. This setup consists of transmitter section, transmission section and receiver section. The component for the transmitter section is *N* pulse pattern generators, MSAC multiplexer, external modulator and continuous wave laser. The number of pulse pattern generator will depend on number of tributary. Each tributary consists of a pulse pattern generator for generating pseudo random binary signal (PRBS). Note that *Tr*1, *Tr*2 and *TrN* represent tributary 1, tributary 2 and tributary *N*, respectively. Each pulse pattern generator has a common clock signal in order to obtain synchronize binary data stream as input signal to the MSAC multiplexer. MSAC multiplexer model implements a conversion process based on the rule. The signal from MSAC multiplexer then modulates the light from a continuous wave laser (CW LD) at 1550 nm, using an external modulator. The optical power of CW LD is fixed at 0 dBm. The external modulator is based on an amplitude modulator (AM) model. Transmission section consists of an optical attenuator and optical fibre. The modulated optical signal is fed into an optical attenuator. In optical attenuator, the signal input electrical field for both polarizations is attenuated. This optical attenuator is used to control the amount of launch optical power from the AM. For this simulation, optical fibre is based on a single mode fibre model, and it is placed after the optical attenuator. The propagation of modulated optical signal in single mode fibre model is based on the Schrödinger equation [5]. The receiver section consists of an optical amplifier, PIN photodiode, electrical low-pass filter and clock and data recovery. An optical amplifier that acts as a pre-amplifier is placed before PIN photodiode in order to boost the signal. Note that the optical amplifier also introduces ASE noise. The received optical signal is then converted to an electrical signal using a PIN photodiode. In this simulation, the signal is corrupted with typical noises in optical system such as shot noise and thermal noise. The PIN photodiode output signal is filtered with a Gaussian low-pass filter (LPF). In order to

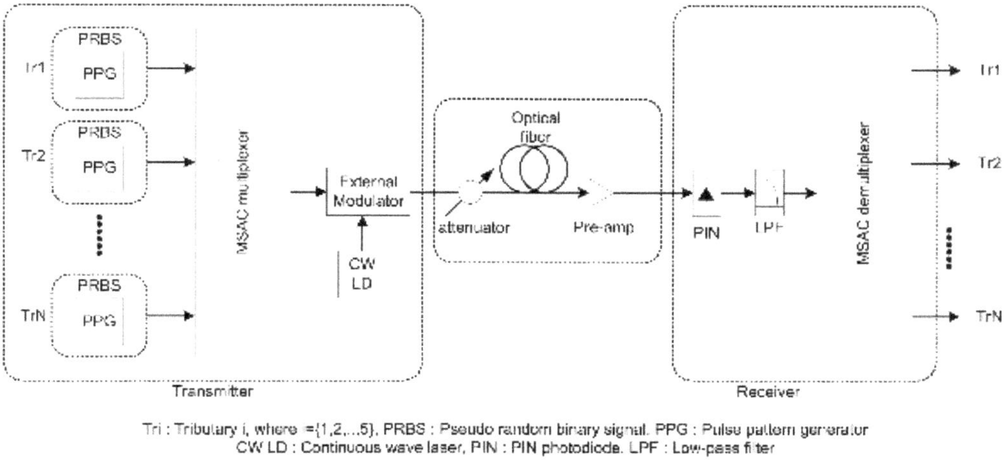

Figure 5. MSAC system setup.

optimize the system performance, the cut-off frequency of filter is set at 0.75 *BW*, where BW is the first null bandwidth of baseband MSAC signal. The filtered electrical signal is fed into the clock and data recovery module to regenerate each tributary data stream. In clock and data recovery module, data recovery process is implemented using MATLAB programming based on the recovery rules of MSAC demultiplexer.

6. Results and discussion

6.1. Optical signal spectrum, spectral width and clock recovery frequency

Figure 6a–c illustrates the optical spectrum of MSAC 3 × 10, 4 × 10 and 5 × 10 Gbit/s, respectively. This spectrum has been observed at the transmitter side (after external modulator) using an optical spectrum analyser.

From this figure, simulated null-to-null spectral width of MSAC 3 × 10, 4 × 10 and 5 × 10 Gbit/s is around 60 GHz. This result shows that spectral width of MSAC remains the same even though N is increasing from 3 to 5. This is because they have similar number of slots in a symbol. In this case, number of slots is three, thus slot period is 1/(3 × 10 Gbit/s). This slot determines the width of main-lobe of optical spectrum. Therefore, increasing aggregate capacity from 30 to 50 Gbit/s using this technique will not affect the required optical bandwidth. Moreover, the modulation speed remains unchanged because of the fixed slot interval.

Another important characteristic of the optical signal spectrum is to visualize the clock information in which the frequency is similar to the symbol rate. Note that impulse or spike in optical signal spectrum means high clock frequency in the signal. As shown in **Figure 6a**, there are seven impulses in the null-to-null spectral width, where f_0 is impulse at optical carrier (1550 nm). Note that the spectral is symmetrical at f_0. Besides that the impulses appear at 10 GHz (f_1) and 20 GHz (f_2) away from f_0 on the right side of the main-lobe. The impulse at 10 GHz (f_1) can be used to recover the clock frequency by a clock recovery circuit in receiver. Note that these impulse frequencies are similar for 4 × 10 (**Figure 6b**) and 5 × 10 Gbit/s (**Figure 6c**).

6.2. Spectral efficiency

In optical communication system based on wavelength division multiplexing (WDM) technology, more than one WDM channel can be propagated in optical fibre. Each WDM channel is separated with channel spacing such as 25, 50, 100 or 200 GHz. Note that spectral efficiency in WDM optical system can be calculated based on formula in [6]. Although the simulation is based on 1550 nm or single channel wavelength, MSAC technique can be implemented over WDM system as well. In order to operate with WDM system, WDM channel spacing must be greater than the spectral width of MSAC signal in order to avoid serious interference between adjacent channels. Since the spectral width of MSAC signal is around 60 GHz, therefore 100 or 200 GHz WDM channel spacing can be used. Assuming that WDM channel spacing is 100 GHz, these are corresponding to 0.3, 0.4 and 0.5 bit/s/Hz of spectral efficiency for 3 × 10, 4 × 10 and 5 × 10 Gbit/s, respectively.

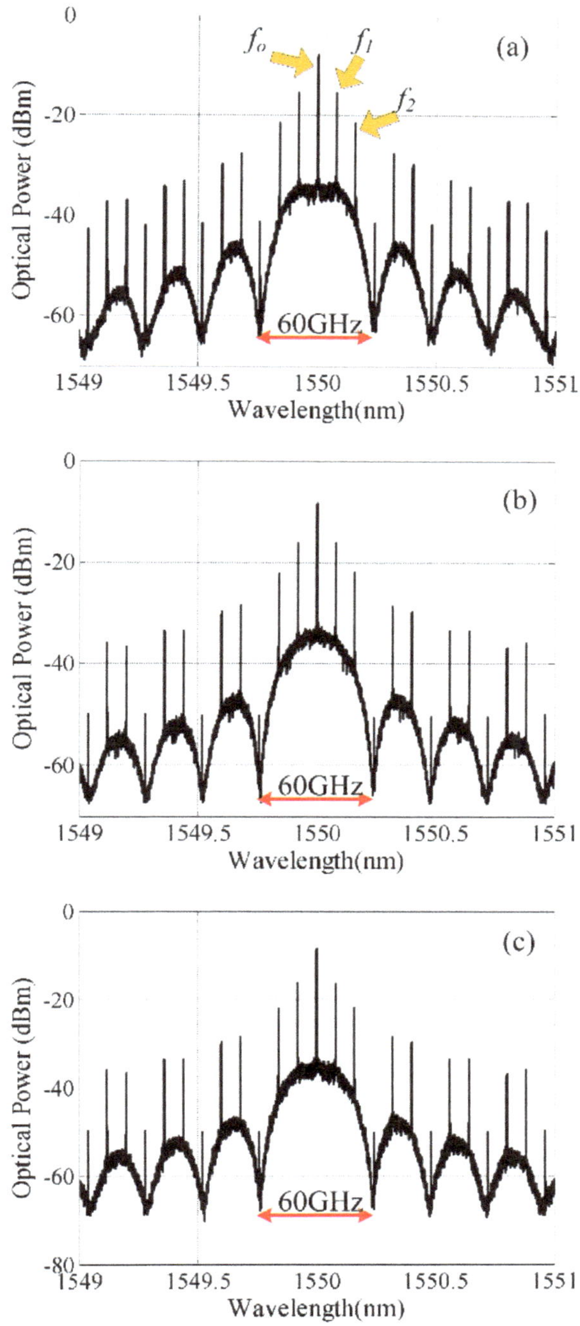

Figure 6. Optical spectrum of MSAC (a) 3 × 10, (b) 4 × 10 and (c) 5 × 10 Gbit/s.

6.3. Received optical power

Bit error rate (BER) estimation in this simulation is based on probability of error method. BER versus received optical power of MSAC at 3 × 10, 4 × 10 s and 5 × 10 Gbit/s has been plotted as in **Figure 7**. Note that this simulation is based on back-to-back setup (optical fibre is not included). As a result, the performance of MSAC system is limited by noises from optical amplifier and PIN

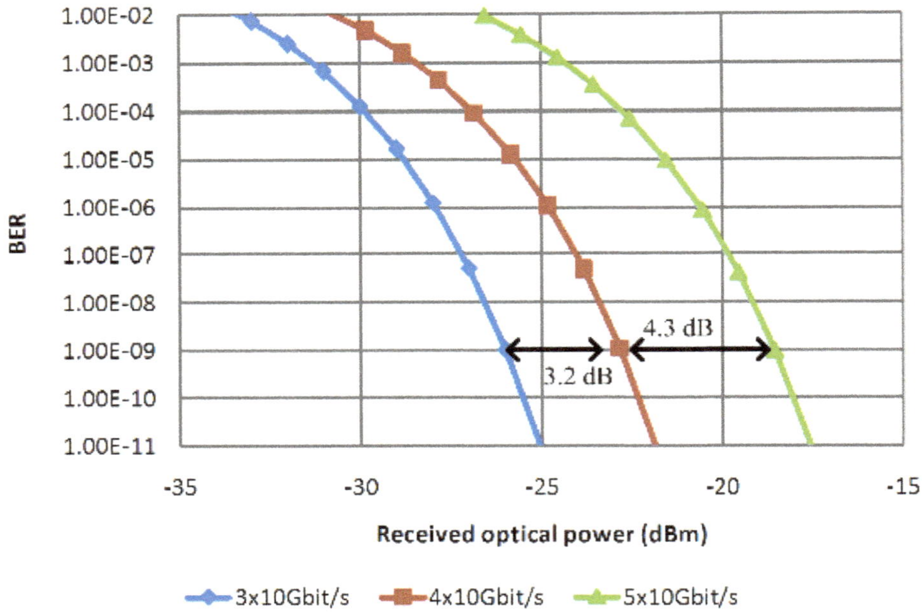

Figure 7. BER versus received optical power of 3 × 10, 4 × 10 and 5 × 10 Gbit/s MSAC.

photodiode. Inter symbol interference is minimized by setting the cut-off frequency of low-pass filter at 0.75 *BW*. From this figure, the required received optical power or receiver sensitivity of MSAC 3 × 10, 4 × 10 and 5 × 10 Gbit/s at BER of 10^{-9} is −26.0, −22.8 and −18.5 dBm, respectively.

This simulation results show that 3 × 10 Gbit/s has the lowest received optical power, whereas 5 × 10 Gbit/s has the highest received optical power. Power penalty of around 3.2 and 7.5 dB is observed for 4 × 10 and 5 × 10 Gbit/s at BER of 10^{-9}, respectively, compared to 3 × 10 Gbit/s. The reason for this penalty is due to the number of signal levels increased. MSAC 3 × 10 Gbit/s uses three signal levels; therefore, it has the lowest received optical power, whereas MSAC 5 × 10 Gbit/s uses six signal levels, thus it has the highest received optical power.

6.4. Optical signal to noise ratio (OSNR)

Figure 8 shows the BER versus OSNR of MSAC at 3 × 10, 4 × 10 and 5 × 10 Gbit/s. From this figure, the required OSNR of MSAC 3 × 10, 4 × 10 and 5 × 10 Gbit/s at BER of 10^{-9} is 25.5, 28.8 and 33.2 dB, respectively. Power penalty due to adding tributary is around 3.3 and 7.7 dB for 4 × 10 and 5 × 10 Gbit/s, respectively. As expected, MSAC system requires more OSNR in order to maintain the BER performance when tributaries increase because the number of levels increases. Note that the variation between power penalty for OSNR and received optical power is small. This indicates that received optical power and OSNR are important parameters for reliable communication.

6.5. Chromatic dispersion tolerance

Chromatic dispersion (CD) is one of the optical fibre impairment in high speed optical communication system. The CD effect in silica fibre will degrade the BER performance due to

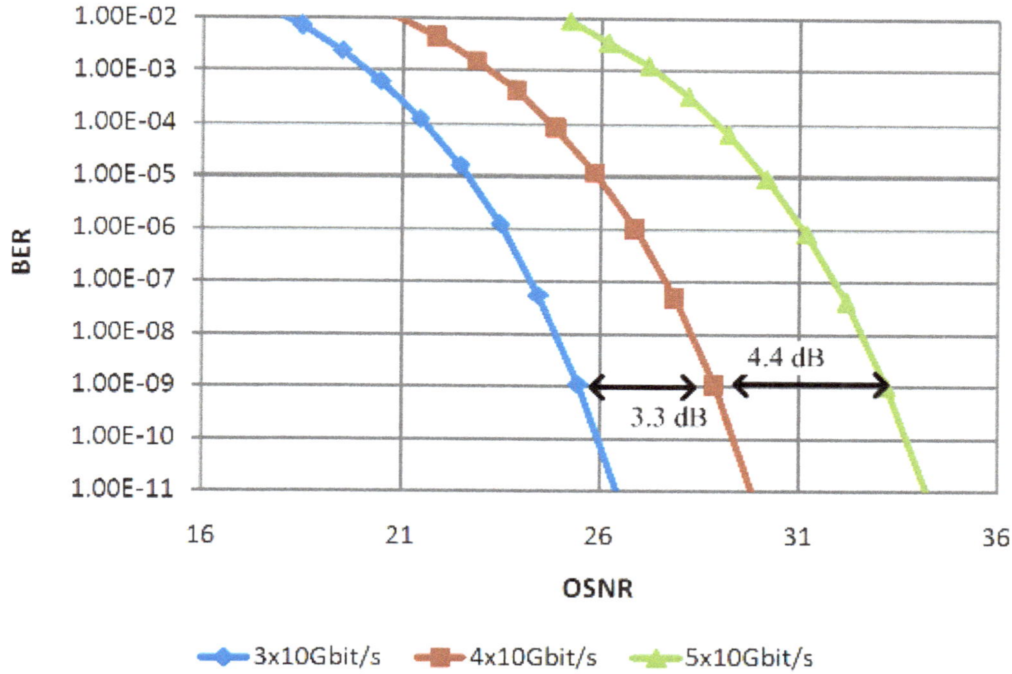

Figure 8. BER versus OSNR of 3 × 10, 4 × 10 and 5 × 10 Gbit/s MSAC.

evolution of signal shape. In order to achieve high performance quality, the accumulated CD must not exceed the allowable CD tolerance. CD tolerance is determined by estimating the amount of CD (positive and negative) at the target BER of 10^{-9}.

In order to determine the chromatic dispersion tolerance, a standard single mode fibre model is included in this simulation setup. The dispersion parameter of optical fibre model is varied from negative dispersion to positive dispersion. Other fibre impairments effect such as attenuation, self-phase modulation (SPM) and non-linear fibre effect are ignored so that the system performance is determined by fibre dispersion only.

Figure 9 depicts the BER versus dispersion of MSAC 3 × 10, 4 × 10 and 5 × 10 Gbit/s. Chromatic dispersion tolerance of MSAC 3 × 10, 4 × 10 and 5× 10 Gbit/s at BER of 10^{-9} is ±164.5, ±149.5 and ±69 ps/nm, respectively. This comparison shows that CD tolerance decreases when the number of signal levels increased. Note that optical power at the highest level for higher number of signal levels of MSAC is high compared to MSAC with small number of signal levels for BER of 10^{-9}. In terms of dispersion mechanism, pulses at higher signal level experience highest energy loss compared to signal at lower level, thus reduces the eye opening and induces a power penalty. Therefore, at similar average power, MSAC with higher number of signal levels losses its CD robustness capability.

6.6. The effect of signal level spacing

In previous work, the separation of signal level or level spacing in MSAC symbol format is equally spaced. The level spacing between level i and level i-1 is Δ. This means that level spacing

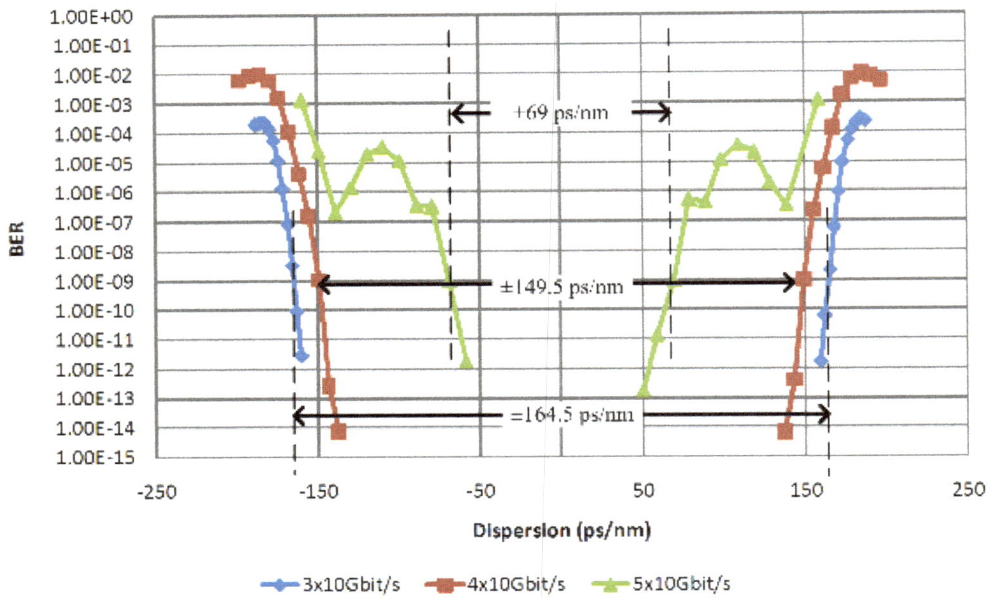

Figure 9. BER versus dispersion of 3 × 10, 4 × 10 and 5 × 10 Gbit/s MSAC.

ratio is 0, 0.5 and 1 for level 0, level 1 and level 2, respectively. This setting is, therefore, known as equal level spacing (ELS). **Figure 10** shows the BER of MSAC against normalized signal spacing of level 1 between 0.1 and 0.6. The optimum level spacing (OLS) is observed at normalized signal spacing of level 1 of 0.31. As expected, higher level spacing for upper level compared to lower level spacing in order to achieve equal probability of error region for each signal level.

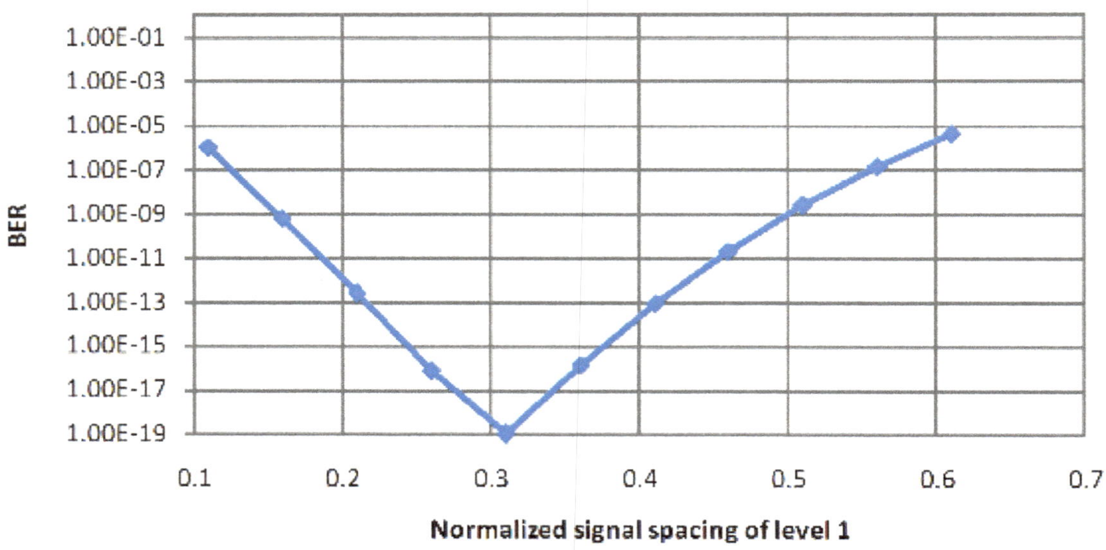

Figure 10. BER of MSAC against normalized signal spacing of level 1 between 0.1 and 0.6.

Figures 11 and **12** show the comparison between equal level spacing (ELS) and optimize level spacing (OLS) of back to back (b2b) BER performance of 3 × 10 Gbit/s MSAC system based on received optical power and OSNR, respectively. From the figure, the received optical powers of −26 and −29.5 dBm are obtained at BER of 10^{-9} for ELS and OLS, respectively. This is an improvement of 3.5 dB when OLS MSAC is adopted at receiver. In term of OSNR, ELS requires 25.5 dB, in contrast OLS requires 21.8 dB. This clearly shows that there is OSNR improvement when OLS method is applied, with improvement around 3.7 dB.

The performance in terms of dispersion tolerance for ELS and OLS method is plotted as shown in **Figure 13**. Based on that figure, ELS and OLS methods are capable of tolerating positive and negative chromatic dispersion of 329 and 311 ps/nm at BER of 10^{-9}. The reduction of 18 ps/nm is observed for OLS method compared to ELS method.

The effect of the signal level is also investigated for the MSAC system with setup of 4 × 10 Gbit/s. In general, the approach to determine the optimum level is similar for previous setup (3 × 10 Gbit/s). Since 4 × 10 Gbit/s MSAC has four signal levels, both signal levels 1 and 2 are adjusted while signal levels 0 and 3 are fixed. It is found that the optimum level spacing is achieved when the normalized signal level 1 and signal level 2 are 0.183 and 0.51, respectively.

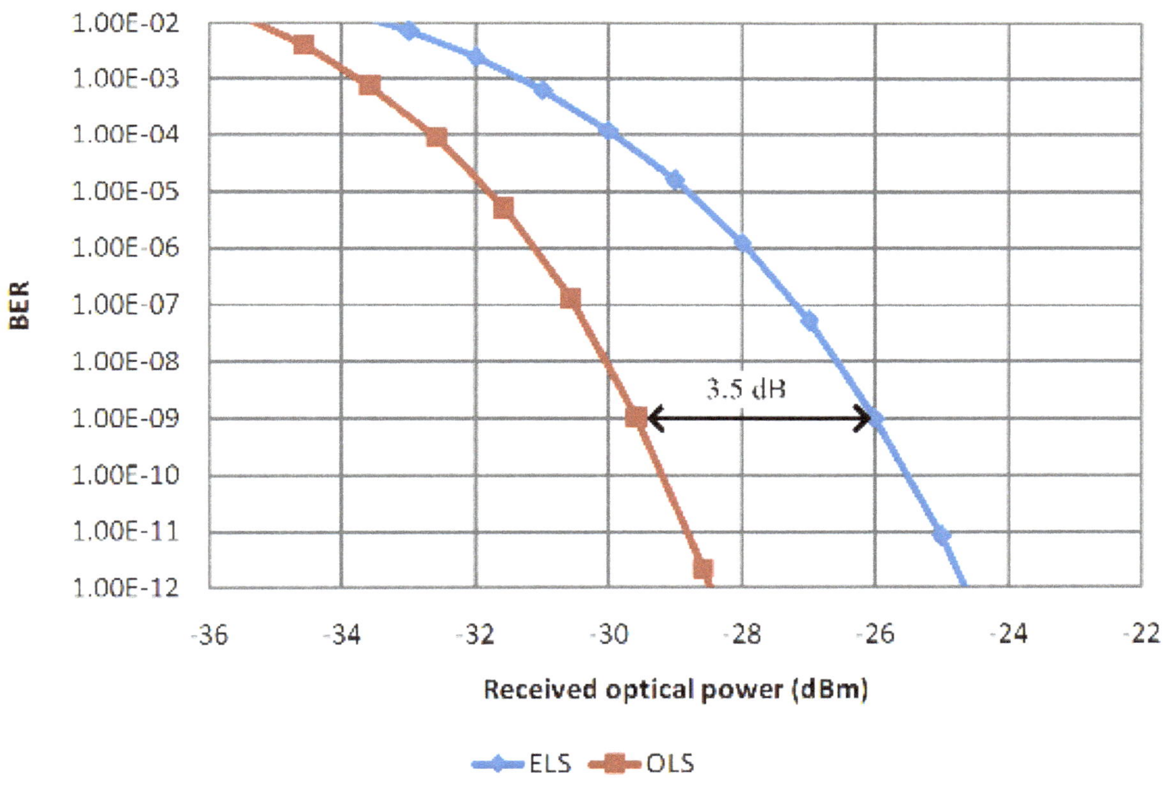

Figure 11. Received optical power comparison between ELS and OLS of b2b 3 × 10 _Gbit/s MSAC system.

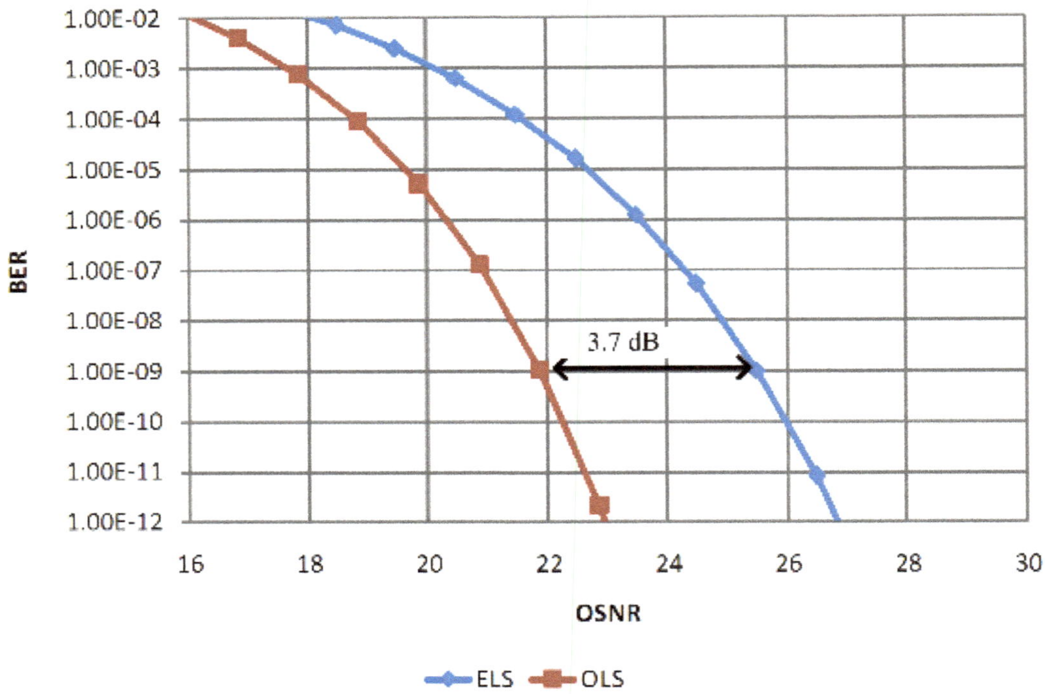

Figure 12. OSNR comparison between ELS and OLS of b2b 3 × 10 Gbit/s MSAC system.

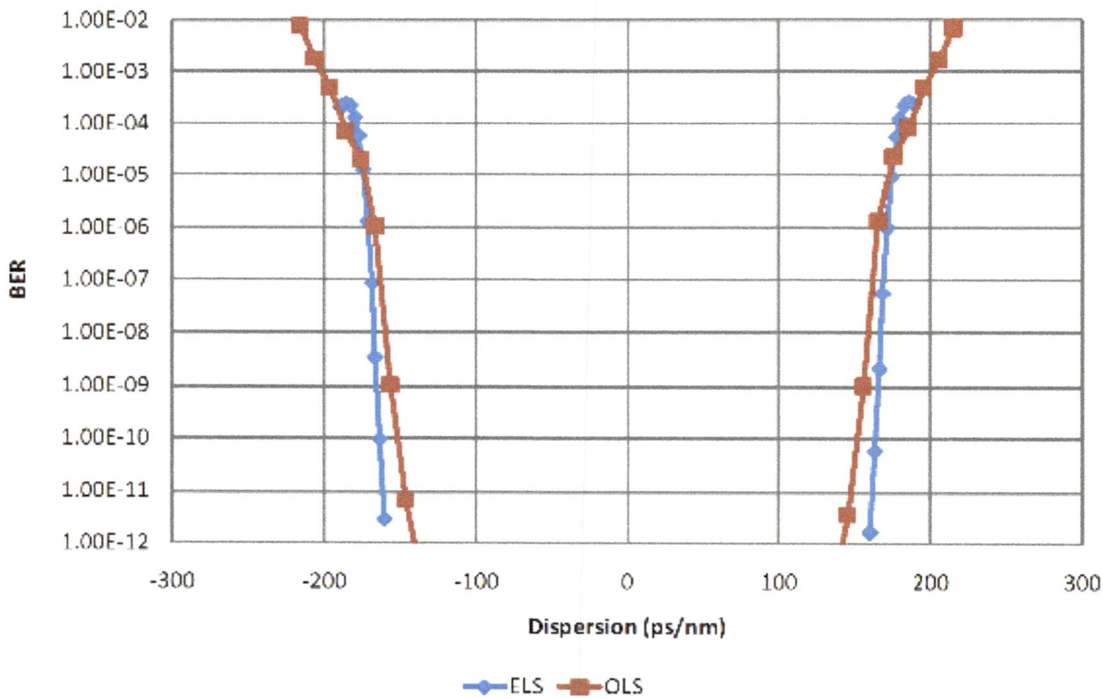

Figure 13. Dispersion tolerance comparison between ELS and OLS of b2b 3 × 10 Gbit/s MSAC system.

Figure 14 shows the comparison between ELS and OLS in terms of the received optical power for b2b 4 × 10 Gbit/s MSAC system. The received optical powers of ELS and OLS MSAC are −22.8 and −27.1 dBm, respectively, at BER of 10^{-9}. The improvement of receiver sensitivity around 4.3 dB is observed. The comparison between ELS and OLS in terms of OSNR is shown in **Figure 15**. Based on this graph, OSNR is 28.8 dB for ELS, whereas 24.5 dB for OLS. This result shows that the OSNR improvement has been achieved with similar amount for receiver sensitivity improvement. **Figure 16** shows the comparison between ELS and OLS in terms of CD tolerance. The CD tolerance is ±149.5 and ±73.1 ps/nm for ELS and OLS, respectively.

6.7. Performance comparison

The performance comparison between other electrical-based multiplexing is made according to transmission capacities, which are 30 and 40 Gbit/s. For this comparison, other versions of DCDM such as DCDM-amplitude distribution controller (DCDM-ADC), DCDM with dual drive Mach Zehnder modulator (DD-MZM), new multiplexed pattern DCDM (NMP-DCDM) and absolute polar DCDM (AP-DCDM) are included. **Table 1** shows the performance comparison between MSAC with various types of DCDM. It is very clear that MSAC with OLS has better performance compared to DCDM-ADC, DCDM with DD-MZM and NMP-DCDM. Note that level spacing for DCDM-ADC was optimized by installing a controller, whereas DCDM with DD-MZM was optimized using a dual drive Mach Zehnder modulator. MSAC with OLS offers better performance in terms of CD tolerance compared to AP-DCDM; however, the receiver sensitivity and spectral width are almost similar. Another advantage of

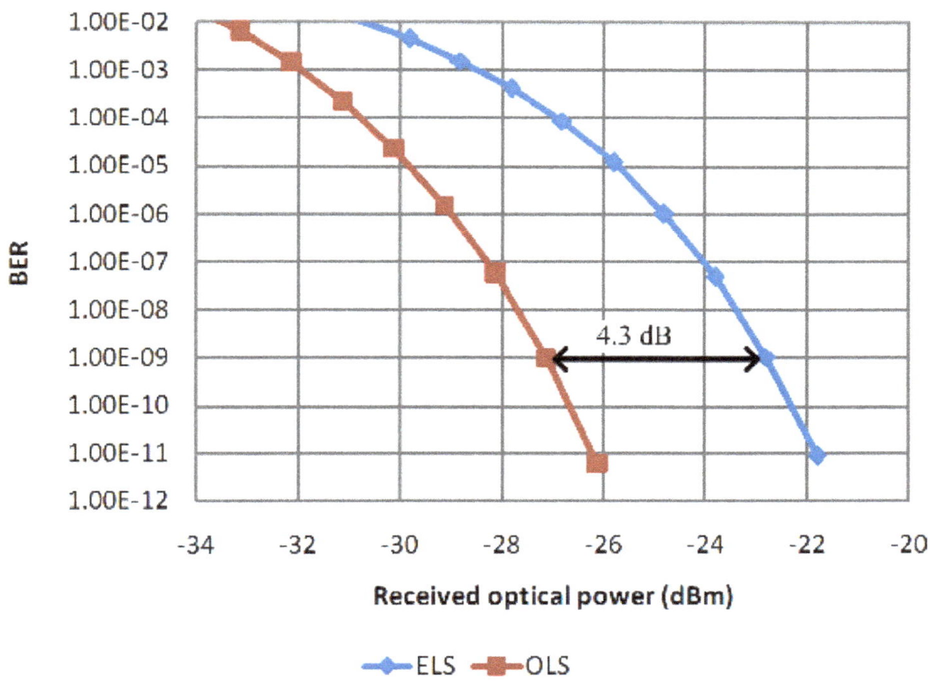

Figure 14. Received optical power comparison between ELS and OLS of b2b 4 × 10 Gbit/s MSAC system.

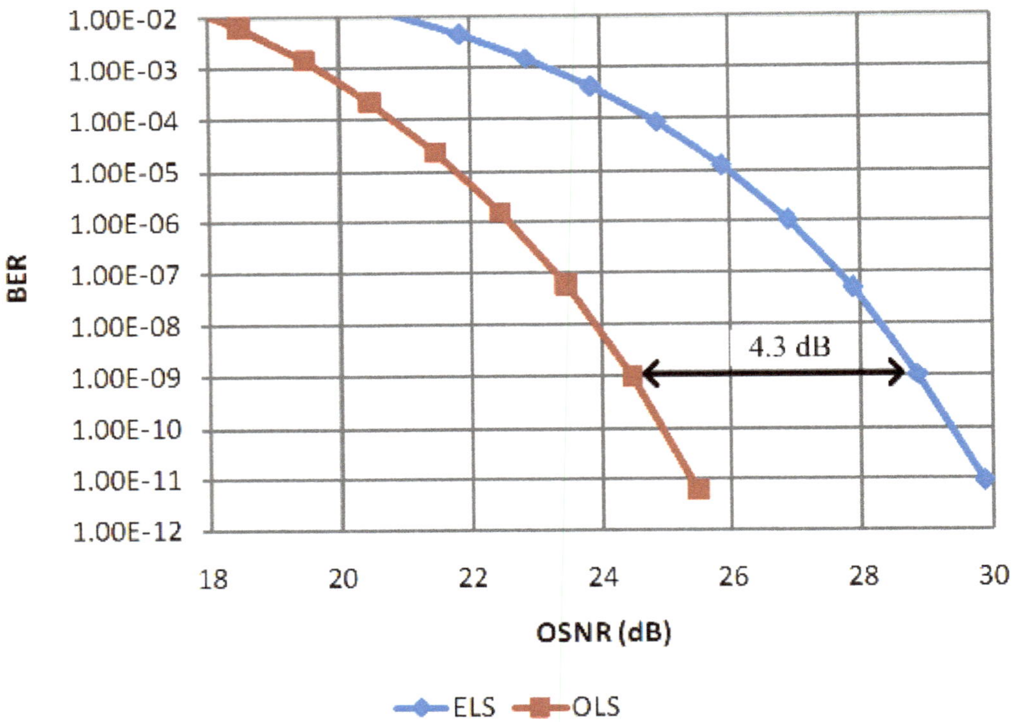

Figure 15. OSNR comparison between ELS and OLS of b2b 4 × 10 Gbit/s MSAC system.

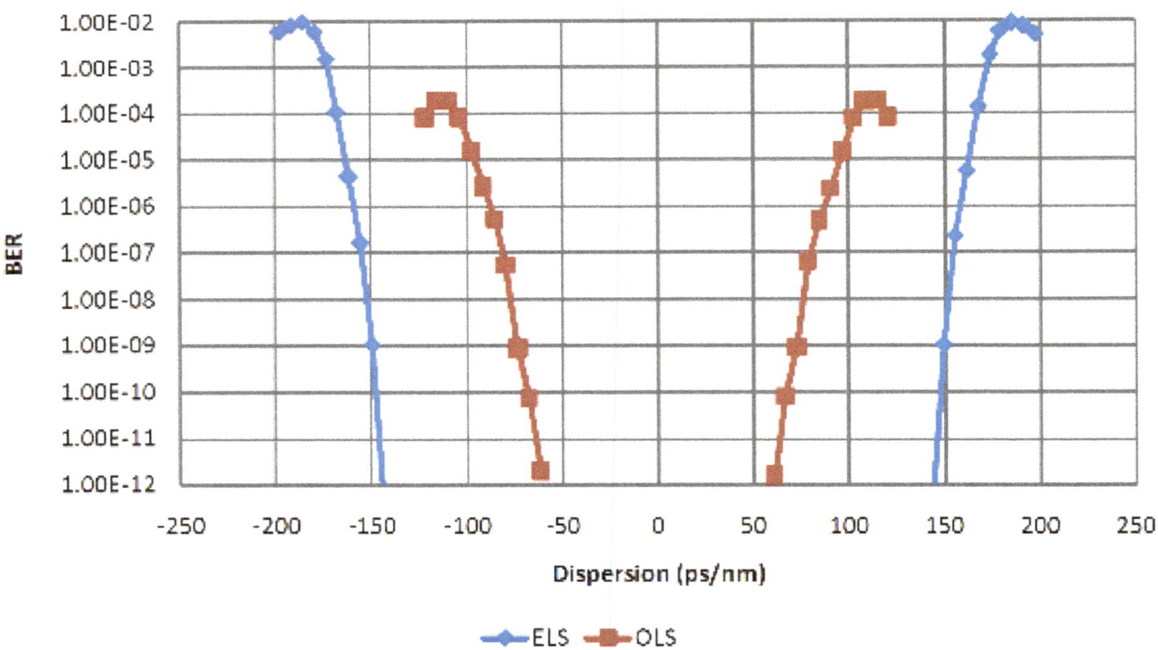

Figure 16. Dispersion tolerance comparison between ELS and OLS of b2b 4 × 10 Gbit/s MSAC system.

Technique	Guard slot	RS BER @ 10^{-9}	OSNR BER @ 10^{-9}	Spectral width (GHz)	CD tolerance (ps/nm)	
DCDM-ADC	Yes	−28.9	23.35	80	±93.5	[7]
DCDM with DD-MZM	Yes	−28.85	23.3	80	−59 to +100	[8]
AP-DCDM	No	−29	22.82	60	±109	[9]
NMP-DCDM	Yes	−28.4	23.5	–	–	[10]
MSAC with OLS	Yes	−29.5	21.8	60	±155	

Abbreviations: RS, receiver sensitivity; DCDM-ADC, DCDM amplitude distribution controller; DD-MZM, dual drive Mach Zehnder modulator; AP-DCDM, absolute polar DCDM; NMP-DCDM, new multiplexed pattern DCDM; NA, not available.

Table 1. Performance comparison between various DCDM version and MSAC for 3 × 10 Gbit/s system setup.

MSAC compared to AP-DCDM is high frequency component at symbol rate. This feature is obtained because MSAC symbols are equipped with guard slot, like DCDM or RZ, therefore, increases the transition density in the modulated signal. As a result, a complex circuit is not required for recovering clock frequency in receiver. This comparison indicates that MSAC provides significant advantages against DCDM and AP-DCDM for multiplexing high speed data stream in electrical domain.

Table 2 shows the performance comparison between DCDM-ADC, AP-DCDM and MSAC for optical transmission system with 40 Gbit/s setup. In general, the performance for all parameters for MSAC with OLS is better than DCDM-ADC. This finding is consistence with the result in **Table 1**. The performance of MSAC is also better than AP-DCDM version with guard slot. The receiver sensitivity for MSAC is inferior when comparing with the optimized level spacing AP-DCDM using DD-MZM, but it has better performance in terms of the spectral width and CD tolerance. Moreover, upgrading MSAC from 3 (3 × 10 Gbit/s) to 4 tributaries (4 × 10 Gbit/s) is achieved without extra spectral width. For ETDM, spectral width of 160 GHz

Technique	Guard slot	RS BER @ 10^{-9}	OSNR BER @ 10^{-9}	Spectral width (GHz)	CD tolerance (ps/nm)	
DCDM-ADC	Yes	−26	26.38	100	±58	[7]
AP-DCDM	Yes	−26.8	25.8	100	±62	[11]
AP-DCDM with DD-MZM	No	−31	NA	80	±68	[12]
MSAC with OLS	Yes	−27.1	24.5	60	±73.1	

NA, not available.

Table 2. Performance comparison between DCDM, AP-DCDM and MSAC for 4 × 10 Gbit/s setup.

is required for RZ format and 80 GHz for NRZ format for 40 Gbit/s system [13]. This means that ETDM is not bandwidth efficient compared to MSAC.

7. Conclusion

The evaluation of MSAC technique in fibre optic-based optical communication system has been carried out through an intensive numerical simulation. From the analysis, it is found that MSAC system performance is certainly better than DCDM-ADC, DCDM with DD-MZM and NMP-DCDM in terms of spectral width, spectral efficiency, receiver sensitivity, OSNR and CD tolerance. Besides that MSAC is also capable of achieving higher bandwidth or spectral efficiency compared to ETDM and AP-DCDM. Note that MSAC system is implemented using simple intensity modulation and direct detection scheme. As a result, the complexity of transmitter and receiver for this system is less than that of coherent scheme or optical time division multiplexing (OTDM) system. This issue is crucial for future high speed optical communication system in metropolitan region. This work has successfully demonstrated the possibility of implementing MSAC as advance multiplexing in the cost efficient high speed optical communication system for metropolitan application.

Acknowledgements

This work has been funded by the Ministry of Higher Education, Government of Malaysia under project FRGS/1/2015/TK04/UTHM/03/5.

Author details

Rahmat Talib and Mohammad Faiz Liew Abdullah*

*Address all correspondence to: faiz@uthm.edu.my

University Tun Hussein Onn Malaysia, Malaysia

References

[1] Lach E, Schuh K. Recent advances in ultrahigh bit rate ETDM transmission systems. Journal of Lightwave Technology. 2006;**23**(12):4455–4467. DOI: 10.1109/JLT.2006.886404

[2] Winzer PJ, Raybon G, Doerr CR, Duelk M, Dorrer C. 107-Gb/s optical signal generation using electronic time-division multiplexing. Journal of Lightwave Technology. 2006;**24**(8):3107–3113. DOI: 10.1109/JLT.2006.878016

[3] Mahdiraji GA, Abdullah MK, Mohammadi AM, Abas AF, Mokhtar M, Zahedi E. Duty-cycle division multiplexing (DCDM). Optics & Laser Technology. 2010;**42**(2):289–295. DOI: 10.1016/j.optlastec.2009.07.007

[4] Talib R, Abdullah MFL, Abdullah MK. Multi slot amplitude coding performance improvement with level spacing optimization. In: Conference Proceeding of the 4th International Conference on Photonics (ICP 2013); 28–30 October 2013; Melaka, Malaysia. New York: IEEE; 2013. pp. 206–208. DOI: 10.1109/ICP.2013.6687115

[5] Agrawal GP. Fiber-Optic Communication Systems. 4th ed. New York: John Wiley & Sons; 2010. DOI: 10.1002/9780470918524

[6] Winzer PJ, Gnauck AH, Doerr CR, Magarini M, Buhl LL. Spectrally efficient Long-Haul optical networking using 112-Gb/s polarization-multiplexed 16-QAM. Journal of Lightwave Technology. 2010;**28**(4): 547–556. DOI: 10.1109/JLT.2009.2031922

[7] Mahdiraji GA. Performance analysis of Duty-Cycle division multiplexing for optical fiber communication systems [thesis]. Serdang: Universiti Putra Malaysia; 2009

[8] Amouzad Mahdiraji G, Abas AF. Improving the performance of electrical duty-cycle division multiplexing with optimum signal level spacing. Optics Communications. 2012;**285**(7):1819–1824. DOI: 10.1016/j.optcom.2011.12.022

[9] Malekmohammadi A, Abdullah MK, Mahdiraji GA, Abas AF, Mokhtar M, Rasid MFA, Basir SM. Analysis of return-to-zero-on-off-keying over absolute polar duty cycle division multiplexing in dispersive transmission medium. Optoelectronics. IET. 2009;**3**(4):197–206. DOI: 10.1049/iet-opt.2008.0050

[10] Dahawü TH, Aljunid SA, Rashidi CBM, Rahman AK. Performance analysis of new multiplexed pattern for duty-cycle division multiplexing over optical communication system. In: 3rd International Conference on Electronic Design (ICED); 11–12 August 2015; Phuket. New York: IEEE; 2016. pp. 170–173. DOI: 10.1109/ICED.2016.7804630

[11] Malekmohammadi A, Mahdiraji GA, Abas AF, Abdullah MK, Mokhtar M, Rasid MFA. Realization of high capacity transmission in fiber optic communication systems using Absolute Polar Duty Cycle Division Multiplexing (AP-DCDM) technique. Optical Fiber Technology. 2009;**15**(4):337–343. DOI: 10.1016/j.yofte.2009.03.001

[12] Malekmohammadi A, Al-Mansoori MH, Mahdiraji GA, Abas AF, Abdullah MK. Performance enhancement of absolute polar duty cycle division multiplexing with dual-drive mach-zehnder-modulator in 40 Gbit/s optical fiber communication systems. Optics Communications. 2010;**283**(16):3145–3148. DOI: 10.1016/j.optcom.2010.04.026

[13] Miyamoto Y, Kataoka T, Tomizawa M, Yamane Y. 40-Gbit/s channel Technologies for optical Transport Network. Optical Fiber Technology. 2001;**7**(4):289–311. DOI: 10.1006/ofte.2001.0359

10

Receiver Performance Improvement in Radio over Fiber Network Transmission

Lisawati Sainawi and Khadijah Ismail

Additional information is available at the end of the chapter

Abstract

Nowadays, wireless demand is raised rapidly since the number of mobile-cellular telephone and broadband internet are springing up. Thus, many base stations (BS) and a lot of money are needed to satisfy the demand. Radio over fiber (RoF) is the solution to this problem since it can operate a lot of BSs that connected to central station (CS) by using optical fiber, as well as offering cost-effective solution. Due to some factors, received signals of RoF tend to be in a low quality. Those factors will lead to some problems such as high value of bit error rate (BER), low value of Q factor, and the receiver could not operate in high data rate network. Wavelength division multiplexing (WDM) network can be one of the solutions of those problems where different signals are transmitted through a single-mode fiber. Bit error rate must be decreased to a certain value, and the Q factor should be increased. The design of RoF will be simulated by using Optisystem software. The performance of RoF's receiver is measured and analyzed based on the obtained BER, value of Q factor and height of opening of eye diagram.

Keywords: radio over fiber, performance, wavelength division multiplexing, receiver, bit error rate, Q factor, eye diagram, optical fiber, eye pattern

1. Introduction

Any form of telecommunication system that having light as the transmission medium is known as optical communication. Optical communication system consists of a transmitter, channel and a receiver. The transmitter will **encode** a message into an optical signal, the channel will carry the signal to its destination, and the receiver will reproduce the message from the received optical signal.

Radio over fiber (RoF) is a technology where RF signal modulates light and then transmitting it over an optical fiber link. Both wireless network and optical are supported by this technology. It is essential for communication system to have high capacity and subcarrier frequency since wireless signal sometimes intend to loss channel at the time of data transmission [1].

RoF is very convenient system since it is less costing and power consumption. This is because RoF lets the electrical signal modulates the optical source and after that the optical signal will travel along the optical fiber to the remote station. When the RF signal is modulated straight to the optical link, the power consumption drops while the antenna side has high frequency radio carriers. The reduction of cost in RoF technology can be explained in two ways. The first one is central station (CS), which provides resources that can be shared by variety of base stations (BS), and secondly, BS only executes simple function. Furthermore, the BS is in a small size and less cost consuming.

Basically, in this technology, central station (CS) is connected with many base stations (BS) by using optical fiber. BS only functions as a converter of optical signal into a wireless signal and vice versa, while at CS, all process involving modulation, demodulation, coding and routing are executed [2]. By using high linear optic link, RoF system distributes the RF signal between CS and BSs. **Figure 1** shows the basic construction of radio over fiber network transmission.

At the transmitter side, data or information from internet or other CS are fed onto modem in the CS during downlink process. Optical signal from the optical source is modulated by the RF signal. After that, the modulated signal will pass through optical fiber toward the BSs. As soon as the signal reaches the BS, photodetector (PD) will function as a detector to detect the modulated optical signal. The PD will also recover the signal before the signal is transmitted through antenna of the BS toward the mobile host. With the similar concept, reverse process is executed for uplink process between mobile host and CS. When the BS receives the signals, optical signal from the LD will be modulated to amplified and transmitted straight toward CS.

In order to determine either the receiver of the RoF in a good quality or not, we can measure and analyze the bit error rate (BER), Q factor value and the eye opening of the resulting result. Bit error rate is the number of received bits of a data stream over a communication channel that has been altered due to noise, interference and distortion orbit synchronization errors. International Telecommunication Union (ITU) has stated that the minimum value of BER of RoF must be below than 10^{-9}. Basically, the value of BER is depending on the measurement time and factors lead to the error. The value of Q depends on the value of BER. There are several ways to determine either our obtained q factor is suitable for our BER value or not. The easiest way is to determine it from BER and Q factor graph. **Figure 2** shows the graph that indicates the relationship between the value of BER and Q factor.

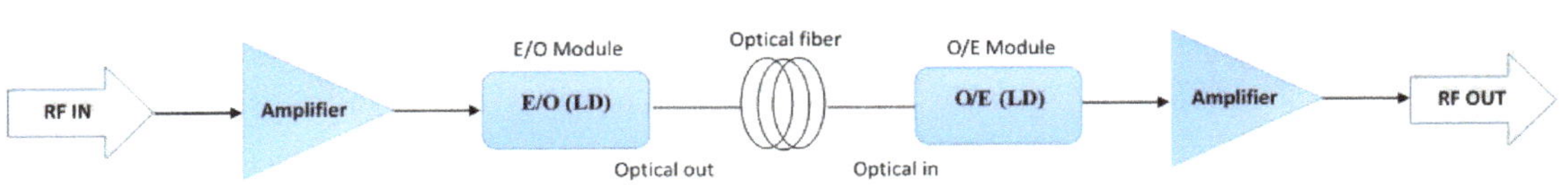

Figure 1. Basic construction of radio over fiber.

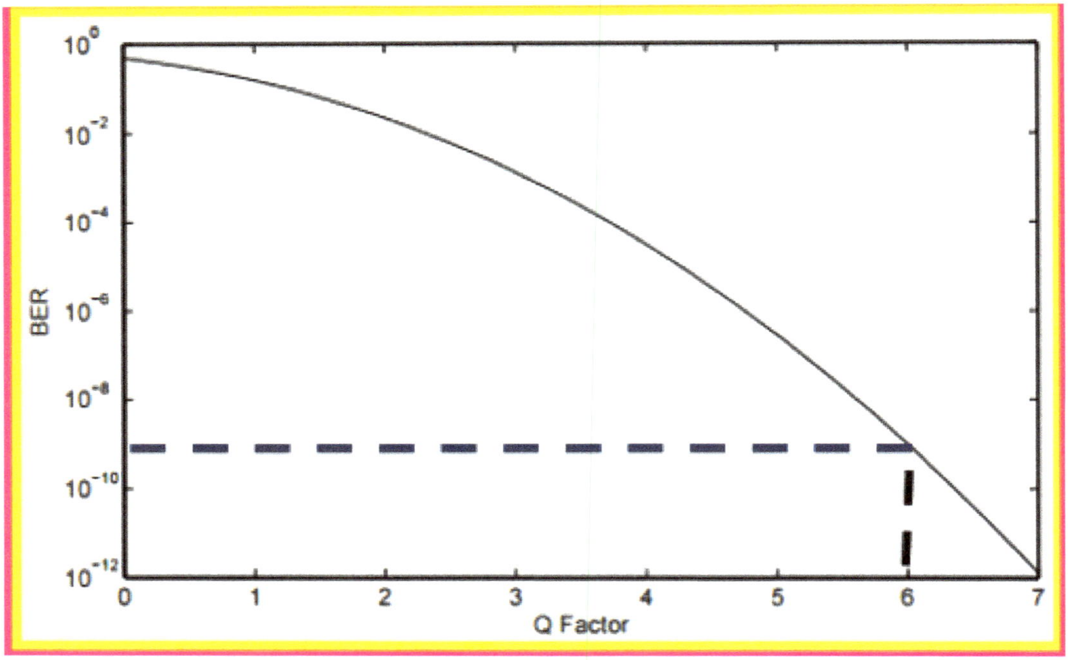

Figure 2. BER versus Q factor.

From the graph, we can see that the value of Q factor increases when the BER decreases. From the graph also, we could see that the value of Q factor for 10^{-9} of BER is approximately 6. Thus, RoF system should have a value of Q factor more than 6 in order to obtain a very good performance of its receiver.

The receiver performance of RoF can also be measured by analyzing the eye diagram of the result after the design has been simulated. Eye diagram analyzer shows multiple traces of a modulated signal to produce eye diagram. This eye diagram is an oscilloscope display where the eye pattern diagram corresponds to minimal signal distortion due to intersymbol interference (ISI) and noise appear in the system [3]. The measurements of eye pattern are made based on time domain. The effect of waveform distortion will appear on the screen of regular BER test equipment. There are many information that could be obtained from eye pattern display. Time interval over which signal that has been reach at the receiver could be sampled without any error causes by ISI could be defined by the width of the eye opening. Noise margin can be determined by looking at the height of eye opening at specified sampling time.

2. Overview of radio over fiber

In 1990, RoF is firstly demonstrated for mobile telephone services or cordless. In this technology, highly linear optic fiber links are used to connect CS and BSs so that RF signal distribution between CS and BSs can be done. Processes involving modulation, demodulation, coding and routing are executed mostly at CS. That means, the only processes occur in BSs only converting optical signal to electrical signal or vice versa. A lot of studies and research have been done just to investigate the limitation and generate new idea to increase the performance of RoF technologies.

Basic RoF consists of all hardware that enable to foist RF signal on an optical carrier at the transmitter side. It also needs fiber optic link to distribute the signal from CS to BSs. At the receiver side, RoF required all the hardware needed to recover the signal from the optical carrier.

2.1. Optical transmission link

2.1.1. Optical fiber

Optical fiber is a platform or a medium to carry information in the form of light from a point to another point. Functioning as a waveguide that will allow the propagation of light, a fiber is a thin filament of glass. One of the advantages of fiber is it provides a path for light with some losses due to the concept of total internal reflection.

In communication system, there are three types of optical fiber, which are step-index multimode, step-index single mode and graded index. Step-index multimode fiber is measured from cladding to the core to cladding as it serves an index of refraction profile that steps from low to high to low. For step index single mode, only one path is allowed for the light to travel in the fiber. Graded index provides large core diameter and higher bandwidth of single-mode fiber.

Optical fiber provides two regions that have low attenuation. First region is at approximately 1300 nm, which has attenuation less than 0.5 dB/km and bandwidth 25 THz. Second region is at approximately 1550 nm, which has attenuation less than 0.2 dB/km and also has bandwidth 25 THz [4]. This combination of two regions will make the total bandwidth as much as 50 THz. Due to the both low attenuation regions, signal loss in data transmission will be very small. Thus, we do not need a lot of amplifiers and repeaters.

2.1.2. Attenuation, dispersion and nonlinearities in fiber

In a fiber optic, attenuation can affect the signal power during the propagation of the signal through distances. Attenuation is a must aspect need to be considered in order to determine the longest distance that signal can go for a given sensitivity of the receiver and the power of the transmitter.

Widening of pulse duration when it propagates through a fiber is known as dispersion. When one pulse starts widening, the pulse is interfering with another pulse besides it. Thus, intersymbol interference (ISI) can happen. Because of this phenomenon, the pulse spacing and the maximum transmission will be limited. There are several types of dispersions and one of them is intermodal dispersion. Intermodal dispersion can happen when several modes of similar signal travels at different velocities through the fiber. Single-mode fiber would not have this kind of dispersion.

Chromatic dispersion is another type of dispersion. Most system will have this type of dispersion since there is no laser able to create a signal that having a single wavelength. Wavelength will be functioning as an index of refraction in dispersive medium. Certain wavelength will travel more fast than other wavelength if the signal that being transmitted has more than one wavelength [5]. Waveguide dispersion might be happen when the propagation of not similar wavelengths depends on characteristics of the waveguide such as indices and shape of fiber core and cladding. Chromatic dispersion is almost 0 in single-mode fiber at 1330 nm. About 1330 nm is also a low attenuation region, and thus, fibers with 0 dispersion can be achieved by using advance techniques such as dispersion shifting.

Nonlinearities may cause an attenuation, distortion and cross-channel interference. Its effect is able to affect the performance of wavelength division multiplexing (WDM) system. In WDM system, nonlinear effects can affect the spacing between similar wavelength channels, limit the maximum power of any channel and able to limit the maximum bit rate.

2.2. Optical transmitter

2.2.1. Optical sources

The most common of light sources used in fiber optic communications are laser diode and light-emitting diode (LED). The benefit of these devices is both have output power for wide range applications. The power also can be directly modulated where the input current is varied to the devices. The efficiency is also high, and they are compatible with the optical fiber.

The diffrence between LED and laser diode is that the laser diodes gives a coherent output where the optical energy is produced in an optical resonant cavity. **Figure 3** shows a basic structure of a laser.

Two mirrors in the laser will form a space between both of them called cavity, a lasing medium that occupied the cavity and a device for excitation. Lasing medium will receive current by the excitation devices and will produce a photon of light. The photon will reflect off the mirrors at both end of the cavity and will go through the medium again.

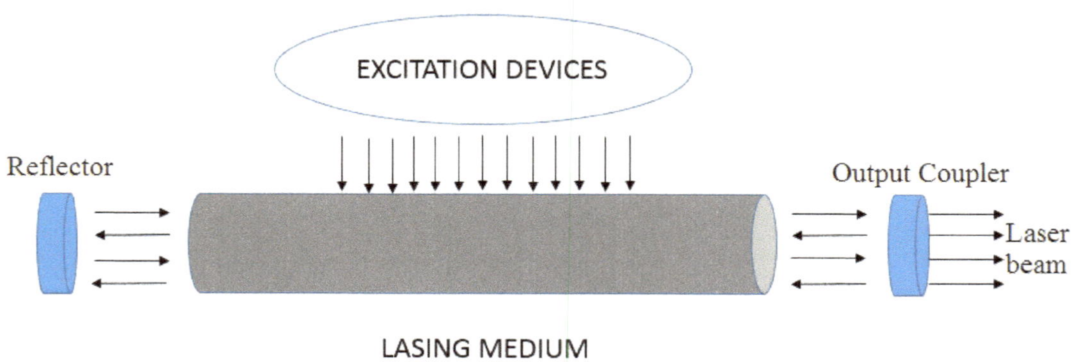

Figure 3. Laser structure.

2.2.2. Optical modulation and line coding

To transmit information across an optical fiber, the data are compulsory to be encoded or modulated onto a laser signal to allow the data being transported through the optical. There is variation of method for modulation. This is including an analog techniques such as frequency modulation (FM), phase modulation (PM) and amplitude modulation (AM). For digital signal, method of modulation includes amplitude shift keying (ASK), phase shift keying (PSK) and frequency shift keying. Among these techniques, binary ASK is the most preferable due to its simplicity. In the systems implementing ASK, the laser is switched on and of to achieved modulation techniques [6]. Current optical communication system also reported on the usage of on-off-Keying (OOK) and DPSK modulations.

To transport digitized information in a communication link, format of transmission the signal must be considered. The signal format is so important since the receiver needs to extract accurately the timing information from the incoming signal. Line coding has several principal functions and one of it is to minimize the errors which causes by noise or any interference effects in the bit stream. The easiest method for encoding data or information is unipolar return to zero (NRZ) code. Logic 1 is represented by a light source that fills the whole bit period, while logic 0 is represented by no pulse transmitted. The process turns the voltage on and off and that's why it is known as (ASK) or on-off keying (OOK). It is essential for NRZ to have a minimum bandwidth and NRZ must be simple to generate and decode.

2.3. Optical receiver

Optical receiver consists of signal processing circuitry, an amplifier and a photodetector [7]. The first thing receiver did when received a signal was converting the optical signal into an electrical signal. After that, the signal will be amplified to an optimum level so that the following process can be done. It is essential to determine and predict the performance of the system based on mathematical models of many receiver stages in order to design a receiver. Also, when designing a receiver, noises and distortions contributed by component in every stage must be considered. Plus, the receiver must have the ability in detecting weak or distorted signals, ability in making a decision on type of received signal and ability to reshaping the distortion signal. This is why it is more complicated compared to process of designing a transmitter. Bit error rate is the most important criteria in measuring RoF system. Other than that are Q factor and the opening of eye diagram.

The height of the eye diagram opening indicated the level of signal distortion. The upper level of the eye represents binary '1' and the bottom level represents binary '0'. Higher eye opening height is desirable, as this indicating that the binary '1' and '0' can be distinguished well. The height of eye opening at a specified time corresponds to the noise margin achieved.

2.3.1. Photodetector

It is compulsory for a receiver to have a device that can interpret the information in the optical signal. Photodetector is a device that can convert the incoming photonic stream into a stream of electrons. When the optical signals reached the receiver, it is in weak and distort condition after undergoes the optical fiber. Thus, photodetector should be sensitive to the emission wavelength range of the optical sources being used. Photodetector also must have an addition of noise to the system, and most importantly, it has to response fast to handle the target data rate. Also, photodetector must be immune to the change of temperature and must be compatible with the physical dimensions of optical fiber. After the photodetector changes the optical signal into an electrical signal, the signal will be amplified and will undergo the threshold device. To determine either the bit is 0 or 1, the presence of light is referred to during the bit duration. It depends on either the electron stream is below or above a certain threshold.

PIN photodiode is the most usable semiconductor photodetector. The structure of this device as shown in **Figure 4** consists of p and n regions. Both regions are separated by intrinsic (i) region [8]. For normal operation, an optimum reverse-bias voltage is supplied across the photodiode to allow the intrinsic region completely depleted of carriers. This will cause the n and p carrier concentrations that become less than impurity concentrations in the region. When a photon flux Φ penetrated into the device, the flux is absorbed.

In PIN photodetector, the light absorption causes the formation of electron-hole pairs. Then, the hole and electron are drifted to the opposite direction, causing the flow of current. More current flows as more light enter the photodetector, which giving rise to the number of electron-hole pairs.

When the energy of an incident photon is higher than or equal to the semiconductor's bandgap energy, the photon excite an electron from valence band to the conduction band by give up its energy. The absorption process will form an electron-hole pairs known as photo carriers. Normally, the photodetector is designed so that those carriers are mainly generated in the depleted intrinsic region. This region has the most absorption of incident light. The large amount of electric field will make the carriers separating between each other and will be collected across the reverse biased junction. Thus, current flow will increase in the external circuit where every carrier pair generated has one electron flowing.

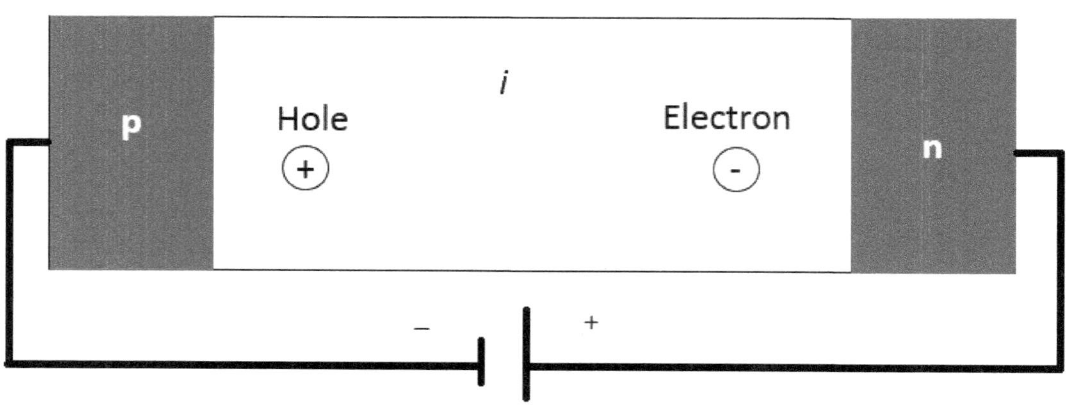

Figure 4. PIN photodiode structure.

2.3.2. Optical amplifier

Optical amplifiers can contribute a lot to long haul or local networks even optical signal still can transmit a long distance without amplifiers. There are several techniques of optical amplification. 1R (regeneration) techniques provide a booster to power up the signal. It does not restore the shape and the timing of signal. It also serves a total data transparency. 2R (regeneration & reshaping) amplification is a technique where the optical signal will be transformed into electrical signal before it directly used to modulate a laser.

3R (regeneration, reshaping, relocking) techniques change the data stream into electronic signal and after that retransmit the signal optically to amplifies the signal [9]. Noise can be eliminated much through the reshaping process where the signal produces back the original pulse shape of each bit. Mostly, reshaping works on digital signal but for some cases it may also works on analog signal but relocking does not work on analog modulated sign.

Basically, there are three main types of optical amplifiers. They are semiconductor optical amplifier (SOA), doped fiber amplifier (DFA) and Raman amplifier. The main function of all amplifiers is to boost the power level of incident light through optical power transfer process. Also, this process can be done through a stimulated emission. Using the concept of laser diode, the mechanism of SOA and DFA is creating the population inversion which is needed for simulated emission. Optical amplifier does not have the ability to generate coherent optical output since it does not have optical feedback which is compulsory for lasing. Thus, optical amplifier can only increase the signal levels.

The amplifier gained an energy from a pump, which is an external source as shown in **Figure 5**. Technically, the external sources are supplying an energy to the electrons in the active medium. This will cause the electrons increased and served a population inversion. The excited electrons will drop to lower level because the electrons are triggered by the incoming photon signal through the process of simulated emission. The output of the signal will be amplified since one incoming photon stimulates a cascaded effect in which way equal energy of photons is emitted by a lot of excited electrons when they hit the ground state.

Advantage of SOA is that this amplifier can be implemented if both signal processing and switching functions were call in optical networks. The drawbacks of SOA is it has rapid gain response that will cause the gain at specific wavelength fluctuate with signal rate for speed up

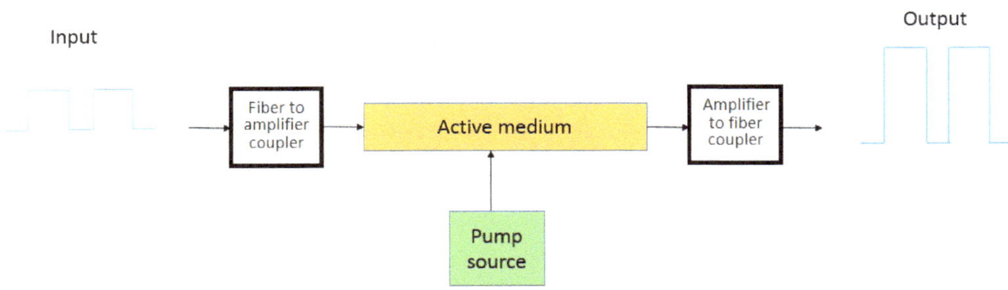

Figure 5. Signal amplification by an optical amplifier.

to some Gb/s. The overall gain also can be affected, and this will cause the signal gain of other wavelengths fluctuate. Thus, cross-talk effect can occur.

For DFA, the length of fiber will be doped with an element that able to amplify the light [10]. The most common element is erbium. The main advantage for this type of amplifier is it can pump the devices at some different wavelengths. DFA also has small coupling loss and low dependence of gain on light polarization. Plus, DFA is immune from interference effects since the gain responses are constant for signal modulations greater than few kilohertz.

Raman amplifier has a transfer of optical power form high power pump wavelength to light wave signals at longer wavelengths. Raman amplification does not need process of population inversion. It is based on stimulated Raman scattering (SRS) which is a nonlinear effect. This effect normally occurs in fibers at high optical powers. Raman gain mechanism was achieved through a discrete amplifier. This type of amplifier can be used in any wavelength band since the gain is based on SRS where the induced transfer of optical power from shorter pump wavelengths to longer signal wavelengths.

3. Overview of wavelength division multiplexing (WDM)

Various methods can be executed to gain full duplex transmission of RoF system and wavelength division multiplexing (WDM) is one of it. WDM has characteristic where the discrete wavelength will form an orthogonal set of carriers that can be separated, routed and switched without interfering with each other [11]. With WDM, many wavelengths are able to be transmitted over a large distance and the downlink and uplink data can be transmitted at the same time through single-mode fiber.

The design of WDM involves some requirements such as selecting various optical sources that has narrow spectral emissions bands. The most straightforward and simplest method can be done by choosing a series of individual lasers that emits at its own specific wavelength. The process of selecting optical sources requires small number of wavelength channel but can be cumbersome for links carrying many wavelengths. In order to implement WDM network, a combination of many passive and active component is need. The difference between passive and active components is passive components and is limited in their ability since passive components did not require an external control to operate, whereas active component can be controlled and has a wide network ability.

WDM contributes simplification to the network. With this method, different wavelengths are allocated toward individually BSs. Thus, the network will be simple and the service upgrades can be done easier. WDM multiplexes optical signals that came from multiple sources, and the signal will be amplified before it is transmitted through optical fiber to under goes demultiplexer so that it can be addressed to each BS as shown in **Figure 6**.

There are two types of WDM architecture: coarse wavelength division multiplexing (CWDM) [12] and dense wavelength division multiplexing (DWDM) [13]. Typically, CWDM systems provide eight wavelengths which separated by 20 nm, from 1470 to 1610 nm. To increase the

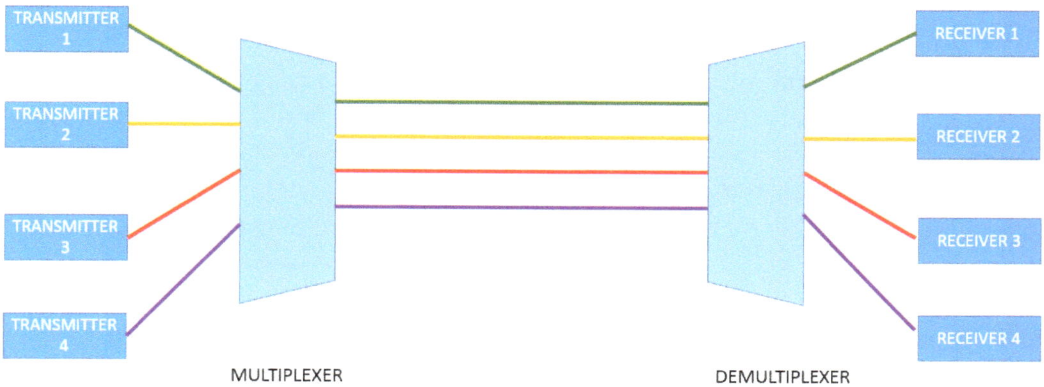

Figure 6. Wavelength division multiplexing.

channel of CWDM to 16, number of wavelength could be increased by using 1310 nm window. One of the advantages of CWDM is that the cost of the optics is 1/3 of the cost of same DWDM optic. This makes that the CDWM is preferable than DWDM. CWDM is able to match the basic capabilities of DWDM but with lower cost and capacity. Typically, CWDM is used for short-range communications. In addition, CWDM equipment is more compact and cost-effective if compared to DWDM designs.

DWDM network systems provide up to 96 wavelengths, which normally has less than 0.4 nm spacing. DWDM is used for long-haul transmission where wavelengths are compacted tightly together. Erbium-doped fiber amplifiers (EDFA) can help the system of DWDM to work over thousands of kilometers. CWDM is not implemented in long haul transmission, where the distance reaches up to thousands of kilometers, causing simpler overall sytem components requirement. This leads to lesser cost implementation, despite of having some limita-tions especially in propagation distance.

4. Methodology

The methodology used in designing RoF system consists of familiarization with Optisystem software, designing system, generate component and simulation object, running the simulation and analyze the data. Optisystem software is used to design, construct and simulate the RoF topology.

Optisystem software is a comprehensive software that provides a platform to plan, test, and simulate optical links in the transmission layer of modern optical networks. Optisystem is also an optical communication system simulation package for the design, testing, and optimization of virtually any type of optical link in the physical layer of a broad spectrum of optical networks, from analog video broadcasting systems to intercontinental backbones.

A system level simulator based on the realistic modeling of fiber-optic communication systems, Optisystem, possesses a powerful simulation environment and a truly hierarchical definition of components and systems. Its capabilities can be easily expanded with the addition of user components and seamless interfaces to a range of widely used tools. Optisystem is compatible with Optiwave's Optiamplifier and OptiBPM design tools. Optisystem serves a wide range of applications, from CATV/WDM network design and SONET/SDH ring design to map design and transmitter, channel, amplifier, and receiver design. Optisystem contains a MATLAB component that enables the user to call MATLAB within its environment to incorporate new components or models into the software. Optisystem uses the MATLAB.dll files to evaluate the MATLAB script in the component to perform the calculations.

Designing RoF system on Optisystem software includes generation of component, simulation object and makes connection between all of them. It is very important to design a very good topology so that the maximum performance of RoF's receiver can be achieved.

5. Simulation

5.1. System design

To send two signals, two continuous wave lasers are used in order to resonate 193.1 and 193.2 THz, respectively, as shown in **Figures 7** and **8**, respectively. To obtain an electrical network that simple and low in speed, a pulse generator is needed. A pseudo-random bit (PRBS) generator is used to operate a Gaussian pulse generator so that baseband signal can be generated. 1 Gbps was set as the bit rate. The PRBS generator indicates the random source of data. To simulate the signal, 1 Gbps per channel was set as the data rate for the simulation.

As the carrier frequency is 2.7 GHz for transmitter 1 and 1.7 GHz for transmitter 2, an amplitude modulator is operated by the 1 Gbps baseband signal. After that, the amplitude modulator will operate the LiNb Mach-Zehnder modulator (MZM)'s port 1. The main function of the amplitude modulator is to translate the baseband signal onto the RF clock. At a frequency of 49.25 MHz for both transmitter, the SCM carrier generator will mix with the RF signal. A 90° hybrid coupler is applied to the signal where it is utilized to separate that the input signal becomes 2 outputs that having 90° phase shift difference between each other. The separated signals are sent to the two arm of M-Z Modulator. At the both transmitter sides, to allow the signal being transmitted along the fiber, the signals change from digital to analog form.

Wavelength digital multiplexer (WDM) is used to multiplex all signals from the transmitter. The bandwidth of the multiplexer was set to 10 GHz. EDFA amplifier with power 10 dBm is used to boost up the power of the optical signal. Then, through an optical fiber, the signals are transmitted. This process can be seen in **Figure 9**.

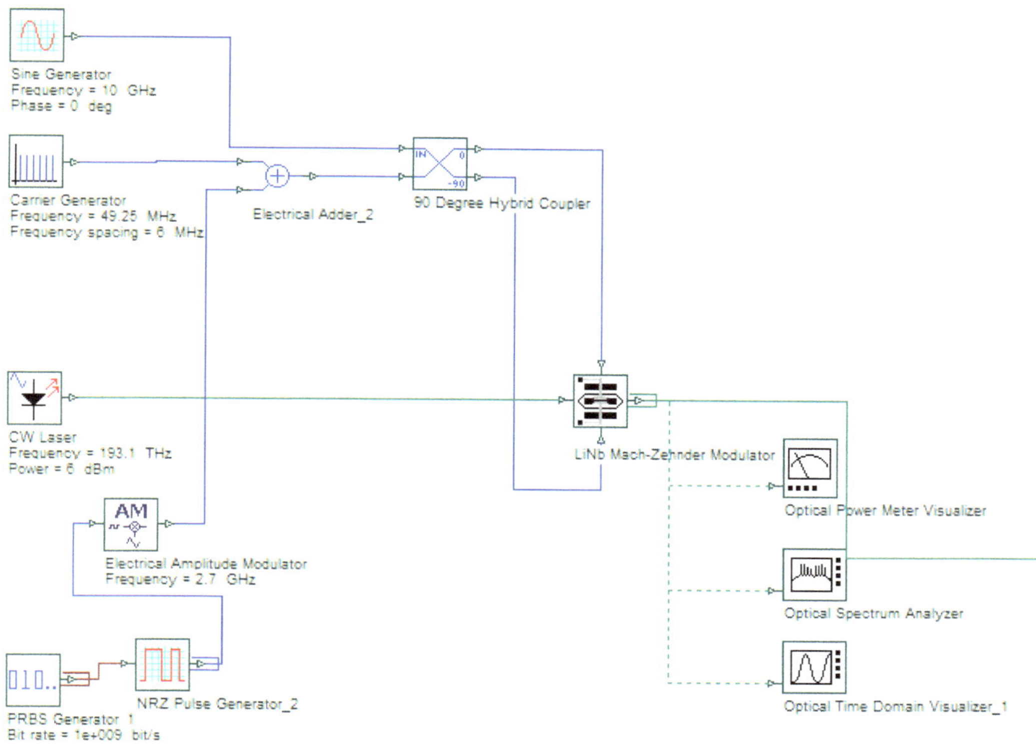

Figure 7. Transmitter 1.

As soon as the signal reaches the receiver side, demultiplexer is used to demultiplex the signals before the signals being distributed to its own receivers. 16 GHz is chosen as the bandwidth of the demultiplexer. At the receiver 1 as shown in **Figure 10**, the signal is once again amplified by using EDFA amplifier with power of 20 dBm. After that, photodetector is used to convert the optical signal into electrical form. Then, bandpass Bessel filter received the signal and filtered the signal. The received signal is amplified by using electrical amplifier with 15 dB gain before demodulated by using demodulator. 3R regenerator technique is applied to the signal before the BER analyzer analyses the received.

Same concept is applied at the second receiver. The EDFA amplifier is used to boost the power of the optical signal after the signal is demultiplexed. Photodetector will transform the signal into electrical form. The converted signal will be amplified by electrical amplifier, and the signal will be passed to the Bessel filter. After the signal has been filtered, demodulator is used to demodulate the signal before the 3R regenerator amplification is applied to the signal. Lastly, BER analyzer will analyze the signal. The configuration of the second receiver can be seen in **Figure 11**.

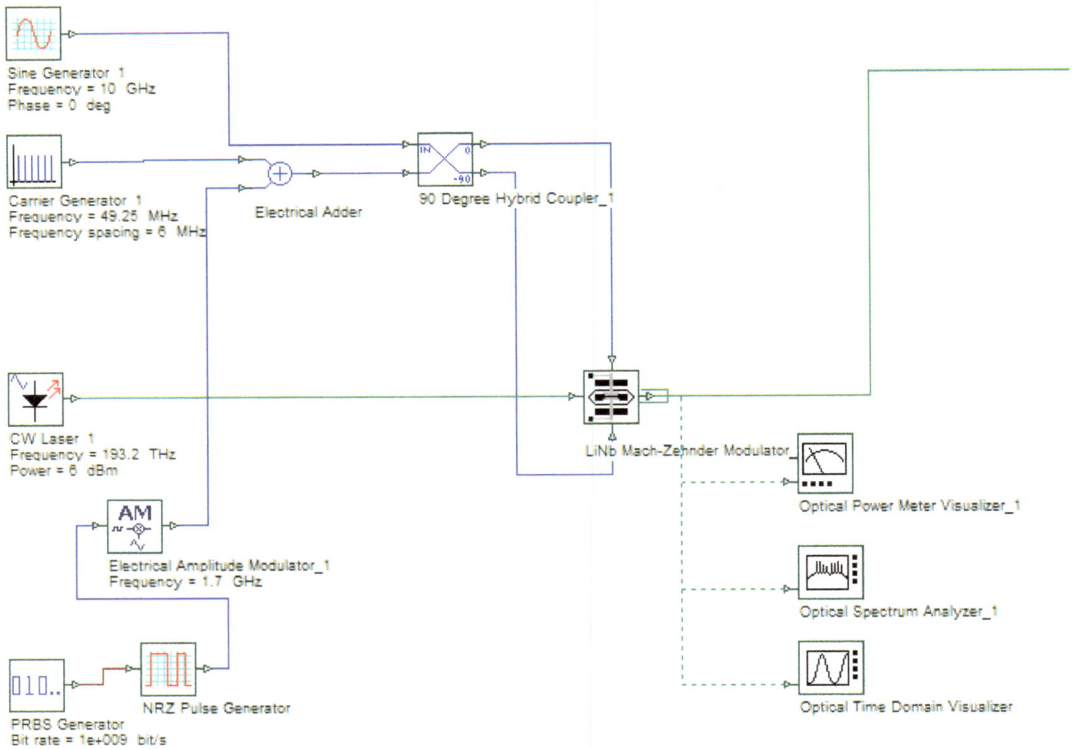

Figure 8. Transmitter 2.

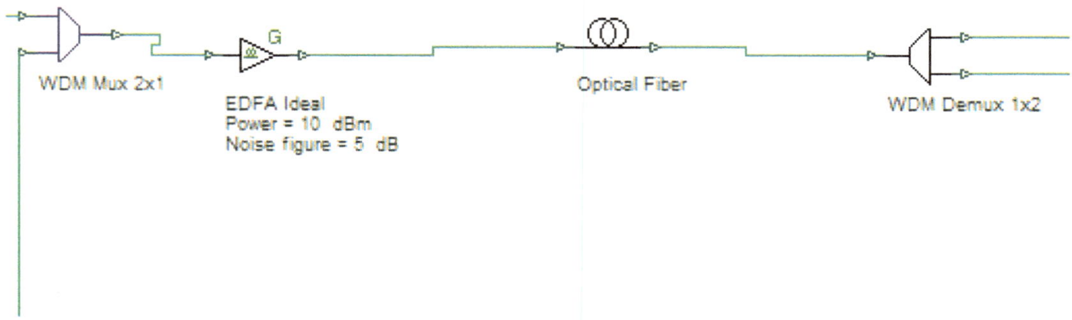

Figure 9. Transmission of data through optical fiber.

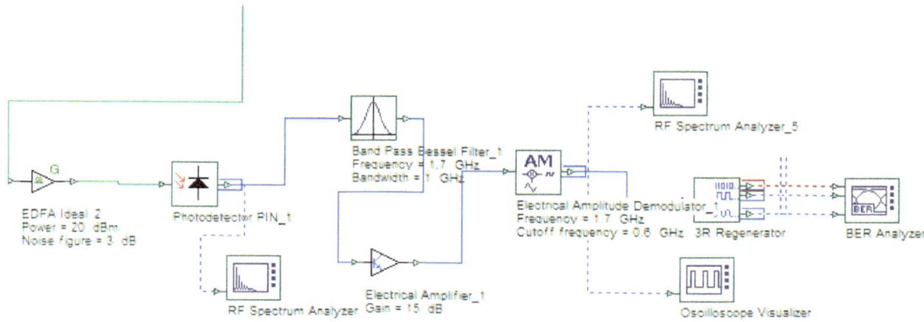

Figure 10. Receiver 1.

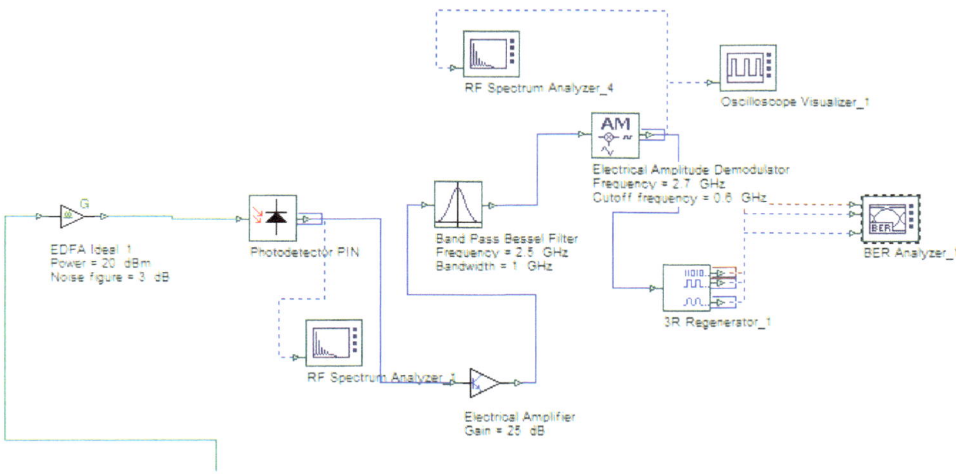

Figure 11. Receiver 2.

5.2. Single parameter optimization (SPO)

SPO can benefit a lot in the simulation. It helps to optimize parameters so we could set a target for the simulation's result. With optimization tools, the software can optimize the fiber length of the EDFA so that a maximum gain could be obtained. It can also calculate the attenuation or the gain in order to get a desired Q factor and minimize the BER by optimizing the fiber length of the system.

After all components had been connected and SPO had been inserted to the topology, the simulation is run to see the result. After the run button is clicked, it will calculate all the calculation of the system. To measure the BER and the Q factor, just simply double click at the BER analyzer component. It will give an eye diagram showing eye opening, BER value and Q factor value. The example of the result is shown as in **Figure 12**.

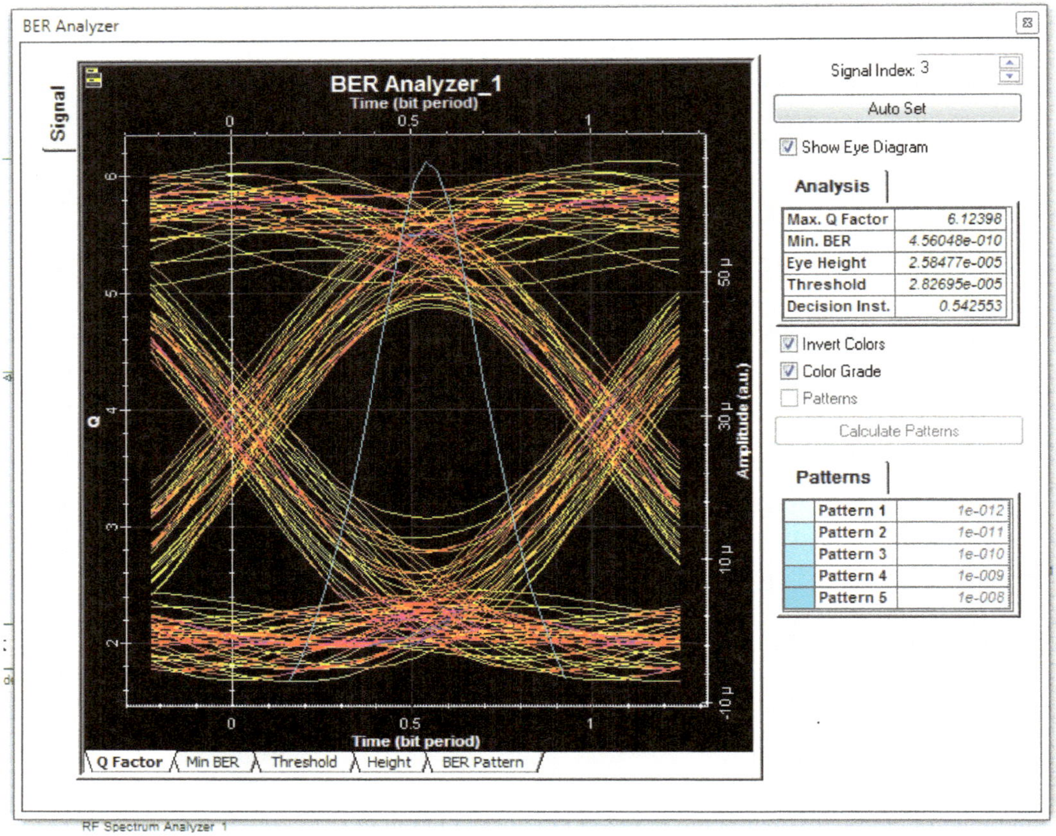

Figure 12. Example of BER analyzer's result.

As shown in the example above, the result tells us that the BER is 4.560^{-10} and the Q factor is 6.124. There is an eye opening in the result of the example which is 2.585^{-5}. The example is a quite good result since the BER is less than 10^{-9} and the Q factor is more than 6. If all the receivers receive an output just like the example above, all objectives had been achieved.

6. Results

After the simulation is done, the performance of the receiver is analyzed by referring the BER analyzer and the eye diagram at the two receivers. All the results can be seen in **Figures 13–16**.

For receiver 1, the BER analyzer shows that the BER for the received signal is 3.54×10^{-20}. The Q factor for this receiver is 9.13. This shows that this receiver has a very good performance since the number of BER is above 10^{-9} and the Q factor is more than 6.

As mentioned before, the height of the eye diagram defines the immunity of the received signal to the noise. The eye diagram for receiver 1 shows an eye opening of 0.0004.

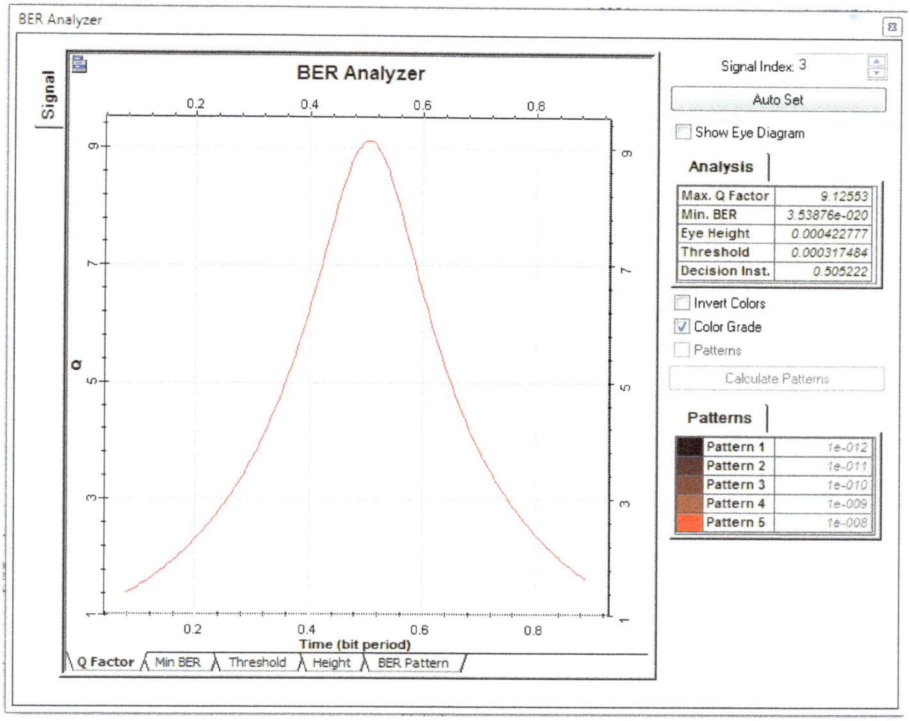

Figure 13. BER result for receiver 1.

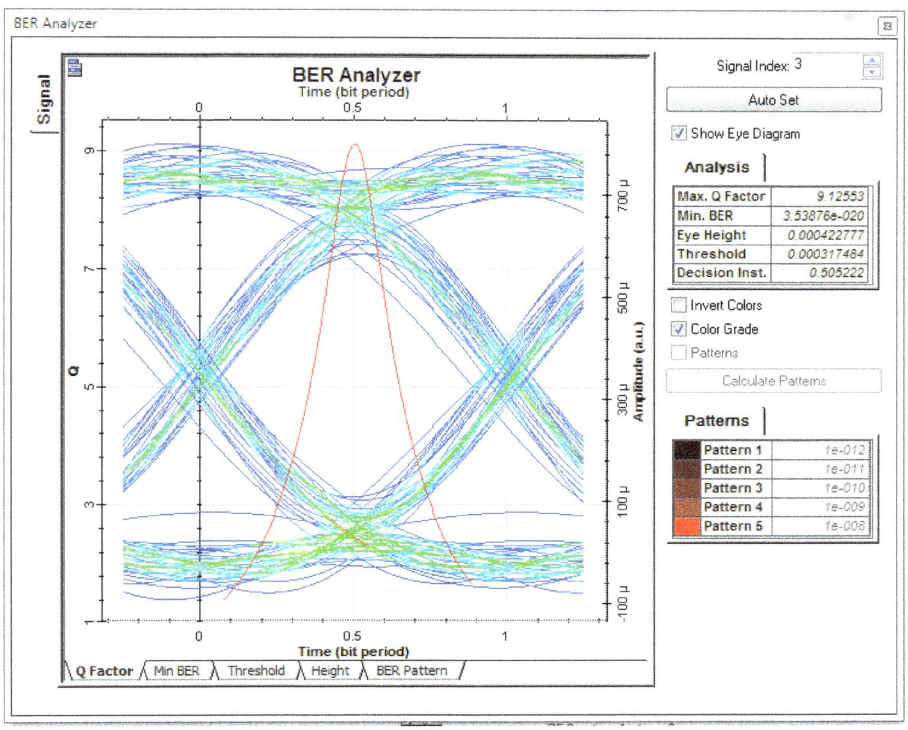

Figure 14. Eye diagram for receiver 1.

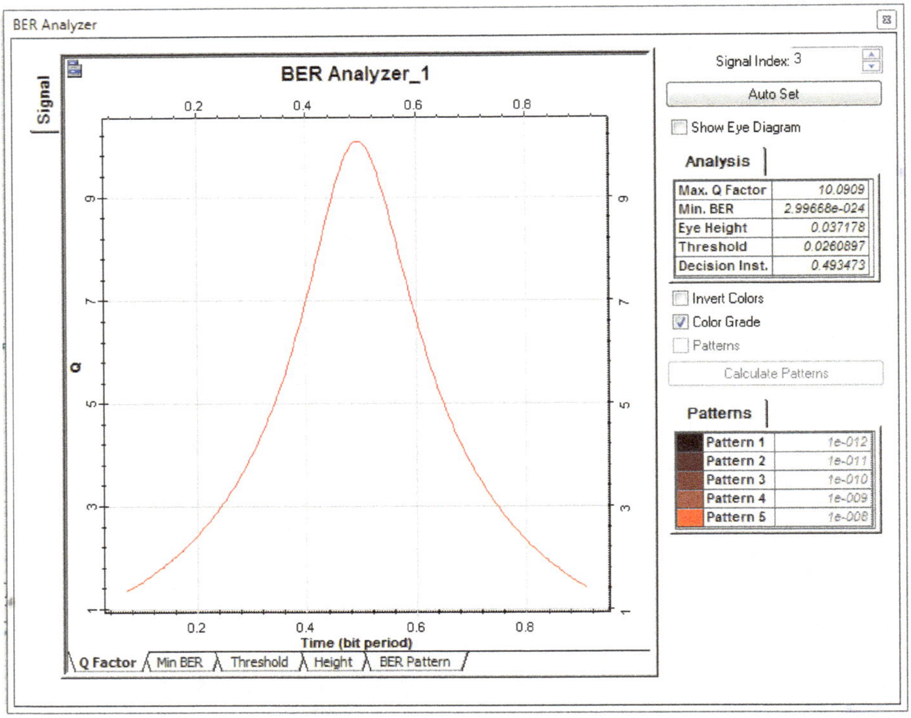

Figure 15. BER result for receiver 2.

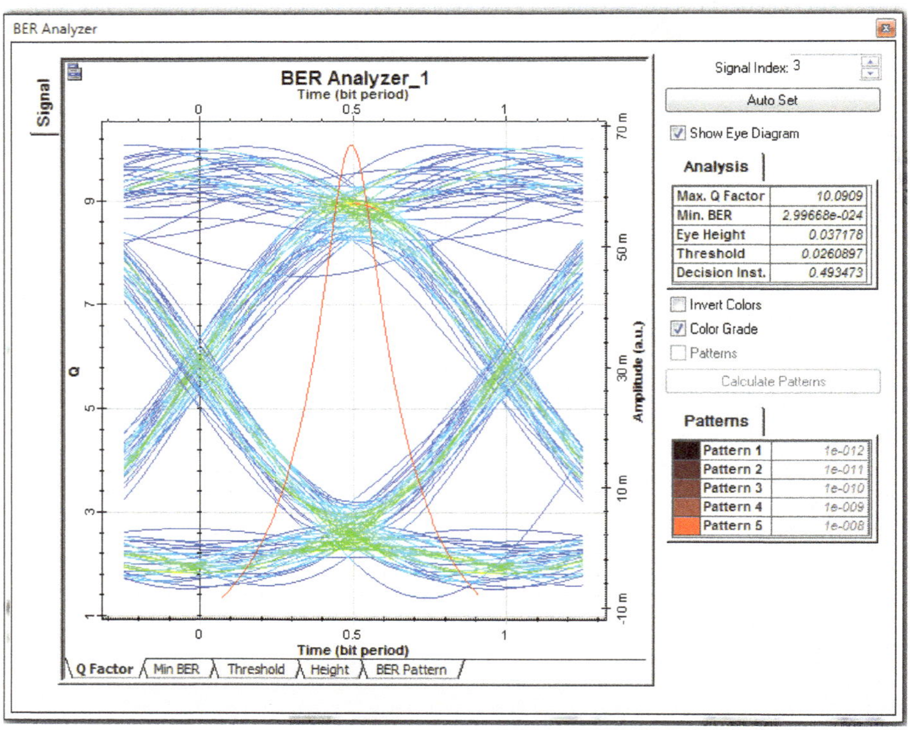

Figure 16. Eye diagram for receiver 2.

For receiver 2, the received signal has BER of 3×10^{-24}. The Q factor for the received signal is 10.09. Thus, the BER analyser shows that this receiver also received a good quality of signal.

The eye opening of the eye diagram is 0.0371. It shows that received signal in receiver 2 has more immunity to noise compared to the received signal in receiver 1.

7. Conclusion

Basically, RoF is a transmission system of analog signal since it distributes a radio waveform from CS to BS direct at radio carrier frequency. A lot of way can be studied and developed to increase the performance of the RoF. This is why RoF is an ideal technology for the future generation. RoF also provides low attenuation and broad bandwidth. It is also immune to electromagnetic interference (EMI), which is a radio-frequency interference where disturbance generated by an external sources that affects an electric circuit by electromagnetic induction, coupling or conduction.

RoF system employing WDM technique is simulated using Optisystem 12.0. WDM is among the best method to be used in RoF. WDM enables transmission of multiple signals through a single fiber over large and can exploit the fiber network bandwidth. Various transmission formats can be supported by various optical channels. Plus, WDM enables the capacity of an optical fiber increase compared to simple point to point link that only enable to carry only single wavelength.

EDFA amplifier is used in this system because it has a lot of advantages. EDFA amplifier provides low noise figure. It also has an independent polarization. Thirdly, the dynamic range is large. The power transfer efficiency of EDFA amplifier is quite large. It also has relatively flat gain. Thus, it is very suitable for long-distance communication.

RoF is a technology that really cost-effective because in this technology, all of the expensive components can be shared by several BS. The installation and maintenance of RoF are easy since the concept of centralized configuration enables all sensitive components being located in safer surrounding. To modulate an optical sources, an electrical signal is used, and the modulated signal will be carried out through optical fiber to the receiver side. When the signal is directly modulated to the optical fiber, power consumption could be decreased, while the frequency of radio carriers is very high at the antenna side. Power consumption also can be decreased by having a simple radio station. All complex equipment and devices are kept at the CS.

RoF also provides operational benefits in terms of operational flexibility. Intensity modulation direct detection (IMDD) technique could be made in order to operate as a linear system. Low dispersion fiber (SMF) with combination with modulated RoF sub carriers (SCM) could be used to achieve linear system. The same radio over fiber network can be used to distribute multi-operator and multi-service traffic which results in huge economic savings.

Author details

Lisawati Sainawi and Khadijah Ismail*

*Address all correspondence to: khadijah_etc@hotmail.com

National Defence University of Malaysia, Kuala Lumpur, Malaysia

References

[1] Prakasam S, Karthikeyan R. OFDM signal improvement using radio over fiber for wireless system. IJCNWC. 2013 June;**3**:287–291

[2] Kim HB and D. Der Ingenieurwissenschaften. Radio Over Fiber Based Network Architecture

[3] Johny J, Shashidharan S, Rof II, and Sing DEU. Design and simulation of a radio over fiber system and its performance analysis; 2017. pp. 655–658

[4] Yunazar F. Modeling and performance analysis of the OFDM scheme for radio over fiber system [thesis]. Malaysia: Universiti Teknologi Malaysia; 2009. p. 74

[5] Iftekhar SF. Radio Over Fiber Transmission by Sub Carrier Multiplexing; August, 2009

[6] Reddy V. Simulation and analysis of radio over fiber (RoF) systems using frequency up-conversion technique. 2016;**133**(12):36–43

[7] Azadeh M. Fiber Optics Engineering. United States: Springer; 2009. p. 374

[8] Keiser G. Optical Fiber Communications. 4th ed. New York: McGraw Hill; 2011. p. 654

[9] Das B, Abdullah MFL, Shah M, Shahida N, and Bukhsh Q. All regeneration for optical communication network using optical 3R regeneration and phase sensitive amplifier. 2015;**8**(8):87–94

[10] Singh YN, Gupta HM, Jain VK. Optical amplifier in broadcast optical network: A survey [thesis]. India: Indian Institute of Technology Kanpur; p. 48

[11] Mohan A and Anisha AP. Performance comparison of radio over fiber system using WDM and OADM with various digital modulation formats. 2015;**4**(2):2013–2016

[12] Basa A, Talukder MZ and Islam MR. Performance analysis and comparison between coarse WDM and dense WDM. 2013;**13**(6)

[13] Robinson S, Jasmine S, and Pavithra R. Investigation on hybrid WDM (DWDM + CWDM) free space optical communication system. 2015 December;**6948**:1187–1192

Permissions

All chapters in this book were first published in OFWC, by InTech Open; hereby published with permission under the Creative Commons Attribution License or equivalent. Every chapter published in this book has been scrutinized by our experts. Their significance has been extensively debated. The topics covered herein carry significant findings which will fuel the growth of the discipline. They may even be implemented as practical applications or may be referred to as a beginning point for another development.

The contributors of this book come from diverse backgrounds, making this book a truly international effort. This book will bring forth new frontiers with its revolutionizing research information and detailed analysis of the nascent developments around the world.

We would like to thank all the contributing authors for lending their expertise to make the book truly unique. They have played a crucial role in the development of this book. Without their invaluable contributions this book wouldn't have been possible. They have made vital efforts to compile up to date information on the varied aspects of this subject to make this book a valuable addition to the collection of many professionals and students.

This book was conceptualized with the vision of imparting up-to-date information and advanced data in this field. To ensure the same, a matchless editorial board was set up. Every individual on the board went through rigorous rounds of assessment to prove their worth. After which they invested a large part of their time researching and compiling the most relevant data for our readers.

The editorial board has been involved in producing this book since its inception. They have spent rigorous hours researching and exploring the diverse topics which have resulted in the successful publishing of this book. They have passed on their knowledge of decades through this book. To expedite this challenging task, the publisher supported the team at every step. A small team of assistant editors was also appointed to further simplify the editing procedure and attain best results for the readers.

Apart from the editorial board, the designing team has also invested a significant amount of their time in understanding the subject and creating the most relevant covers. They scrutinized every image to scout for the most suitable representation of the subject and create an appropriate cover for the book.

The publishing team has been an ardent support to the editorial, designing and production team. Their endless efforts to recruit the best for this project, has resulted in the accomplishment of this book. They are a veteran in the field of academics and their pool of knowledge is as vast as their experience in printing. Their expertise and guidance has proved useful at every step. Their uncompromising quality standards have made this book an exceptional effort. Their encouragement from time to time has been an inspiration for everyone.

The publisher and the editorial board hope that this book will prove to be a valuable piece of knowledge for researchers, students, practitioners and scholars across the globe.

List of Contributors

Yossef Ben Ezra
MER Cellos, Holon, Israel
Department of Electrical Engineering and Electronics, Holon Institute of Technology (HIT), Holon, Israel

Boris I. Lembrikov
Department of Electrical Engineering and Electronics, Holon Institute of Technology (HIT), Holon, Israel

Xuelin Yang, Adnan A.E. Hajomer and Weisheng Hu
State Key Laboratory of Advanced Optical Communication Systems and Networks, Shanghai Jiao Tong University, Shanghai, China

Benjamin Dingel
Photonics Inc., Painted Post, NY, USA
Ateneo Innovation Center, Ateneo De Manila University, Quezon City, Philippines

Nicholas Madamopoulos
Department of Aeronautical Science, Hellenic Air Force Academy, Dekeleia, Greece
Department of Electrical Engineering, The City College of City University of New York, USA

Andru Prescod
ManTech International Corporation, Arlington, VA, USA

Edward M.D. Fisher
Institute of Digital Communications (IDCOM), The School of Engineering, The University of Edinburgh, Scotland, UK

Mitsuru Kihara
NTT Access Network Service Systems Laboratories, Nippon Telegraph and Telephone Corporation, Hanabatake, Tsukuba, Ibaraki, Japan (He is currently with Osaka Electro-Communication University)

Danilo Bórquez-Paredes and Alejandra Beghelli
Science and Engineering Faculty, Universidad Adolfo Ibañez, Chile

Ariel Leiva
Electrical Engineering School, Pontificia Universidad Católica de Valparaíso, Chile

Konstantinos Manousakis, Panayiotis Kolios and Georgios Ellinas
Department of Electrical and Computer Engineering, KIOS Research Center for Intelligencez Systems and Networks, University of Cyprus, Cyprus

Rahmat Talib and Mohammad Faiz Liew Abdullah
University Tun Hussein Onn Malaysia, Malaysia

Lisawati Sainawi and Khadijah Ismail
National Defence University of Malaysia, Kuala Lumpur, Malaysia

Index